POPULATION ECOLOGY

OF THE

Cooperatively Breeding
Acorn Woodpecker

MONOGRAPHS IN POPULATION BIOLOGY

EDITED BY ROBERT M. MAY

POPULATION ECOLOGY

OF THE

Cooperatively Breeding
Acorn Woodpecker

WALTER D. KOENIG

RONALD L. MUMME

PRINCETON UNIVERSITY PRESS

PRINCETON, NEW JERSEY

1987

Library of Congress Cataloging in Publication Data will be found
on the last printed page of this book
ISBN 0-691-08422-X (cloth)   0-691-08464-5 (pbk.)
This book has been composed in Linotron Baskerville
Clothbound editions of Princeton University Press books are printed on
acid-free paper, and binding materials are chosen for strength and durability.
Paperbacks, although satisfactory for personal collections, are not usually
suitable for library rebinding
Printed in the United States of America by Princeton University Press,
Princeton, New Jersey

To Fanny Hastings Arnold and Frank A. Pitelka;
also to Fanny and Lucia.

Contents

CONTENTS

CONTENTS

CONTENTS

Acknowledgments

Long-term studies usually have long lists of people to whom the investigators are indebted. In our case, chief among these is Frank Pitelka, who was (and continues to be) a constant and valuable source of ideas, encouragement, and support, not only as a graduate adviser but also as a friend and colleague. We also owe a great debt to Fanny Hastings Arnold, whose continuing generosity makes Hastings Reservation an outstanding study site, and to Michael and Barbara MacRoberts, who began the study in 1971, collected the first three years of data, and who were unfailingly generous in helping us carry on where they left off in 1974. Other colleagues who have collaborated with us include Alan de Queiroz, Rocky Gutiérrez, Susan Hannon, Philip Hooge, Nancy Joste, Jill Marten, Mark Stanback, and Bob Zink; their help has aided the project immensely.

Since 1978 we have benefited from the help of a small army of field assistants: Sue Bonfield, Janet Conley, Vasiliki Demas, Alan de Queiroz, Monica Farris, Melissa Fleming, Denise Fruitt, Larry Hanks, Katy Heck, Philip Hooge, Nancy Joste, Gwen Moore, Sandy Nishimura, Leslie Nitikman, Mary Peacock, Sandy Spon, Molly Williams, and Byron Wilson. Others who aided directly or indirectly with fieldwork include Andy Griffin, Michael MacRoberts, and Pam Williams.

The staff and residents of Hastings Reservation have aided us in innumerable ways over the years. In particular, we thank John Davis, Jim Griffin, and Suzanne Schettler of the staff, and Steve Albano, Shorty Boucher, Bill Carmen, Joanne Daly, Carol Davis, Steve Dobson, Andy Griffin, Joan Griffin, Rocky Gutiérrez, Carol Patton, Jim Patton, Mark Reynolds, and others with whom we were fortunate to live and work at Hastings during the course of our work. In addition, the senior author would like to thank the Pattons for continuing to provide a hotel

ACKNOWLEDGMENTS

facility for him in Berkeley. We are also indebted to Robert Noyce for his permission to conduct research on the land adjacent to Hastings Reservation.

We wish to thank our friends and colleagues in the Museum of Vertebrate Zoology and the Department of Zoology, University of California, Berkeley, who fostered us through graduate school and beyond. Foremost among them include our academic siblings Tom Custer, Glenn Ford, Russ Greenberg, Pete Myers, Steve Pruett-Jones, Van Remsen, Mark Reynolds, Steve West, Pam Williams, and Dave Winkler. Other Berkeley friends and colleagues who facilitated our work include Judy Gradwohl, Melinda Pruett-Jones, Dave Wake, and the staff of the Museum of Vertebrate Zoology. We especially thank Janet Conley, Beth Schwehr, Kim Sullivan, and Pam Williams for their help and continuing friendship.

We are happy to acknowledge the intellectual stimulation provided by the following students of social systems and cooperative breeding: Jerram Brown, John Craig, Nick Davies, Steve Emlen, John Fitzpatrick, Patty Gowaty, Dave Ligon, Robert May, Marty Morton, Paul Sherman, Bill Shields, Pete Stacey, Sandy Vehrencamp, Jeff Walters, and Glen Woolfenden. We owe a special debt to Glen and John for proving that a book such as this one could actually be written.

An appalling number of errors, statistical travesties, grammatical insults, and outright lies were pointed out by those colleagues who were generous enough to review early drafts of all (A) or part (chapters in parentheses) of the manuscript; they include Jerram Brown (9), Bill Carmen (4), Janis Dickinson (1–2), John Fitzpatrick (A), Patty Gowaty (8), Susan Hannon (11), Katy Heck (A), Henry Horn (A), Lois Krieger (A), Dave Ligon (5–7), Kevin McGowan (1–4), Bert Murray (10), Frank Pitelka (A), Pete Stacey (1–4, 14), and Mark Stanback (A). We sincerely thank them; the manuscript is greatly improved as a result of their efforts. We would be happy to attribute those errors that remain to them as well, but alas, they are ours. *All* ours.

ACKNOWLEDGMENTS

We also thank George Barrowclough for help in calculating effective population size, Jerry Brown for urging us to distinguish "nonbreeders" from "helpers," Bill Carmen for helping with the acorn censuses, Jim Griffin for allowing us to use his acorn crop data, John Marzluff for following up on the *Time* magazine story, Judith May from Princeton University Press for encouragement, Fernando Ortiz-Crespo for finding and translating the record of Herrera's encounter with acorn woodpeckers, Kathy Ralls for prompting us to calculate coefficients of relatedness, and Peter Stacey for sharing data from his study of acorn woodpeckers in the Southwest. Bonnie Bowen, Jerram Brown, John Craig, Steve Emlen, John Fitzpatrick, Patty Gowaty, Gustavo Kattan, Rolf Koford, Dave Ligon, Uli Reyer, Bill Shields, Pete Stacey, Tobbe von Schantz, and Glen Woolfenden supplied us with unpublished manuscripts or information on their unpublished work. Drawings were executed by Gene Christman.

Finally, we are grateful for the continuing support of the National Science Foundation, most recently through grant BSR84-10809, as well as earlier graduate fellowships to the authors and grants DEB78-08764 and DEB81-09128 to F. A. Pitelka. Additionally, in 1981–82 RLM was supported by a Betty S. Davis memorial fellowship made possible through the generosity of Fanny Hastings Arnold. Other financial support was provided by the Museum of Vertebrate Zoology, University of California, Berkeley.

To all these individuals and institutions we express our sincere thanks.

Walt Koenig
Ron Mumme
Jamesburg, California
July 1986

POPULATION ECOLOGY

OF THE

Cooperatively Breeding
Acorn Woodpecker

CHAPTER ONE

Introduction

If I were a Californian boy, I think I should spend my time in trying to find out more about this wise woodpecker, concerning which much remains to be discovered.

Fanny Hardy Eckstorm, 1901

1.1. HISTORICAL PERSPECTIVE

If Fanny Hardy Eckstorm were alive today, she doubtless would be pleased to discover that many biologists and natural historians, Californian and otherwise, have heeded her advice concerning the acorn woodpecker (*Melanerpes formicivorus*). Ever since this species was observed and described by Spanish explorers visiting Central America near the turn of the sixteenth century, its behavior—particularly its unique habit of storing acorns in specialized storage trees or granaries—has impressed ecologists and explorers alike. No doubt the various tribes of native Californian Indians and Central American Mayas and Aztecs were similarly impressed (see, for example, Margolin [1978]).

Despite considerable interest in and controversy over the food habits of acorn woodpeckers throughout much of the nineteenth century (see MacRoberts, 1974), it was many years before an equally unusual feature of their biology came to light: their extreme sociality. Acorn woodpeckers live in groups of up to fifteen individuals, all of whom help rear the young of a single nest, a phenomenon now known as cooperative (or communal) breeding. The group-living habit of this species was mentioned by several early American ornithologists, including Henshaw (1876) and Bendire (1895).

1

In what may be one of the earliest papers focusing specifically on the phenomenon of cooperative breeding, Harriet Williams Myers (1915) of Pasadena, California, reported observations of at least three adults caring for the young of a fall nest. Somewhat later, Frank Leach of Walnut Creek, California, confirmed Myers's findings in a paper titled "Communism in the California woodpecker" (Leach, 1925). Shortly thereafter, E. Michael, a resident of Yosemite Valley, observed two males mating in sequence with a single female (Michael, 1927), thereby providing the first evidence for mate-sharing (the presence of more than one breeder of the same sex within a single social unit) within acorn woodpecker groups or, for that matter, in any species of bird.

Most prominent of these earlier students of the acorn woodpecker was William E. Ritter, a California zoologist who, among other accomplishments, founded the Scripps Institution of Oceanography in La Jolla. Beginning about 1908 and continuing for over three decades, Ritter devoted considerable time to field observation and philosophic contemplation of the acorn woodpecker's behavior. His first papers (Ritter, 1921, 1922, 1929) were devoted primarily to acorn storage. But after he retired in 1923, he expanded his efforts, synthesizing his own and other workers' observations in a monograph (Ritter, 1938) titled *The California Woodpecker and I: A Study in Comparative Zoology—in which are set forth numerous facts and reflections by one of us about both of us.*

Although many useful facts of natural history, some portending insights of recent work, are reported in his treatise, Ritter's book is so rambling and florid that the modern-day reader may become quickly exasperated. In an obituary for Ritter, another eminent California zoologist, Francis B. Sumner (1944), comments admiringly on this work, but adds, "The reader may fail to detect any great degree of coherence at all points in the narrative."

About the same time as Ritter's later work, Alexander Skutch (1935) authored the first full-length discussion of the phenomenon of avian "helping"—parentlike behavior directed toward nondescendent young (the defining feature of cooperative breeding systems). Skutch's extensive work in Costa Rica presaged the still unexplained observation that cooperative breeding is more common in tropical areas. Thus, by the 1930s helping behavior and cooperative breeding were well known, and the acorn woodpecker was known to be one of the few temperate-zone species exhibiting this unusual social organization. Nonetheless, little new information on acorn woodpeckers was published for over thirty years following Ritter's book.

In retrospect, the apparent lack of interest in the acorn woodpecker following Ritter's account is surprising. Enough information on the unusual social organization and cooperative behavior of this species was available to suggest that it posed potential problems to Darwinian selection; that is, how could parentlike behavior directed toward nondescendent offspring be favored by natural selection? (Darwin himself mentions cooperation among animals as being a problem for his theory.) Among the first workers to recognize this difficulty was V. C. Wynne-Edwards, who in his controversial and now classic *Animal Dispersion in Relation to Social Behaviour* (1962) discussed the social behavior and acorn-storing habits of the acorn woodpecker at length.

In keeping with his ideas concerning the evolution of social behavior in general, however, Wynne-Edwards proposed that the function of cooperative behavior in this (as well as many other) species is to allow individuals to assess the food supply and adjust their reproductive rate to avoid overexploitation of the available food resources ("epideictic displays"). With respect to the acorn woodpecker, he concluded that acorn storage presented "the perfect example of an epideictic rite, combining as it does a sampling of the food-supply, a territorial symbol (the tree), and social competition" (Wynne-Edwards, 1962:325).

As we will discuss in Chapter 4, it is indeed true that the reproductive rate of acorn woodpeckers is intimately entwined with the acorns they store. Nevertheless, Wynne-Edwards's theory for the evolution of cooperation and higher forms of sociality has been rejected by the general scientific community. Notable among his many critics was G. C. Williams (1966), who argued strongly in favor of individual selection and for resorting to group selection, central to Wynne-Edwards's theory, only when necessary. In so doing, Williams also found cause to remark on the acorn woodpecker. Drawing on W. D. Hamilton's (1964) then recently published theory of inclusive fitness, he wrote:

> Of the multiple birds that attend a single nest, some are manifestly immature and may represent older offspring helping with the rearing of their younger brothers and sisters. I would predict that this would be the case, and that the societies of the California woodpecker, of the social insects, and of all other such organized groups, will be found to be based almost entirely on family relationship. (Williams, 1966:201–202).

As it turns out, this generalization is not completely true: at least some such societies appear *not* to be family based (e.g., the Galápagos hawk [Faaborg et al., 1980]; pied kingfishers in poor habitats [Reyer, 1980]; house sparrows in Mississippi [Sappington, 1977]; acorn woodpeckers in New Mexico [Stacey, 1979b]). However, in the case of acorn woodpeckers in California, as well as the vast majority of other cooperatively breeding species, Williams was manifestly correct: groups nearly always consist of extended families. The significance of the close degree of relatedness within groups of cooperative breeders—specifically, the extent to which kin selection (Hamilton, 1964; Maynard Smith, 1964; Michod, 1982) or indirect selection (Brown and Brown, 1981b) has been crucial to the evolution of these species—is still a matter of considerable controversy, and one that we will discuss in Chapters 12 and 13.

Williams's book was published at a time when the work of Hamilton (1964), Crook (1965), Lack (1968), and others synergistically forged the new field of behavioral ecology—a field focusing on the ultimate causes of behavior. Investigations of ultimate causation frequently necessitate not only detailed field observations but also long-term information on marked individuals. Thus, it is not surprising that concurrent with the emergence of behavioral ecology as a discipline, a number of intensive long-term studies of marked bird populations were initiated. Among these were several involving cooperatively breeding species, including the work of Jerram and Esther Brown and their colleagues on the Mexican jay (Brown, 1970, 1972; Brown and Brown, 1980, 1981a; Trail et al., 1981), Glen Woolfenden and John Fitzpatrick (1978, 1984) on the Florida scrub jay, Steve Emlen and his colleagues (Emlen, 1981; Hegner et al., 1982; Emlen and Wrege, 1986) on the white-fronted bee-eater, and Michael and Barbara MacRoberts (1976) on the acorn woodpecker. It is this last study, begun in October 1971 and continued by us since July 1974, that forms the cornerstone of the work we present here.

With the appearance of E. O. Wilson's monumental *Sociobiology* (1975), interest in cooperative social systems expanded. Today is an undeniably exciting time for study of cooperative breeding: many studies begun in the mid-1970s and earlier are coming to fruition, producing a large amount of diverse and exciting information on a variety of species. Combined with numerous recent theoretical advances in the field, these studies promise to provide an outstanding overview of not only the diversity in but also the evolutionary causes and consequences of cooperative breeding.

1.2. THE ACORN WOODPECKER

The acorn woodpecker ranges throughout California, the American Southwest, and the mountains of western Mexico through Central America and northern Colombia. Within this

area acorn woodpeckers are closely associated with oak woodlands. In California and many other parts of their range, they
live in family groups of up to fifteen individuals of both sexes
and all ages (MacRoberts and MacRoberts, 1976).

Recent work by Stacey (1979b) and ourselves (Mumme et
al., 1983a; Koenig et al., 1984) has produced a coherent picture of the breeding system of this unorthodox species. In our
population at Hastings Natural History Reservation in central
coastal California, groups consist of 1 to 4 (rarely more) breeding males, 1 or 2 (rarely 3) breeding females, and 0 to 10 nonbreeding offspring from prior nests. Groups containing virtually
all permutations of these 3 components may be found simultaneously in the population. Thus, groups of a single breeding
male and female (a monogamous pair) coexist in the population with groups containing 1 breeding male and 2 breeding
females (a polygynous trio), groups with 3 breeding males and
1 breeding female (a polyandrous quartet), and groups with 3
breeding males and 2 breeding females (a polygynandrous
quintet); in addition, such groups may contain offspring from
prior years who do not breed. The range in group composition
in this population is illustrated in Figure 1.1.

Each group defends an all-purpose territory, including one
or more granaries for acorn storage, and produces only a single
nest at a time, regardless of size or composition. Thus, up to
three females may lay eggs in a single nest (Koenig and
Pitelka, 1979; Mumme et al., 1983b), a situation that parallels
the nesting habits of anis (Davis, 1940a, 1940b, 1941; Vehrencamp, 1977, 1978). In addition, however, up to four males
may mate with and, presumably, contribute genetically to the
offspring in the communal nest (Stacey, 1979b; Mumme et al.,
1983a, 1985; Joste et al., 1985; see also section 2.2). Traditional dyadic associations (pairing behavior) do not occur
among these multiple breeders, thus contrasting with group
organization both in the anis and in Mexican jays, where several pairs breed concurrently within a social unit (Brown,
1970; 1972; Brown and Brown, 1981a).

BREEDERS: males   and   females   plus   NONBREEDING HELPERS

FIGURE 1.1. Group composition of acorn woodpeckers at Hastings Reservation, California. Following the postjuvenal molt in the autumn, males have red on the crown abutting the white above the bill and females have a black band separating the white forehead from the red crown. Groups may contain any combination of one to four breeding males (either brothers or a father and his sons) sharing one to three females (usually sisters but sometimes a mother and her daughter). Pairing among breeders is absent, and the mating system is polygynandrous. In addition to the breeders there are up to ten (usually fewer) nonbreeders of either sex. Nonbreeders are offspring of the breeders born in prior years.

Genetic relatedness within these social units is high. Breeders are almost invariably closely related to one another *within each sex;* thus, male breeders are generally siblings or a parent and one or more offspring, while the same is true for females. Nonbreeding helpers are virtually always the product of prior nests of the breeders, and thus, with one exception, everyone is closely related to everyone else within a group. The exception is the two sexes of breeders in a group: breeding males are generally unrelated to the breeding females, and thus close inbreeding is usually, although not invariably, avoided (Koenig et al., 1984).

Some of the most intriguing aspects of this system involve the means by which the social organization of groups is perpetuated from one generation to the next. Offspring born into a group remain there as nonbreeders until they either disperse and become breeders elsewhere (Koenig et al., 1983; Hannon et al., 1985), inherit and breed in their natal territory, or die. Territorial inheritance occurs following the death of all breeders of the *opposite* sex in a group (Koenig and Pitelka, 1979; Koenig et al., 1984), an event that creates a reproductive vacancy. For example, a male nonbreeder may inherit and breed in his natal territory following the death of his mother (as well as aunts who may have bred the year he was born). Such heirs may then share mates with their parent of the *same* sex; hence, the male just described may subsequently breed along with his father. Such a system illustrates well an apparent incest taboo acting to decrease the degree of close inbreeding in the population below levels that would be attained by indiscriminate mating within groups. It also results in a degree of genetic continuity within groups as a territory is handed down from one generation to the next. The consequences and significance of this phenomenon are discussed in Chapter 9.

Offspring of the same sex as that of a breeding vacancy (e.g., the female nonbreeders in the group discussed above) do not fare as well: because of the incest taboo, vacancies are generally filled by unrelated birds following contests, which we call "power struggles," involving up to twenty or more nonbreeders (Koenig, 1981a; Hannon et al., 1985). If only a single nonbreeder of the sex being replaced is present in the group, he generally leaves or is forced out of the group by the new immigrants, as discussed in Chapter 3. (The result is more complex if more than one such nonbreeder is present; see Hannon et al., 1985).

Groups of acorn woodpeckers thus are highly variable assemblages of related and unrelated individuals. Many additional details concerning the social behavior of this species are available in our earlier papers, cited where relevant throughout the text, as well as in the parallel work of Peter Stacey and his co-workers in the Southwest (Stacey and Bock, 1978; Stacey 1979a, 1979b, 1982; Stacey and Edwards, 1983; Stacey and Ligon, 1987). A recent review synthesizing salient features of studies in both California and the Southwest is available in Stacey and Koenig (1984).

1.3. THE MAJOR QUESTIONS

Ever since Darwin (1859), cooperation and apparent altruism, such as shown by cooperative breeders, have been topics of significant interest. However, behavioral ecologists have failed to reach a consensus concerning the evolution of these phenomena. A primary reason for this is purely logistical: in order to adequately test hypotheses for the evolution of cooperation and altruism, data are needed on the causes, context, ontogeny, and ecological consequences of cooperative behaviors, as well as on the ecology of the population in which the behavior occurs. The acquisition of such data sets requires long-term studies (Fitzpatrick and Woolfenden, 1981).

In the case of the social behavior shown by acorn woodpeckers and other cooperatively breeding species, there are two basic, although not unrelated, questions, each involving several parts. First, why live in groups? Two important aspects of this general problem are: why do offspring remain as nonbreeders up to several years in their natal groups? (Or, alternatively, why do their parents tolerate them for so long?) And in the case of cobreeders of the same sex within groups: why do birds share mates rather than breed on their own?

The second basic question is: once groups have formed, why do group members feed young that are not their own, as well as engage in other cooperative behaviors? And by doing so, how do they contribute to their own fitness? In other words, do helpers gain by helping? And if so, how and by how much?

These two general questions—concerning the factors promoting group existence and the fitness gains of helpers—focus on different (but overlapping) aspects of the biology of cooperative breeders: the first, emphasizing ultimate factors, is more ecological and demographic, whereas the second, emphasizing proximate factors, is more behavioral and demographic. It is the first question—why do acorn woodpeckers live in groups?—that forms the primary focus of this book. However, because of their joint concern with demography, the results we present are also critical to resolving questions about helping behavior. In part this is because nondispersal and group living are obligate prerequisites for subsequent helping behavior by nonbreeders. Hence, the fitness consequences of helping behavior must be taken into consideration when focusing on either of these two phenomena. Conversely, the mere act of nondispersal may have important fitness consequences to helpers by influencing the survivorship of other (related) group members, as we show for acorn woodpeckers in Chapter 12. Thus, despite their apparent distinctness, the questions "why live in groups?" and "why do helpers help?" are closely interrelated, and neither is likely to be resolved without reference to the other.

Why Live in Groups?

The currently favored hypothesis for why cooperative breeders live in groups is that of "habitat saturation" or "ecological constraints" (Selander, 1964; Brown, 1974, 1987; Stacey, 1979a; Koenig, 1981a, 1981b; Koenig and Pitelka, 1981; Emlen, 1982a). The critical feature of habitat saturation is that nonbreeders are ecologically forced to remain in their natal groups as a result of some constraint, usually but not necessarily space related. For example, we have suggested that in the acorn woodpecker granaries may constitute the critical resource (Koenig and Pitelka, 1981; see also Stacey, 1979a and Chapter 4). According to this hypothesis, nonbreeding helpers are "making the best of a bad job" (Dawkins, 1980) and would do better to breed on their own if they were able to overcome the constraint that keeps them from doing so. Analogous reasoning can be applied to the phenomenon of mate-sharing: although individuals may have higher fitness if they are able to breed on their own, ecological constraints may restrict them from doing so and result in them making the best of a bad job by forming coalitions with others, as do male lions, *Panthera leo* (Bygott et al., 1979; Koenig, 1981c).

Our favorite analogy for this hypothesis is that of postdoctoral students (Koenig and Pitelka, 1981; Woolfenden and Fitzpatrick, 1984). Although no one denies that researchers obtain valuable experience as postdoctoral students, there are nonetheless few Ph.D.'s who would turn down a tenure-track academic position for a postdoctoral appointment, and a majority are likely to abandon a postdoc immediately upon receiving such an offer. Thus, even though direct academic benefits accrue through postdoctoral experience, it is nonetheless "making the best of a bad job." Obtaining a tenure-track job is the ultimate goal, and the one perceived as more valuable to total lifetime academic fitness.

Although few recent authors have seriously questioned the importance of this hypothesis as a major, if not the exclusive, ecological feature selecting for group formation and setting the stage for cooperative breeding (for a recent exception involving acorn woodpeckers, see Stacey and Ligon, 1987), at least two alternatives exist. The first, termed the "skill hypothesis" by Brown (1987), proposes that young of cooperative breeders delay breeding and remain in their natal groups because they have not yet attained the skill necessary to breed independently. A second alternative, which we call the "hard life hypothesis," is that conditions are so poor that more than two individuals are necessary to obtain enough food, or to protect against predators, to raise any young whatever. One apparent example of this phenomenon is the stripe-backed wren, in which pairs are virtually never successful and the reproductive advantages of having two or more helpers is substantial (Rabenold, 1984; 1985; see also section 14.2). Thus, the ecological factors leading to cooperative breeding are not a closed issue, and they continue to pose important questions for behavioral and evolutionary ecologists.

If ecological constraints are paramount in explaining group formation, there remains the question: what is the mechanism by which group living and cooperative breeding become a strategy superior to dispersal and attempted independent breeding? In other words, what are the specific conditions leading to ecological constraints so severe that group living results?

Koenig and Pitelka (1981) suggested that cooperative breeders live under conditions in which there is a relatively high proportion of "good" compared to "marginal" habitats. Under such conditions, the production of surplus offspring in "good" areas will regularly outstrip the ability of the few marginal habitats to accommodate them, and faced with the option of dispersing to "poor" habitat where fitness is by definition low, selection will be strong for young to try to remain in their natal groups until a breeding vacancy can be found.

The importance of this idea can be appreciated by realizing that most territorial species produce on average more potential recruits than "slots" available through adult attrition (see section 8.8). (Note that the actual number of recruits filling breeding vacancies in a population is quite likely to equal adult attrition on a long-term basis, but in order for this to occur there must have been at least as many *potential* recruits, usually first-year birds, to begin with.) It is thus not sufficient to focus on the "surplus" of offspring over reproductive vacancies to explain cooperative breeding (see, for example, Brown, 1987): we propose that the majority of demographically sound, territorial species have some such surplus much, if not all, of the time. Thus, all such populations are to a greater or lesser extent ecologically constrained, and the hypothesis of Koenig and Pitelka (1981), or an alternative, is necessary in order to explain how it is that some more severe constraints operating on co-

operative breeders lead to group living, whereas those operating on other species (which, after all, are in the vast majority)
do not.

Why Help?

Why and how do helpers help raise young not their own? To
the degree that these problems require behavioral and physiological data on the extent and specific effects of helping behavior,
we defer a discussion of them to future papers. However, the
demographic context and consequences of helping behavior are
critical to an understanding of this phenomenon, thus we will
be addressing the evolution of helping behavior from this standpoint. For a thorough review of helping behavior in general,
as well as the specific hypotheses discussed below, see Brown
(1987).

Briefly, major hypotheses include the following:

1. Kin selection. Since the papers of Hamilton (1964) and
Maynard Smith (1964), the hypothesis that selection acts on
individuals to increase their inclusive fitness, and thus that aid
rendered to nondescendent kin can be as selectively advantageous as that rendered to direct offspring, has become one of the
most important concepts of behavioral ecology (e.g., Brown,
1974, 1978, 1987). Nonetheless, there are few data to support
unambiguously the importance of kin selection (through non-
descendant young), in species other than eusocial insects (see,
for example, Markl, 1980).

Kin selection has frequently been uncritically applied to the
phenomenon of cooperative breeding, leading many authors,
ourselves included, to emphasize alternatives (e.g., Woolfenden
and Fitzpatrick, 1978, 1984; Ligon and Ligon, 1978a; Koenig
and Pitelka, 1981). Discussion of this phenomenon is further
complicated by semantic difficulties, some of which are discussed in detail by Brown (1987). Here we will generally employ the useful distinction of "indirect fitness" (that accruing

through production of nondescendent kin) and "direct fitness" (that accruing through production of offspring) made by Brown and Brown (1981b).

We (Koenig and Pitelka, 1981), like many others (e.g., Woolfenden and Fitzpatrick, 1984; Brown, 1987), maintain that consideration of kinship has little to do with the resolution of the first of the questions raised above, namely, why individuals form groups rather than disperse. However, the fact that the social structure of most cooperative breeders, and certainly the acorn woodpeckers in our population, are based on family units renders the potential importance of kin selection in the evolution of group behaviors undeniable. We agree with Woolfenden and Fitzpatrick (1984) that the critical feature is the "*relative* importance of the direct versus the indirect components to inclusive fitness" (italics theirs). Thus, an important goal of this book is to begin analyses aimed at determining the relative importance of kinship to the social behavior of the acorn woodpecker and to quantify indirect and direct fitness components of helping behavior.

2. Group selection. Widely rejected during the late sixties, group selection has made something of a comeback in the models and data of D. S. Wilson (1975, 1977, 1980), as well as the work of Wade (1979, 1980; see also Boyd and Richerson, 1980), indicating that the dichotomy between kin selection and group selection is artificial. As such, it is not now productive to attempt to separate their effects, particularly in a system such as the one we are dealing with here, in which relatedness within groups is very high.

Critical to models of group selection is the degree of genetic differentiation among groups (Wade, 1980; Slatkin, 1981). Quantification of such differentiation, as indexed by between-group differences in mean fitness, has recently been performed for macaques by Silk (1984), and for pinyon jays, some lineages of which contain helpers, by J. Marzluff (personal communication). A similar analysis for acorn woodpeckers is beyond the scope of this book.

3. Reciprocity. Trivers (1971) was the first to stress the potential for aid-giving behaviors to evolve when altruistic acts can be repaid in kind, especially those which cost the donor relatively little but benefit recipients a great deal; an elaboration of this idea is presented by Axelrod and Hamilton (1981). Such reciprocity has been proposed to be important in the cooperative breeding systems of green woodhoopoes by Ligon and Ligon (1978a; see also Ligon, 1983), stripe-backed wrens by Wiley and Rabenold (1984), Mexican jays by Brown and Brown (1980; see also Caraco and Brown, 1986), and pukeko by Craig (1984).

Much of the data critical for testing the importance of this hypothesis (for example, the relatedness and subsequent reproductive bias among mate-sharing siblings, as well as the timing and frequency of sibling dispersal related to the degree of help) is beyond the scope of this treatise. For a critical review of several of these proposed examples of reciprocity described in cooperative breeders, see Brown (1987) and Koenig (in press).

4. Individual selection. Under this heading lie all hypotheses suggesting that the apparent altruism in cooperative breeding is illusory and instead serves primarily or exclusively to increase the donor's direct fitness. Potential direct benefits to helpers that are independent of genetic relatedness or reciprocity include obtaining access to resources that would otherwise be unavailable to them (Craig, 1976), gaining experience that could aid them later in life (e.g., Skutch, 1961; Rowley, 1965; Lack, 1968; Fry, 1972; Ricklefs, 1975), improving their social status (Zahavi, 1976), and increasing their chances of future breeding (Woolfenden and Fitzpatrick, 1978, 1984), at least on high-quality territories (Stacey and Ligon, 1987).

Regardless of the ultimate evolutionary factors that have led to the evolution of helping behavior, once helpers exist there is every reason to expect that they will take advantage of whatever benefits they can, direct or otherwise. Consequently, except in cases where helpers are permanently nonreproductive (such as the sterile castes of eusocial insects), we would be surprised if helpers in any species did *not* obtain some direct fitness benefit as a result of helping. This hypothesis is thus likely to be unrejectable. Consequently, as before, we feel that the most fruitful approach is to attempt to assess the relative impact of helping behaviors on direct and indirect fitness, as has recently been accomplished by Rowley (1981), Reyer (1984), and Rabenold (1985). Such a procedure does not necessarily answer the question of which ecologic settings and population processes cause the evolution of helping behavior, but it does potentially solve the problem of how the behavior affects various components of individual fitness and clarifies questions of adaptive significance. A more precise partitioning of selection may only become possible with long-term data far exceeding what is currently available.

1.4. ORGANIZATION

This book is primarily a discussion of the population ecology of the acorn woodpecker, with emphasis on its unusual group-territorial, cooperative social system. After a brief summary of procedures and methods (Chapter 2) and an introduction to the overall features of the population (Chapter 3), we discuss at length the critical role that acorns play in the demography of this species (Chapter 4), focusing on the use of acorns and the phenomenon of acorn storage. Next we present chapters on basic demographic features, including reproduction (Chapters 5 and 6), survivorship (Chapter 7), and the fate of offspring (Chapter 8), along with a discussion of the demographic consequences of territorial inheritance by group offspring (Chapter 9). Next we synthesize the data from these chapters by considering fitness on a lifetime basis, both for the population as a whole (in the form of a synthetic life-table analysis) and for individuals (in the form of lifetime reproductive success; Chapter 10). This is followed by a discussion of population regulation (Chapter 11). We then return to the social behavior of this species and confront directly some of the evolutionary questions raised by the phenomena of helping-at-the-nest and mate-sharing (Chapters 12 and 13). We summarize our conclusions and describe some of the remaining questions in Chapter 14.

In a few cases, the analyses we present update those presented earlier (e.g., Koenig, 1981b), but we do not attempt review of all prior work on the social behavior of acorn woodpeckers. Instead, our goal is to present a comprehensive analysis of the population ecology of the Hastings population and to apply these data to some of the more pressing questions concerning the evolution of cooperative breeding in this species. We also do not attempt to duplicate any of the several excellent recent reviews of cooperative breeding, either in birds (e.g., Brown, 1978; Emlen and Vehrencamp, 1983), mammals (Gittleman, 1985; Bekoff and Byers, 1985), or both (Riedman, 1982; Emlen, 1984; Brown, 1987).

Research on the population ecology of a species whose social organization is as complex as the acorn woodpecker is in many ways an arduous affair. Consequently, it has been inevitable that many relevant analyses have been deferred, and we make no claim to have answered, or even to have attempted to answer, all the important questions. We console ourselves by maintaining that good projects raise as many questions as they answer. Whether this adage is true or not, ours has certainly fallen into this category. But then, the fun is in the quest, and the quest continues.

1.5. SUMMARY

Cooperative behavior in general, and cooperative breeding in particular, have a rich theoretical history, largely because of the apparent paradox such behavior presents for traditional Darwinian natural selection. One of the earliest cooperative breeders identified and studied extensively was the acorn woodpecker, extraordinary not only for its highly social tendencies but also for its habit of storing vast numbers of acorns in specially modified trees known at storage trees or granaries.

Acorn woodpecker societies consist of up to four breeding males (usually siblings but sometimes a father and one or more sons), one to three breeding females (again sisters or a mother and daughter), and up to ten offspring from prior seasons. The mating structure of any particular group depends on its composition; the population as a whole is best described as polygynandrous or simply as exhibiting mate-sharing while also including nonbreeding helpers-at-the-nest. Despite the possibility of inheritance of the natal territory by both male and female group offspring, a system of incest avoidance assures that the breeding males are generally unrelated to the breeding females, and thus inbreeding is uncommon.

There are two major questions, not entirely distinct, concerning the evolution of cooperative breeding. First, why do groups form? And once group living has evolved, how and why has cooperative nest care (helping) evolved? The currently reigning hypothesis to answer the first of these questions is that ecological constraints act to restrict the available opportunities for offspring to disperse and breed independently. The precise benefits derived from helping behavior are generally controversial. In particular, it may be unfeasible to separate completely the effects of helping on direct versus indirect fitness, at least in a species such as the acorn woodpecker in which the entire social structure is family based. It is, however, possible to estimate the relative impact of helping behavior, or other cooperative behaviors such as mate-sharing, on different components of fitness. It is this latter approach that we will follow.

CHAPTER TWO

Procedures and Methods

2.1. THE STUDY SITE

Our study area is located on and adjacent to Hastings Natural History Reservation, a 900-ha tract located at the northern edge of the Santa Lucia Mountains in the upper Carmel Valley of Monterey County, California. Elevation ranges from 450 m to 920 m. Summers are hot and dry, whereas winters are generally cool and wet; snow is infrequent and does not occur every year. Additional details on climate and history of Hastings and the surrounding area may be found in MacRoberts and MacRoberts (1976), Williams and Koenig (1980), Griffin (1983), and Steinbeck (1952).

Seven plant communities are represented at Hastings (Griffin, 1974, 1983). Those inhabited extensively by acorn woodpeckers include foothill woodland, savanna-grassland, and riparian woodland (see Figures 2.1 and 2.2); most territories also contain areas of mixed evergreen (broad sclerophyll) forest intervening between adjacent territories (MacRoberts and MacRoberts, 1976). Dominant tree species include five species of oaks: valley oak (*Quercus lobata*), coast live oak (*Q. agrifolia*), blue oak (*Q. douglasii*), black oak (*Q. kelloggii*), and canyon live oak (*Q. chrysolepis*). These oaks all occur in the higher elevations, such as on the Arnold site (see Chapter 3); however, the majority of groups on lower elevation areas have only three species of oaks (valley, blue, and live) readily available to them.

The five species of oaks belong to the three subgenera of *Quercus* (see Table 4.3) and vary considerably in their biology. The two species of live oaks are evergreen, and the others are deciduous. More important, the acorns of canyon live oak and

19

FIGURE 2.1. Representative foothill woodland and oak-savanna at Hastings Reservation. The majority of acorn woodpecker territories in this area contain habitat of these types.

FIGURE 2.2. An area of riparian woodland used by acorn woodpeckers at Hastings Reservation. This habitat type includes valley oaks, live oaks, and California sycamores. About 35% of territories contain extensive areas of this habitat (MacRoberts and MacRoberts, 1976).

black oak require two years to mature, whereas those of the other three species mature in a single year. This basic difference in maturation rates may be important in producing a lack of synchrony among the crops of the various oaks species at Hastings (see Chapter 4).

Other common trees include California sycamores (*Platanus racemosa*) and willows (*Salix* spp.) in riparian areas and California buckeye (*Aesculus californica*) and madrone (*Arbutus menziesii*) in the mixed evergreen forests. There are only a few individual pine trees in the study area, although two species of pines are common only a few kilometers away.

Areas used by acorn woodpeckers are generally fairly open, with few shrubs or other understory but a dense ground cover of grasses and forbs. Some territories are located on relatively flat savanna, others include steep hillsides. More detailed descriptions of these communities are given in MacRoberts and MacRoberts (1976).

2.2. FIELD AND ANALYTICAL PROCEDURES

Data Included in the Analyses

Work on the Hastings acorn woodpecker population was begun in October 1971 by M. and B. MacRoberts. Between that time and July 1974, a period encompassing 3 breeding seasons, they made observations on 24 groups and color-banded about 150 individuals (MacRoberts and MacRoberts, 1976). Beginning in July 1974, Koenig "inherited" the project under the direction of F. A. Pitelka. Fieldwork was conducted by Koenig between 1974 and 1979, at which point Mumme joined; fieldwork was subsequently performed by both Koenig and Mumme in 1980 and 1981, and primarily by Mumme between 1982 and 1984; Koenig "returned" in 1985. The basic data set on which the majority of the demographic analyses here are performed thus runs through the 1982 breeding season. However, data and observations subsequent to 1982 (through 1986) are

21

included for a few analyses (e.g., lifetime reproductive success and density dependence of reproduction) when of special interest. These additional data were added only to confirm results based on the smaller data set; we never added data in the hopes of making the outcome of a statistical test more to our liking.

During the evolution of the project many new groups have been added to those originally banded and censused by the MacRobertses; thus, not all groups discussed were monitored for the identical length of time. In total, we followed birds from 35 groups; 18 of these were censused by the MacRobertses starting in 1972 and were followed during the entire 11 breeding seasons covered through 1982. When appropriate, we have made use of the data collected and presented by the Mac-Robertses (1976), adding them to our own data from subsequent years. Thus, some tests include data beginning in 1972 and others include data only from 1975, the first year we monitored the population. Sample sizes therefore vary depending on when we first deemed sampling procedures and data to be comparable to that collected in later years.

Banding and Censusing

As in other studies focusing largely on individual behavior, the most critical component of our field techniques involved banding birds. Through 1982, over 800 birds had been banded, 447 as nestlings. Our goal was always to find *all* nests produced by groups under observation and to band all fledglings they produced. With few exceptions, we succeeded; in particular, we are confident of having found all successful nests between 1977 and 1982 and having found the vast majority of nests that survived more than a few days but ultimately failed. Because acorn woodpeckers are cavity nesters, this is no simple feat. All nests followed throughout the study occurred in natural cavities.

Once a nest was found, we climbed up and opened it so as to examine and monitor the contents. Opening holes generally consisted of sawing a wedge-shaped piece of wood from below the nest cavity entrance large enough to allow us to reach the bottom. Wedges were subsequently nailed back and patched with glazing compound. Birds readily accepted such interference without abandoning or otherwise modifying their behavior. Nests were subsequently checked at irregular intervals depending in part on their accessibility. Young were banded at 20 to 25 days of age; unless we had reason to suspect that the nest failed subsequent to banding, the number of young "fledged" was taken to be the number banded at that time. Actual fledging takes place at about 30 days after hatching.

We employed several techniques to capture adults. During periods when acorns were being stored or when birds were flying along predictable paths, birds could be captured by use of mist-nets placed in open areas inside tree canopies. A few individuals were caught while inside the nest cavity feeding young; typically, we set up a rectangular trap made of flexible material that was pinned beside the cavity and connected to the ground by a string, which we pulled after the bird entered the hole, thereby shutting the entrance and trapping the bird inside. A similar technique was frequently employed at roost holes to catch birds after they had retired for the night. This latter technique often succeeded in yielding several birds roosting communally and was the most frequent method we employed to capture adults. Our goal was to band all birds in the population. We were never entirely successful, but in general about 95% of the population of up to 150 birds present at a given time were banded.

Once captured, birds were weighed, measured, banded, and bled. Blood was frozen and used for electrophoretic examination of population structure and parentage (Mumme et al., 1985). For banding, we used three colored plastic leg bands and one U.S. Fish and Wildlife Service band. Color bands were a single solid color, two colors adjacent to each other ("split" bands), or two colors alternating several times ("narrow split" bands); approximately 100 different kinds of bands were used during

23

the study. Very few such bands were ever lost by individuals, and only three birds (<0.5%) lost legs, apparently as a result of injuries or infection resulting from their bands. No other abnormalities in behavior or survivorship were noted that were likely to have been a result of banding; however, in 1975 only, many nestlings were fitted with plastic wing-streamers, which may have had a detrimental effect on their long-term survivorship.

Colors assigned to particular birds were unique, and a special effort was made to make them distinctive within groups. Thus, cohorts, within either years or groups, were not banded in any consistent fashion.

Group censuses, conducted at irregular intervals but generally at least every two to three months, consisted of observing the territory (usually the granary or, during the breeding season, the nest) from a vantage point until all group members had been seen. Acorn woodpeckers are generally quite wary; thus, several censuses were usually necessary before we could be confident of group membership. (Unfortunately, we do not enjoy the benefits of a tame population, such as the scrub jays studied by Woolfenden and Fitzpatrick [1984].) Censuses were conducted from a blind or other concealed area and involved searching for birds and identifying their bands. Individuals observed were identified from "field lists" of group members and recorded in permanent field notes.

When birds disappeared, the date of disappearance was taken to be the midpoint between the date last observed and the date when they were first noted as missing. This usually resulted in an uncertainty of two months or less over the precise date of disappearance.

Determination of Group Composition: Age and Sex

We determined group composition as well as individual survivorship and fates of fledglings from our census records. Age categories we use are as follows:

1. Juvenile. A bird prior to its postjuvenal molt, which occurs in late summer or autumn of its first year. Birds cannot be externally sexed while juveniles; all have crown plumage similar to older males (Spray and MacRoberts, 1975).

2. First-year. A bird from the time of fledging until its first spring, and thus including the juvenile period. Following the postjuvenal molt, first-year birds are sexed based on the presence (female) or absence (male) of a black band on the head separating the red crown from the white on the forehead, and can often be aged by criteria described in Koenig (1980a).

3. Adult. Any bird known to have fledged prior to the current or most recent spring. Thus, all fledglings in year x (even any that may have fledged in the autumn) graduate to adults in April of year $x + 1$.

Individual birds are referred to by their sex followed by a unique number assigned to it when it is banded; e.g., ♂140. Where relevant, the age category of the bird may be appended; hence "A♂140" indicates a male who was an adult at the time under consideration, and "F♀193" indicates a first-year female. Juveniles are indicated by a "j" (e.g., j672).

Determination of Group Composition: Breeders and Nonbreeders

Determination of the breeding status of individuals, critical to many of the analyses performed here, is more difficult. In general, birds that immigrated into groups are classified as breeders. Group offspring are classified as nonbreeders unless all breeders of the opposite sex have disappeared and have been replaced by unrelated individuals, at which point nonbreeders may "inherit" their natal territory. Thus, in most cases, our classifications of breeding status are based on prior history rather than direct observation and are meant to represent *potential*, not necessarily actual, status. For example, the classification of a male as one of four breeding males in a group is meant to convey our judgment that the bird was both capable of and in a position to attempt reproduction; such an individual may or may not have successfully sired offspring during any particular year (see Koenig et al., 1984; and below). This definition of what constitutes a breeder assumes that males attempting copulation behave as breeders, even if, genetically, they do not succeed. We believe this assumption to be plausible, but unfortunately there are currently few data with which to test it.

The criteria by which we determined the status of individuals in groups are ultimately based on several lines of evidence discussed at length in Koenig et al. (1984) and therefore not repeated here. We are, with minor exceptions involving only a few individuals ($<5\%$ of the population), confident that these criteria accurately depict the reproductive status of birds in our population. Nonetheless, our determination of breeding status is usually indirect (copulations, for example, are almost never observed, and we have seen relatively few eggs actually being laid by known females). Furthermore, exceptions to the above guidelines, resulting in close inbreeding, are known to occur (see Koenig et al., 1984). Such exceptions were taken into consideration whenever possible. Nevertheless, these problems, compounded by our inability thus far to determine reproductive bias among cobreeding males (see Mumme et al., 1985), are an important limitation of this study, which should be borne in mind by the reader.

Reproductive bias among cobreeders, particularly males, is a difficult issue, which permeates many of the analyses throughout the book and to which we will return in Chapter 13. The problem is as follows: within a group, some individuals are "attempting" to breed (the breeders) and others are not (the nonbreeding helpers). However, depending on the degree of reproductive bias within the group, a particular "breeder" may or may not successfully sire offspring in any one year (or even for his entire "breeding" career). How do we classify such an individual?

Evolutionarily and demographically, such a thwarted "breeder" is indistinguishable from the "nonbreeding helpers" living in the same group: neither contribute genetically to the group's nest. (This similarity is stressed by Vehrencamp's [1979, 1983] discussions of "despotic versus egalitarian societies" in which nonbreeding helpers are modeled as birds kept from breeding because of the high degree of bias imposed by the dominants in the group.) Behaviorally, however, the two are separable (see, for example, Mumme et al., 1983a), and it is this behavioral distinction that we use to justify our classification system.

However, the fact remains that we usually have little choice but to define breeders and nonbreeders based on indirect evidence, oblivious of the effect that reproductive bias may have on the conclusions that follow. Worse yet, we are generally forced to assume *equal* parentage among cobreeders, particularly for the analyses of lifetime reproductive success in Chapters 9 and 10. We have some evidence to support this critical assumption for females (Mumme et al., 1983b; Mumme, 1984; see also Chapters 10 and 13), but none for males. Ideally, of course, we wish to know the parents of all eggs, or at least all fledglings, in the population. Unfortunately, the difficulties in attaining this goal are currently all but insurmountable even for "apparently" monogamous species (see Gowaty and Karlin, 1984), and we have certainly fallen far short of it ourselves, despite our attempts using starch-gel electrophoresis (Mumme et al., 1985).

It now appears that a new generation of genetic techniques, including genetic "fingerprinting" using hypervariable minisatellite regions of DNA (Jeffreys et al., 1985), may ultimately render the goal of complete and unambiguous genealogies attainable in field studies of species with even the most complex breeding systems, thereby engendering what we predict will be a virtual revolution in behavioral ecology. However, until that time, we must emphasize our ignorance concerning precise reproductive roles, much less reproductive bias, of birds in our

27

population. This shortcoming is made more visible by the complex breeding system of our population, but is in fact shared by all existing studies of avian sociobiology (Sherman, 1981; Gowaty and Karlin, 1984; Gowaty, 1985).

Assessment of Food and Territory Quality

We monitored several of the major food items eaten by acorn woodpeckers during part or all of the study and also conducted vegetational surveys of the territories themselves. Food items that we measured were acorn stores (done by direct counts) and relative abundance of flying insects. The latter was measured by a series of nine "yellow pan" traps (yellow dishpans filled with soapy water) and by several Malaise traps, one large (3 m) and four small (2 m). Details and discussion of these trapping techniques can be found in Southwood (1978).

For territories, we measured the number of oak species present and the total basal area of oaks. Further details on methodology for both our vegetation and our insect sampling may be found in Chapter 4.

For quantification of the acorn crop itself we attempted several methods, most of which were ultimately unsatisfactory. We finally settled on an annual sampling of 250 marked individual oak trees distributed throughout the study area (Hannon et al., 1987). Unfortunately, we did not begin this procedure until 1980, and thus we can only use these data in a preliminary way. Instead, we are forced to take our measure of crop abundance primarily from a small series of acorn traps run annually throughout the duration of the study by J. R. Griffin. Again, further details may be found in Chapter 4.

Estimating Fitness

For the majority of our analyses, we use offspring fledged (or their equivalent) as the currency with which to compare the direct and indirect (Brown and Brown, 1981b) fitnesses of in-

28

dividuals pursuing particular behavioral strategies. Given the difficulty of obtaining reproductive opportunities in acorn woodpeckers, a better estimate of fitness might be the number of breeders produced (or "breeder equivalents") rather than fledglings (Woolfenden and Fitzpatrick, 1984; Fitzpatrick and Woolfenden, in press). However, dispersal and other complications render such a measure unavailable from our current database. For additional discussion of this important issue, see Woolfenden and Fitzpatrick (1984: 308).

Experimental Procedures

Although the majority of the data discussed here are descriptive, we also performed various experimental manipulations to complement and extend our demographic studies. One such experiment involved the erection of artificial granaries in various parts of the study area. Artificial granaries were composed of sections of fallen granaries (more recently we have used large pieces of styrofoam). Three such granaries were constructed during the study to test the hypothesis that, on a proximate basis, granaries limit the number of groups present in an area. Results from this experiment are discussed in Chapter 11.

We also performed preliminary experiments manipulating group composition. These experiments, designed to test hypotheses concerning incest avoidance and reproductive competition, were performed primarily on groups adjacent to the main study area and were expanded in 1983 in collaboration with S. J. Hannon. Results of these expanded efforts may be found in Hannon et al. (1985, 1987). Additional experiments are currently being conducted but are beyond the scope of this book.

Statistics

A diversity of statistical techniques have been employed to analyze the data discussed here. Most of the more complex analyses were performed using spss (Nie et al., 1975; Hull and

Nie, 1981), with the exception of the BMDP log-linear analyses used to examine survivorship (Dixon, 1983). Hand calculators and microcomputers were used to perform many of the simpler analyses. We have restricted ourselves to nonparametric procedures for most univariate analyses under the assumption that most of the data with which we are concerned are unlikely to be normally distributed. F-tests, which are relatively insensitive to modest violations of normality (Cochran, 1947; Donaldson, 1968), are used for most of the parametric multivariate analyses we employ. Values are usually reported as means $\pm$ standard deviations. When not otherwise specified, the degrees of freedom for χ^2 tests is 1. P-values ≤ 0.05 are considered significant.

In all cases, we have employed two-tailed tests when possible. We recognize that this violates the standard procedure of employing one-tailed tests when there is a priori reason to predict that a difference will go in a particular direction. However, at least with the types of data we are dealing with, we rarely feel confident that the answer can go in only one direction (acorn woodpeckers have surprised us too many times).

Opposing this conservatism is the overall large number of tests we perform and the instances in which we apply more than a single test to the same data set. Both these procedures will tend to increase the probability that we will find "significant" differences due to chance (type I errors). Our reliance on two-tailed tests thus offsets this tendency, resulting in (we hope) a relatively constant probability (~ 0.05) of committing such errors.

2.3. TERMINOLOGY

Cooperative versus Communal Breeding

We refer to species (such as the acorn woodpecker and Florida scrub jay) in which more than two birds care for the young at a single nest as "cooperative breeders." There appears to have been no generally applied name for this phenomenon (other than the explicit definition of "helpers" provided by Skutch [1961]) until Lack (1968) discussed it at length in a chapter of *Ecological Adaptations for Breeding in Birds* titled "Co-operative Breeding." (The term "communism," used by Leach [1925] to describe group living in the acorn woodpecker, was never widely accepted, despite being used again by Skutch [1935].) Since that time, both "cooperative breeding" and "communal breeding" have generally been used synonymously. A brief review of the use of these terms in the literature is provided by Brown (1978).

Of these two terms, our preference is to use "cooperative breeding." We use "communal" as a generic term designating activities performed jointly by several individuals; i.e., "communal care of young" to refer to the aid that numerous acorn woodpeckers render to young, only some of which may be their own.

We have two reasons for this decision. First, we believe the definition of "cooperation" ("the act of working or operating together to one end; joint operation; concurrent effort or labor" [Webster's Unabridged Twentieth Century Dictionary, 2d ed.]) is appropriate for the phenomenon in question. Unfortunately, there has been considerable misunderstanding concerning the prior use of this term. Brown (1978), Rowley (1981), and Dow (1982), for example, justify their preference for using "communal breeding" on the erroneous basis that Hamilton (1964) preempted the term "cooperation" to designate a particular fitness relationship (specifically, the relationship in which both the recipient and donor benefit; see the fitness space diagrams presented in Brown and Brown [1981b] and Brown [1987]). In fact, he did not; rather, Hamilton (1964:44) used "cooperation" in precisely the traditional, nongenetic sense implied by the dictionary definition given above.

Second, the term "communal" has and continues to be used for a wide variety of diverse phenomena in the ornithological literature; thus, papers with "cooperative breeding" in the title

31

are generally, as expected, about species with helpers-at-the-nest, whereas those containing the words "communal breeding" may concern anything from species with helpers to those in which numerous pairs build a single structure for a nest.

The phenomenon of mate-sharing, that is, several birds of the same sex sharing breeding status and mates, has also been given numerous names in the literature. Among these are "wife-sharing" (Maynard Smith and Ridpath, 1972), "communal breeding" or "true communal breeding" (Vehrencamp, 1977; Koenig and Pitelka, 1979, 1981; Skutch 1985), "promiscuity" (Stacey, 1979b), "joint-nesting" (Brown, 1978), "cooperative polyandry" or "cooperative polygamy" (Faaborg and Patterson, 1981; Mumme et al., 1983a), and "polygynandry" (Brown, 1983; Koenig et al., 1984; Davies, 1985; Joste et al., 1985). Skutch (1935), the first to refer specifically to this phenomenon, used the terms "complete cooperation or communism." Neither of these terms has been used more than passingly since 1935, and thus we find little basis for choosing any one of the terms applied subsequently over the others besides personal preference and general uniformity. Here we refer to males sharing breeding status within a group as "mate-sharing" males and females laying eggs together in the same nest as "joint-nesting" females; either mate-sharing males or joint-nesting females may be referred to as "cobreeders."

Breeders and Helpers

Skutch (1961:220) defined "helpers" to include "a bird that assists in the nesting of an individual other than its mate, or feeds or otherwise attends a bird of whatever age that is neither its mate nor its dependent offspring." Hence, only the breeding male and female in acorn woodpecker groups containing no additional breeders could *not* be considered to be helpers, as otherwise both breeders and nonbreeders alike are presumably assisting in raising young that are not their own (as well as, in the case of parents, those that are their own).

32

PROCEDURES AND METHODS

In prior publications (e.g., Koenig et al., 1984), we have referred to the breeders within groups (whether "helpers" *sensu* Skutch [1961] or not) as "breeders," and nonbreeding helpers as "helpers." So as to reduce ambiguity, here we follow Skutch's definition more closely and divide groups into "breeders" and "nonbreeders" (or "nonbreeding helpers"). This terminology better reflects the functional classifications that are of greatest interest to the goals of this monograph; that is, whether individuals are or are not breeding. In general, we are *not* directly concerned here with whether breeders are feeding their own young or the young of others (which, because of the mating system, is not possible for us to determine in most cases; see section 2.2), or whether all nonbreeders are physically "helping at the nest." Such alternatives are, of course, relevant to a wide variety of questions concerning the evolution of helping behavior, but are outside the scope of this book and will be considered in detail in subsequent publications.

2.4. SUMMARY

The study site is located at Hastings Natural History Reservation in the Santa Lucia Mountains of Monterey County, California. Acorn woodpeckers are quite common in the habitats represented on the reservation, including foothill woodland, riparian woodland, savanna-grassland, and mixed broad-sclerophyll forest. Five species of oaks are common in the study area, although many groups have only three species readily available to them. The habitat is rugged but open, allowing relatively easy access to most granaries and territories.

Beginning with the work of the MacRobertses in 1971, acorn woodpeckers from 35 groups at Hastings have been caught and given unique color-band combinations to facilitate individual recognition. We also found all nests attempted by groups and banded all nestlings. Nests were exclusively in natural cavities.

33

During the period covered here, we banded over 800 birds and examined over 175 nests.

The primary field procedures consisted of banding all birds, following the histories of all nests, and censusing birds at irregular intervals. These procedures allowed us to follow the histories of a large number of individuals and to determine group composition. From these data, in turn, we calculated survivorship, reproductive success, and dispersal. Additional procedures assessed stored acorns, the acorn crop itself, flying insect biomass, and vegetation characteristics of territories. Experimental manipulations carried out during the period covered here consisted primarily of adding several artificial granaries and performing preliminary group manipulations on groups adjacent to the main study area.

The goal of this book is to present an integrated set of demographic and ecologic data for the acorn woodpecker, focusing on what these demographic processes can tell us about the evolution of cooperative breeding. Our study is a continuing one, however, and we thus make no promises about presenting definitive answers. The acorn woodpecker has surprised us in the past, and we trust that it will continue to do so in the future.

CHAPTER THREE

Overview of the Population

3.1. GRANARIES AND THE DISTRIBUTION OF TERRITORIES

The central feature of acorn woodpecker territories at Hastings Reservation and throughout much of this species' range is the storage tree or granary. With few exceptions, each territory contains at least one storage tree (the "primary" granary) and often one or more auxiliary granaries with fewer storage holes. Acorn woodpeckers drill small holes in granaries and store acorns that are harvested directly off trees in the fall and early winter (see Figure 4.1). Storage holes are reused year after year, and as a result, holes accumulate with time, eventually to be lost as the tree rots, burns, or falls. Numerous tree species have been recorded as granaries, including pines, oaks, firs, redwood, sycamore, and various exotics. Acorn woodpeckers also use utility poles, fence posts, and to the chagrin of many California homeowners, the wood trim of buildings.

At Hastings Reservation, where no pines are available, granaries are predominantly valley oaks, *Quercus lobata* (78% of 113 granaries tallied, MacRoberts and MacRoberts, 1976) and blue oaks, *Q. douglasii* (15% of granaries), followed by lesser numbers of sycamores, *Platanus racemosa* (4% of granaries), and willows, *Salix laevigata* (1% of granaries). Acorns of all available oak species are stored.

The importance to our population of granaries and the stored acorns they contain cannot be overemphasized. Other workers have noted apparent geographic or local changes in group size depending on both the acorn crop (Roberts, 1979a; Trail, 1980) and the local oak community (Bock and Bock, 1974; Roberts,

1979a). At Hastings Reservation, acorn crop failures also lead to marked changes in the social behavior within groups and ultimately to severe population crashes as birds exhaust their stores and abandon their territories (e.g., Hannon et al., 1987; see also Chapter 11). The acorn crop also has pronounced effects on reproductive success (see Chapters 4 to 6).

This strong dependence on acorns and granaries is one of the most characteristic features of the Hastings population and, based on the distribution of granaries, is apparently representative of a considerable fraction of the acorn woodpecker's range not only in California but in Mexico (e.g., MacRoberts and MacRoberts, 1976; Koenig and Williams, 1979), Central America (Skutch, 1969; Howell, 1972; Stacey, 1981), and Colombia (Miller, 1963). However, there are a growing number of observations of birds both in the United States and in Central America that may not use granaries and do not appear to store acorns or other similar kinds of nuts.

The most detailed study of such a population is that by Stacey and Bock (1978) in southeastern Arizona. Correlated with a lack of storing behavior, birds in this population usually live in monogamous pairs and are migratory. Such populations are probably not genetically distinct from more "typical" acorn-storing birds, as groups with large granaries are found only a few kilometers away from the migratory population. Nonetheless, the persistence of such populations indicates that the dependence of acorn woodpeckers on granaries is not obligatory.

More recently, Kattan (MS) has observed several populations of acorn woodpeckers in Colombia. One area studied in some detail, located at Finca Merenberg in the Cordillera Central, consists of groups of five to ten resident individuals despite an absence of granaries or acorn storage. Apparently other food sources provide a year-round supply without the necessity of acorn caching.

Additional information concerning this and other nonstoring but still group-living populations will be of considerable interest: their very existence suggests that granaries, which are a critical feature of the social behavior of acorn woodpeckers at Hastings (see Chapters 4, 11, and 12), are not an indispensable component of the system, at least under some ecological conditions. Obviously, an understanding of why group living in a nonstoring population occurs—despite the extreme ecological contrasts with birds in California—would be especially prized.

At Hastings, the acorns of all five common oak species (see Chapter 2) are used extensively; not all five are available on most territories, however. Besides the short-term storage of insects (MacRoberts, 1970), acorn woodpeckers also have been recorded storing pine nuts (MacRoberts and MacRoberts, 1976; Stacey and Jansma, 1977), cultivated nuts (Ritter, 1929, 1938; Bent, 1939), and various odd items such as Douglas fir (*Pseudotsuga menzeisii*) cones (W. D. Koenig, personal observation) and stones (e.g., Ritter, 1938). Neither pines nor cultivated nuts are available at Hastings. Stones, on the other hand, are plentiful, but we have never observed birds storing them.

Granaries serve as the focal point for both the activities of group members and the territory itself. Although territorial boundaries may vary slightly from year to year (such boundaries are often difficult to determine as territories frequently do not abut each other), granaries change ownership only rarely, and then do so abruptly following abandonment and subsequent recolonization. We are thus able to maintain long-term records on individual territories, defining them by their associated granary or granaries. As a result, the pattern of territory occupancy can be quantified directly by the proportion of territories occupied in any one year.

The distribution of the 35 territories we followed is presented in Figure 3.1. No more than 26 of these 35 territories were occupied simultaneously during the breeding season; the minimum number occupied at one time was 16. Most are either along or adjacent to watercourses or in old fields with scattered oaks on the tops of hills. Groups did not occur in chaparral or mixed deciduous forests (MacRoberts and MacRoberts, 1976;

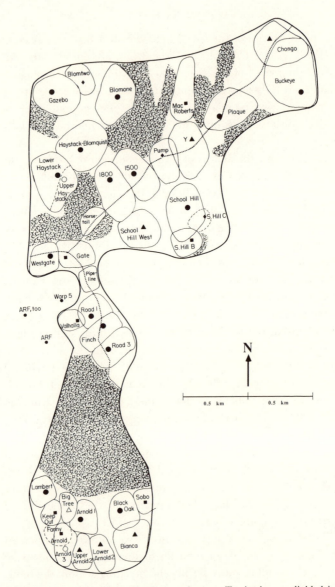

FIGURE 3.1. A map of the Hastings study area. Territories are divided into two groups: the larger set of Hastings groups and the smaller set of groups on the Arnold Field (located at the southern end of the map). Approximate territorial boundaries are shown; stippled areas are chaparral or mixed deciduous forest, habitats generally unsuitable for acorn woodpeckers. Symbols for territories represent territory quality (see Table 3.1). The total size of the study area is 300 ha.

these areas are marked with stippling in Figure 3.1). A brief introduction to the history of each focal group is presented in Table 3.1; we will refer to groups by the names used in this table and in Figure 3.1.

We categorized territory quality based on their occupancy record. Those occupied during all spring breeding seasons are listed as "type 1" territories. "Type 2" territories are those occupied at least 75% of breeding seasons but unoccupied at least one spring. "Type 3" territories, which we generally combine with type 2 territories for comparative purposes, were occupied at least one spring but unoccupied over 25% of the years they were watched. "Type 4" territories, of which there were only two, were never occupied during the spring. Neither of these type 4 territories contained more than a few storage holes, and they were occupied only temporarily. Excluding type 4 territories, as well as two others on which we erected artificial granaries, we followed 33 territories intensively during the study. Successful breeding is known to have occurred on 30 of these, is suspected in one additional territory, and is known not to have occurred (although attempts were made) on the final two territories.

The granary facilities of four territories (Upper Haystack, Arnold 3, Big Tree, and MacRoberts) were lost catastrophically during the study—two through fire and two by falling. These territories are considered to have existed only until the loss of their granary, and are unoccupied and ignored in the occupancy record after that time. This procedure follows from the emphasis we, and the birds, place on granaries as a resource of critical importance in defining territories. Finally, the occupancy record of group Fanny Arnold is considered to have started following the loss of the primary granary at Arnold 3. This was done because these two territories apparently could not exist contemporaneously, given the high degree of spatial overlap between them. For further details on these territories, see Chapter 11.

TABLE 3.1. Brief Synopsis of Acorn Woodpecker Focal Territories at Hastings Reservation

Group Name	Code[a]	Type[b]	Occupancy Record	First Observed	First[c] Censused
Buckeye	3	1	Continuous	Fall '73	Feb. '79
Chongo	1	2	Abandoned winter '73 to spring '74 and winter '78 to fall '81	Fall '72	
Plaque	2	1	Continuous	Fall '71	
MacRoberts	—	3	Not occupied fall '71 to fall '73; abandoned spring '74 to fall '74 and Jan. to Sept. '77; granary fell summer '78, not occupied subsequently	Fall '71	
Y	4	2	Abandoned fall '78 to spring '79 and summer '79 to fall '81	Fall '71	
Pump	5	4	Occupied winter '73–74. Artificial granary erected Mar. '76; reoccupied fall '77; granary fell Dec. '77, unoccupied subsequently	Fall '71	
1500	6–6a	1	Continuous	Fall '71	
1800	7–7a	1	Continuous	Fall '71	
Haystack-Blomquist	11	1	Continuous	Fall '72	Spring '74
Lower Haystack	10	1	Continuous	Fall '72	
Upper Haystack	9	1[d]	Occupied continuously spring '72 to summer '74. Granary fell summer '74, unoccupied subsequently	Spring '72	
Blomone	—	1	Continuous	Winter '78–79	
Blomtwo	—	2	Occupied all springs except '79	Spring '77	Spring '81
Gazebo	—	1	Continuous	Spring '77	Spring '79
Horsetail	—	—	Artificial granary	Sept. '78	
Gate	8–8a	1	Continuous	Dec. '71	
Westgate	—	3	Occupied fall '74 to fall '75 and Dec. '79 to Oct. '82	Dec. '71	

			Artificial granary	Sept. '78	
Pipeline	—				
Road 1	12	1	Abandoned summer '72 to spring '73 only	Summer '72	Spring '73
Valhalla	—	3	Occupied May to Dec. '79 only	Spring '75	
Finch	13	1	Abandoned summer '72 to spring '73 only	Summer '72	Spring '73
Road 3	14	1	Abandoned summer '72 to spring '73 only	Summer '72	Spring '75
School Hill	15–15c	1	Continuous	Feb. '72	
School Hill B	15a	3	Occupied Sept. to Nov. '72 and May to Dec. '75 only	Feb. '72	
School Hill C	15b	4	Occupied Sept. to Nov. '72 only	Feb. '72	
School Hill West	—	2	Abandoned summer '77 to spring '79	Spring '75	Spring '77
SOBO	—	3	First occupied winter '80–81	Summer '72	
Black Oak	18	1	Continuous	Summer '72	
Arnold 1	19	1	Continuous	Summer '72	
Big Tree	20	2	Abandoned Sept. '78 to summer '79, when the granary burned; not reoccupied	Summer '72	
Lambert	21	1	Abandoned Sept. '78 to Apr. '79 only	Summer '72	
Keep Out	—	3	First occupied fall '74; abandoned summer '78 to spring '82	Winter '72–73	
Arnold 3	22	1d	Continuous until granary burned summer '79; not reoccupied	Summer '72	
Fanny Arnold	—	2e	First occupied fall '80; reoccupied part of Arnold 3's old territory	Summer '79	
Upper Arnold 2	23	2	Abandoned summer '76 to spring '77 and summer '77 to fall '79	Summer '72	
Lower Arnold 2	—	2	First occupied fall '74; abandoned Jan. to Apr. '78 and June '78 to May '79	Summer '72	
Bianca	24	2	Abandoned Nov. '78 to May '79 and Aug. '79 to Oct. '81	Summer '72	

a Code is the group number assigned by MacRoberts and MacRoberts (1976), appendix 2.

b Type 1 = occupied all breeding seasons.
Type 2 = occupied 75 to 99% of breeding seasons.
Type 3 = occupied 1 to 74% of breeding seasons.
Type 4 = never used for breeding.

c Date of first census is same as that for first observed unless indicated.

d Type of territory determined by occupancy record until loss of granary.

e Type of territory determined by occupancy record starting from the dissolution of Arnold 3.

In all, 18 of the 35 territories (51%) were occupied continuously or at least during all breeding seasons during the study. Eight (23%) were occupied at least three-fourths but not all breeding seasons, and 7 (20%) were occupied at least one breeding season but less than three-fourths of the springs they were observed. Additional observations on the occupancy record of groups in the study area are given in section 3.4.

The temporal record of territory occupancy by groups, plotted on a quarterly basis, is shown in Figure 3.2. Plotted are the proportion of all territories occupied and the proportion of type 2 and 3 territories occupied. Occupancy of all territories ranges from a low of 57% in January 1979 to a high of 96% from October 1974 to October 1975. Occupancy of type 2 and 3 territories varies more dramatically, from a low of 0% in January 1979 to 92% in July–October 1975.

3.2. POPULATION SIZE

Residents

Our data allow us to estimate the total number of resident acorn woodpeckers in an area of approximately 300 ha on and adjacent to Hastings Reservation beginning in April 1973. These data are plotted at quarterly intervals in Figure 3.3. During this time, the maximum population size was 156 birds (following breeding in April 1975) and the minimum 48 (prior to breeding in April 1979). Population size during the breeding season ranged from 113 (1976) to 48 (1979). Thus, the population during the breeding season has varied by a factor of 2.4 during the 10 seasons on which we report here (coefficient of variation [hereafter C.V.] = 26%). This coefficient of variation is generally within the normal range of values calculated for a variety of avian populations and summarized in Ricklefs (1973); it is considerably greater than the 4% coefficient of variation reported in breeding population size for the cooperatively

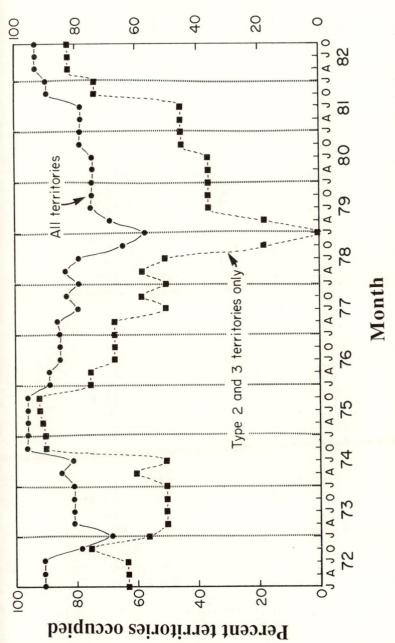

FIGURE 3.2. Territory occupancy for all territories and for territories not occupied all years (type 2 and 3 territories) plotted quarterly.

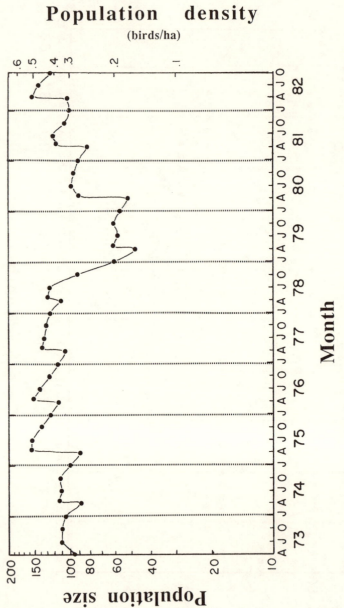

FIGURE 3.3. Population size and density plotted logarithmically at quarterly intervals, April 1973 to October 1982.

breeding Florida scrub jay by Woolfenden and Fitzpatrick (1984).

The larger fluctuations in population size observed during the study were the result of variation in both reproduction and survivorship and generally correlated with major variations in the size and quality of the acorn crop (but see section 4.6). For example, the rise in 1976 followed two excellent acorn crops in the autumns of 1974 and 1975, and the crash in the 1978–80 period followed a regional, near total crop failure in autumn 1978. This failure resulted in territory abandonment by a sizable proportion of the population during winter 1978–79 and poor reproductive success by the few remaining individuals in spring 1979. Further discussion of the relationship between population ecology and the acorn crop is presented in Chapters 4 and 11, as well as in Hannon et al. (1987). Overall, the population was static: 96 birds comprised the breeding population in 1973 compared to 103 in 1982 (a factor of 1.1 over the 1973 figure). Thus, demographic analyses spanning the years 1973 to 1982 are based on a population that increased very slightly during the course of the study.

Floaters

A potentially important component of the population not included in Figure 3.3 is that of nonresident floaters. Although almost all acorn woodpeckers are members of a group, there is usually a small number of nomadic individuals or floaters present in the population at any one time. However, in many cases it is ambiguous whether these nomads are actually home-based and merely engaged in extensive exploratory forays searching for reproductive vacancies, or whether they have in any real sense severed ties with their natal group. Although only birds in the latter category can be considered floaters in the traditional sense, birds in both categories are engaged in nomadic behavior and are thus, at least from the population standpoint, functionally similar.

We know of four ways in which individuals may become floaters. First, nonbreeders may become floaters when the breeder(s) of the same sex as a nonbreeding helper dies. Such birds usually neither inherit their natal territory nor breed under such circumstances, apparently because incest avoidance prevents them from breeding with their parent of the opposite sex (Koenig and Pitelka, 1979; Koenig et al., 1984). The reproductive vacancies are subsequently filled by unrelated immigrants, following which the group offspring may either leave or be forced from the group (Hannon et al., 1985).

With no other territory to which they can go, such birds become nomads, intruding onto a series of territories in search of a reproductive vacancy to fill. We have documented such floaters on four occasions ($\male$22, $\female$522, $\female$587, and $\male$584). They apparently may wander for at least several months before finally moving into another territory or, in the case of $\female$587, returning to their natal territory. The histories of these floaters are summarized in Table 3.2.

Second, floaters may be former resident breeders that have left or been forced out of their territory by immigrants. One such case occurred when group Fanny Arnold was reduced to a mother (A$\female$652) and two of her (formerly nonbreeding) sons. Rather than the nonbreeders leaving, as occurred in the four cases mentioned above, A$\female$652 left and floated for a period of two months before occupying a vacant territory nearby.

A second case of a breeder becoming a floater is provided by A$\female$478, of group Blomone, who was evicted by a coalition of three unbanded females that participated in a power struggle precipitated by the death of the male breeder in the group. A$\female$478 subsequently floated for a short time before moving to a nearby vacant territory. This example is the better of two cases of eviction that we have documented, but we have recorded at least one other instance (not recorded in Table 3.2) of resident breeders leaving their territory, which may have been the result of similar evictions. Such cases all involved females; the possibility of such eviction by potential immigrants may explain in

TABLE 3.2. Case Histories of Known Floaters

Bird	Time as a Floater	History
♂22	3 months	Left natal group (Y) after replacement of his father by four new ♂♂ in April 1976. Seen wandering at School Hill in May. Joined group Plaque in late July 1976.
♂571	3 months	Left natal group (Gazebo) following the replacement of his mother in April 1981. Seen at Lower Haystack (where he roosted several times) in May. Returned to Gazebo in early July 1981.
♂584	3 months	Evicted from his natal group (Blomone) by three immigrant ♂♂ in late April 1981. Seen wandering in various unoccupied areas. Joined a group adjacent to the study area in July 1981.
♀522	?	Left natal group (Plaque) following the replacement of her mother by two new ♀♀ in April 1980. Not seen subsequently.
♀587	5 months	Evicted from natal group (Lower Haystack) when three females replaced her mother in April 1981. Wandered for the next several months, and eventually returned to Lower Haystack in September 1981.
♀652	2 months	Left group Fanny Arnold, where she was a breeder, in September 1981 after her mate died, leaving her alone with two sons. Seen in a power struggle at Upper Arnold 2 and eventually moved to group Lambert in November 1981.
♀478	1 month	Evicted by a coalition of three ♀♀ in April 1981 during a power struggle precipitated by the disappearance of her mate. She wandered for several weeks and then occupied the nearby territory of Blomtwo in May 1981.

part the tendency for females (but not males) to participate in power struggles precipitated by the death of the opposite-sexed breeders in a group (Koenig, 1981a).

A third route to becoming a floater in an acorn woodpecker population involves individuals that leave their natal group without the incentive of having new unrelated birds of the same sex joining the group. ♂571 (Table 3.2) fits into this category, as he left his natal group following the replacement of his mother and wandered for several months before returning home. In addition, we have numerous examples of nonbreeders who leave their natal group for extended periods of time only to return at a later date, but whose activities in the interim are unknown. We suspect that some of these individuals filled reproductive vacancies and subsequently abandoned their new territory following an unsuccessful breeding attempt. However, it is likely that some wander for extensive periods of time, possibly even longer than the maximum of five months for the birds discussed in Table 3.2, in search of vacancies until they either return home, find a vacancy and join a new group, or die.

A fourth route to becoming a floater occurs during years of acorn crop failures. Individuals with no access to acorns in the autumn because of a crop failure generally abandon their territory. The fate of such birds is usually not known, but at least some become nomads until they are able to find an unoccupied area in which to settle. Usually such individuals are not detected until they do settle, at least temporarily, onto a vacant territory. Thus, the frequency of this phenomenon is not known. Because spatial variability in the acorn crop is often great, it is likely that at least some of the population over a wide geographic area finds itself in such a position in any one year, and in some years, such as autumn 1978 in central coastal California, crop failure may be widespread and the subsequent numbers of nomadic wandering acorn woodpeckers quite large. In years when the acorn crop fails locally, a substantial fraction of the population may even become "migratory," moving to areas where the crop is better and then returning to their original territories the following spring (MacRoberts and MacRoberts 1976; Hannon et al., 1987). Such behavior parallels that of the Arizona population studied by Stacey and Bock (1978).

Because of the difficulty in detecting or even recognizing floaters, we have no way of objectively estimating their absolute numbers. Many and perhaps all nonbreeders engage in at least some exploratory wandering (Mumme and de Queiroz, 1985).

48

It seems unlikely, however, that more than a small fraction (less than 5%) of the population could be individuals lacking a strong group affiliation, at least during the breeding season. In contrast, during years when the acorn crop is poor, a considerable fraction of the population may be forced out of their home groups to become nomadic floaters, at least until some interim settlement can be located (Hannon et al., 1987). In any case, the true population size in the study area is certainly slightly larger than that graphed in Figure 3.3 due to the existence of an indeterminate number of such individuals.

3.3. GROUP SIZE AND COMPOSITION

Acorn woodpecker groups consist of a variable number of individuals of both sexes and of several ages, as well as a variable number of breeders and nonbreeding helpers. With such a plethora of criteria with which to categorize group composition, we briefly summarize group characteristics, beginning with group size and working through the various ways that individuals within groups may be classified.

Overall Group Size

The maximum group size encountered among acorn woodpecker groups at Hastings during the study was 15 birds, achieved 5 times, generally following the fledging of large broods in the spring. Breeding groups contained between 2 and 12 adults. The distribution of group size during the 11 years of the study is graphed in Figure 3.4. Mean ($\pm$SD) group size was 4.4 $\pm$ 2.4 adults. The size of the group in which the average woodpecker lives during the breeding season ("typical group size" [Jarman, 1974]) was 5.7 $\pm$ 2.8 adults. The most common group size was a simple pair, whereas most birds lived in groups of 4. Nearly all groups, with only 5 known exceptions (2.4%),

49

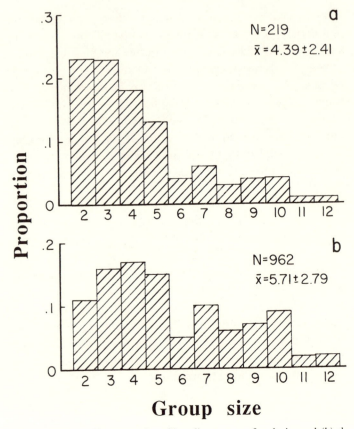

FIGURE 3.4. (a) The proportion of breeding groups of each size and (b) the proportion of individuals living in breeding groups of each size. Mean ± SD and sample sizes are given.

contained at least 1 male and 1 female at all times except for interim periods (which were usually quite short but in some cases lasted up to several months) between the death of the only male or female resident and its replacement by birds from outside the group; 4 of the 5 unisexual groups persisted through a breeding season. Case histories of these 5 groups are detailed in Table 3.3.

TABLE 3.3. History of Unisexual Groups

Group	Composition	Duration	History
Finch	A♂63 A♂66 A♂67 A♂282 F♂386	1 yr (1 BS)	A♀232, the breeder and only ♀ here in 1977, disappeared in April 1978, leaving numerous ♂♂. The group forewent breeding in 1978 and remained without a ♀ until April 1979, when A♀293 joined the group and bred.
School Hill	A♂86	7 months (0 BS)	During August 1978, the year of an acorn crop failure, all birds disappeared except for A♂86. He remained the sole group member until joined by A♀260 in April 1979.
Keep Out	F♂705 F♂706	3 months (1 BS)	These two siblings were evicted from their natal group (Fanny Arnold) in February 1982 and moved here in late March. They remained the only birds here until June, when they were joined by A♀567.
Upper Arnold 2	A♀128 A♀230	1 year + (1 BS)	When first observed in spring 1975, these birds were the only two present in this group. (Both had probably moved here in late summer 1974, but whether a ♂ was with them at that time is unknown.) A♀230 disappeared in September 1975, leaving A♀128 by herself until April 1976, when she was joined by F♂205.
Bianca	A♂123 A♂126	3 months (1 BS)	These two birds, probably siblings from Arnold 3, moved here in May 1979 and remained through the breeding season despite the lack of a ♀. Both abandoned in late August and founded a new group nearby (Upper Arnold 2).

NOTE: BS = breeding season.

51

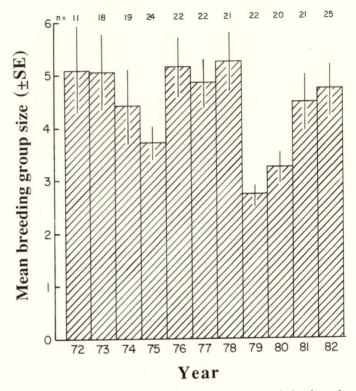

FIGURE 3.5. Breeding group size (mean ± SE) for each year during the study. Differences among years are significant (ANOVA, $F_{10,214} = 2.3$; $p < 0.05$).

Mean breeding group size varied significantly between 2.7 in 1979 and 5.2 in 1978 (ANOVA, $F_{10,214} = 2.3$, $p < 0.05$; Figure 3.5). Mean group size suffered markedly following the widespread crop failure in autumn 1978, but otherwise hovered in the range of 4.5 to 5.3 birds per breeding group. Also varying considerably was the proportion of groups consisting of a pair only; this percentage varied from a low of under 8% in 1972 and 1977 to a high of 50% in 1979, the year following the poor 1978 acorn crop. As shown below, annual variation in group size reflects changes in all fractions of the population, but especially that of the nonbreeders.

Breeding Status

The mean number of breeding males and females per group during the breeding season is 1.84 (C.V. = 54%) and 1.19 (37%), respectively. Interyear variation in the number of nonbreeders per group is even greater, with means (C.V.) of 0.70 (180%) for males and 0.49 (176%) for females. Thus, annual variation in average group composition is considerable, especially among nonbreeders.

Any particular group may contain up to 4 breeding males (rarely more) and up to 3 breeding females (usually 1 or 2). The frequency distribution of the number of male and female breeders per group is given in Figure 3.6. We will explore the consequences of mate-sharing and joint-nesting in Chapter 13; here we note only that the observed frequency of the number of breeding males and breeding females in groups during the

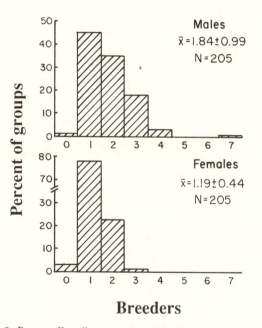

FIGURE 3.6. Percent of breeding groups containing a given number of breeding males and females. Mean ± SD and sample sizes are given.

TABLE 3.4. Observed (Expected) Percentage of Groups with a Given Number of Breeding Males and Females

Number of Breeding ♀♀	Number of Breeding ♂♂			
	1	2	3+	Total
1	40 (37)	25 (29)	13 (14)	78 (80)
2	5 (8)	9 (6)	7 (3)	21 (18)
3	0 (1)	0 (1)	0.5 (0)	0.5 (2)
Total	46 (47)	34 (35)	20 (17)	100

NOTE: Data presented are observed and expected percentages, excluding unisexual groups. Expected percentages are calculated using a Poisson distribution and the observed means for males and females independently, subtracting one in each case in order that observed values begin at 0 rather than 1. Total may not add up to 100% due to rounding errors. $N = 205$ groups.

breeding season is distributed independently of one another and that each follows a Poisson distribution (Table 3.4; for males, $\chi^2 = 4.3$, df = 3, ns; for females, $\chi^2 = 3.7$, df = 2, ns; expected values are calculated using a mean of one less than the observed values such that the "0" class corresponds to a single male or female). The only possible exceptions are that groups with two females and two or three males are slightly overrepresented, whereas those with two females and a single male are somewhat less frequent than expected.

Sex Ratio

The sex ratios of the breeding and nonbreeding fractions of the population are graphed in Figure 3.7. The sex ratio of breeders is fairly constant and is always biased toward males by about a three-to-two margin (mean percent males among breeders = 60.4). The sex ratio of nonbreeders is more variable, ranging between being all females (between July 1979 and January 1980, when only a single nonbreeder was present

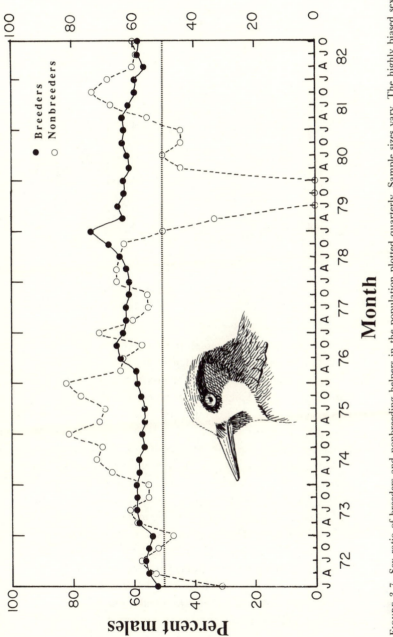

FIGURE 3.7. Sex ratio of breeders and nonbreeding helpers in the population plotted quarterly. Sample sizes vary. The highly biased sex ratio among nonbreeders in 1979 was due to the existence of only a single female nonbreeder at that time.

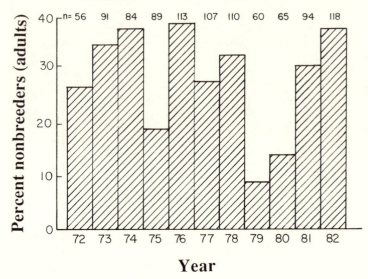

FIGURE 3.8. Percent of the adult population consisting of nonbreeders during the breeding season, 1972–82.

in the study area) and 81% males (mean percent males = 54.3). One reason for this slight, although insignificant, bias toward males among nonbreeders is the tendency of males to remain as nonbreeders slightly longer than females, a finding discussed further in Chapter 8. More important, the sex ratio of young produced appears to be slightly biased toward males, as discussed in section 8.2 when we consider the sex ratio of offspring.

Figure 3.8 graphs the proportion of adult nonbreeders in the population during the breeding season for each year. On average, 25% of adults in the population were nonbreeders. Including first-year birds, 39% of the population consisted of nonbreeding helpers. Thus, a substantial fraction of the population is not in a position, socially, to breed in any given year.

The percent of adult nonbreeders varies considerably among years, ranging from 9% in 1979 to 39% in 1976. Overall, 45% of groups contained at least one nonbreeder during the breeding season, and the mean number of nonbreeders per breed-

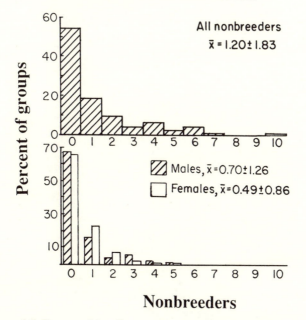

Nonbreeders

Figure 3.9. Percent of breeding groups containing the number of non-breeders given on the x-axis (top). Percent of breeding groups containing the number of nonbreeders of each sex given on the x-axis (bottom). Data are $\bar{x} \pm$ SD.

ing group was 1.20 (Figure 3.9). Male nonbreeders ($\bar{x} =$ 0.70 group^{-1}) outnumbered female nonbreeders ($\bar{x} = 0.49$ group^{-1}). However, the proportion of groups lacking a male nonbreeder is very similar to the proportion of groups lacking a female nonbreeder (Figure 3.9).

The observed frequency of nonbreeder males and females in groups is given in Table 3.5. In contrast to breeders, male and female nonbreeders are not distributed randomly compared to the corresponding Poisson distributions. For both sexes, there are more groups with no nonbreeders and fewer groups with 1 nonbreeder than expected. Groups with 2 male nonbreeders are also underrepresented, while those with 3 or more non-breeders are more frequent than expected. These differences are significant, particularly for males (lumping groups with 3

57

TABLE 3.5. Observed (Expected) Percentage of Groups with a Given Number of Nonbreeding Males and Females

Number of Nonbreeding ♀♀	Number of Nonbreeding ♂♂				
	0	1	2	3+	Total
0	54 (30)	10 (21)	1 (7)	3 (2)	68 (60)
1	10 (16)	7 (11)	1 (4)	4 (1)	22 (31)
2	2 (4)	1 (3)	2 (1)	2.5 (0)	8 (8)
3+	1.5 (1)	0.5 (0)	0 (0)	2.5 (0)	4.5 (1)
Total	67 (50)	16 (35)	5 (12)	12 (8)	100

NOTE: Data presented are observed and expected percentages, excluding unisexual groups (see Table 3.4). $N = 205$ groups.

or more nonbreeders of either sex: for males $\chi^2 = 87.2$, df = 3, $p < 0.001$; for females $\chi^2 = 18.5$, df = 3, $p < 0.001$). Thus, groups tend to either have no nonbreeders of a given sex or several nonbreeders of that sex more often than expected. The reasons for this are probably related to the coalitions of 2 to 4 siblings formed for dispersal (Koenig et al., 1984). Such sibling units are more successful at filling reproductive vacancies than singletons (Hannon et al., 1985); hence, assuming that nonbreeders are "trying" to disperse as soon as possible (Koenig, 1981a; see also Chapter 8), nonbreeders are likely to accumulate in groups until successful dispersal takes place, an event that becomes more likely as additional nonbreeders are available to form sibling coalitions. Successful dispersal then effectively eliminates all nonbreeders of that sex.

An alternative explanation for the tendency for groups to have nonbreeders of the same sex more frequently than expected is that groups adjust the sex ratio of young so as to result in unisexual broods. Although this particular pattern of sex ratio manipulation—whereby a higher proportion of unisexual broods would be produced than expected—has yet to be documented in any avian species, data from two species, the red-cockaded woodpecker and the green woodhoopoe, suggest that an overall bias in the sex ratio of nestlings produced by at least a subset of females may occur in some cooperative breeders (Gowaty and Lennartz, 1985; Ligon and Ligon, MS; see also Emlen et al., 1986). This mechanism has the advantage of directly endowing young with a sibling support group to aid in subsequent dispersal attempts. Given the possible fitness benefits of such a strategy, the sex ratio of offspring is worth investigating in detail, an analysis presented in Chapter 8. Here we simply note that we have no evidence that such sex ratio manipulation occurs, although our sex ratio data are not sufficient to completely rule out the possibility.

Territory Quality

Finally, group composition varies with the quality of the territory. Territories that are always occupied during the breeding season (type 1) are larger and have more breeding males and more nonbreeders than those that are intermittently occupied (types 2 and 3; Figure 3.10). This in part is because type 2 and 3 territories are, by definition, recolonized more frequently; since new groups rarely contain nonbreeders (Koenig et al., 1984), such territories tend to have fewer nonbreeders overall than groups on permanently occupied territories. The difference in the number of breeding males cannot be explained in this way, however, and may reflect a real difference in the tendency for male sibling units to fill reproductive vacancies in better territories. Such a pattern is suggested by the contests over breeding vacancies (Koenig, 1981a), which involve more contestants and last longer on territories with greater numbers of stored acorns (Hannon et al., 1985). Thus, power struggles on high-quality territories generally are won by large sibling units (Hannon et al., 1985).

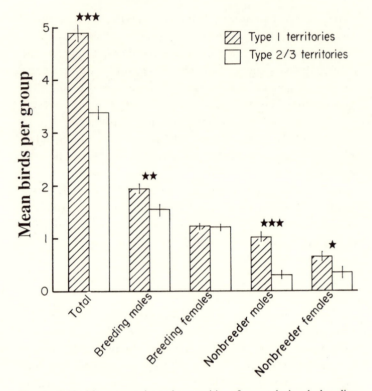

FIGURE 3.10. Mean group size and composition of groups during the breeding season of type 1 territories (those occupied all breeding seasons) and type 2 and 3 territories (unoccupied at least one breeding season). Data are $\bar{x} \pm$ SE. Starred comparisons are significantly different by a t-test; *** $p < 0.001$, ** $p < 0.01$, * $p < 0.05$.

3.4. DISCUSSION

A summary of the population and group characteristics described in this section is given in Table 3.6. Also listed in this table are several measures of annual variability around the mean values. Several points are worthy of discussion:

1. Territory occupancy is generally very high: 83.8% of all territories are occupied at any one time, and 18 of 35 territories (51%) were occupied all breeding seasons. In Figure 3.11, these

TABLE 3.6. Variation among Group and Population Characteristics

Variable	Mean $\pm$ SD[a]	C.V.	Individual[b]			Mean[b]	
			Min.	Max.	Max./Min.	Min.	Max.
Percent territory occupancy (all territories)	83.8 (74–92)	11	—	—	—	57	96
Percent territory occupancy (type 2/3 only)	58.4 (48–91)	25	—	—	—	0	92
Population size	104 $\pm$ 26	25	48	156	3.3	—	—
Population density (birds ha^{-1})	0.35 $\pm$ 0.09	25	0.16	0.52	3.3	—	—
Population size during the breeding season	88 $\pm$ 23	26	48	113	2.4	—	—
Group size during the breeding season (birds group^{-1})	4.39 $\pm$ 2.41	55	2	12	6	2.73	5.24
Percent of groups a pair only	23.5 (11–40)	35	9	50	—	—	—
Adult males per group	2.55 $\pm$ 1.67	65	0	8	—	1.71	3.33
Adult females per group	1.70 $\pm$ 1.05	62	0	7	—	1.10	2.45
Male breeders per group	1.84 $\pm$ 0.99	54	0	7	—	1.52	2.20
Female breeders per group	1.19 $\pm$ 0.44	37	0	3	—	0.71	1.40
Male nonbreeders per group	0.70 $\pm$ 1.26	180	0	6	—	0.00	1.27
Female nonbreeders per group	0.49 $\pm$ 0.86	176	0	5	—	0.05	0.94
Percent nonbreeders (adults only)	25.0 (15–37)	25	—	—	—	2	39
Percent nonbreeders (all, including first-year)	39.4 (28–51)	18	—	—	—	9	58
Percent males breeders)	60.4 $\pm$ 4.0	7	—	—	—	52	74
Percent males (nonbreeders)	54.3 $\pm$ 19.3	36	—	—	—	0	81

NOTE: Values are calculated at quarterly intervals except for breeding season data, which are taken from April, prior to the fledging of young.

[a] Means and standard deviations for percentages calculated by arcsin transformation.

[b] "Individual" values are for individual birds or individual groups; "mean" values are yearly means.

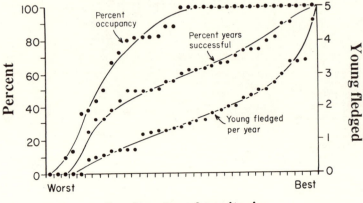

Rank order of territories

FIGURE 3.11. Habitat gradient slopes for acorn woodpeckers at Hastings Reservation, 1972–82. Plotted are the percent of years a territory was occupied, the percent of years individual territories were successful, and the mean number of young fledged per year by birds occupying territories. In all cases values are ranked from worst to best on the *x*-axis.

data are arranged in order of increasing number of breeding seasons each territory was occupied.

The shape of the percent occupancy curve in Figure 3.11 is reminiscent of that predicted by Koenig and Pitelka (1981) to be expected if group living is "forced" upon young by ecological constraints (Koenig and Pitelka, 1981; Emlen and Vehrencamp, 1983). The key to this hypothesis is the existence of relatively few "marginal" territories compared to "optimal" territories, thereby resulting in the production of many young compared to the marginal areas available for colonization. Under these conditions, young that have no place to move, except to areas where survival and reproduction are unlikely, will benefit by remaining as nonbreeders in their natal groups until a breeding opportunity elsewhere can be found. This hypothesis will be discussed in further detail in Chapter 12 when we consider the ecological factors selecting for delayed dispersal and delayed breeding. Figure 3.11 presents data relevant to the hypothesis using three measures of territory quality:

percent occupancy, percent years territories are successful, and the mean number of young fledged per year on individual territories.

Unfortunately, the data in Figure 3.11 cannot be construed as a test of the Koenig and Pitelka (1981) model. This is because the model is only useful in a comparative context, and there are few data from other species (ideally nonsocial and other social populations of woodpeckers) with which the data in Figure 3.11 may be compared.

With respect to the territory occupancy data, there is the additional problem of the time dependence of type 1 territories as defined here. Not surprisingly, the proportion of territories always occupied declines with time (Table 3.7); ultimately, long-term changes in habitat, granaries, and food resources result in the elimination of all type 1 territories. Thus, as reported by MacRoberts and MacRoberts (1976), all twenty-one territories they watched were occupied all three years of their study; however, after the eleven years considered here, the proportion of type 1 territories has declined significantly to only 55% of all territories occupied at least one season.

TABLE 3.7. Occupancy Record of Territories at Hastings and at Water Canyon, New Mexico

	Territories Always Occupied	Total Territories	Percent Always Occupied	χ^2
Hastings (11 years)	18	33	55	12.2***
Water Canyon (11 years)	2	26	8	
Hastings (first 3 years)	21	21	100	7.0**
Water Canyon (first 3 years)	12	19	63	

NOTE: Water Canyon data from P. Stacey (personal communication). Includes only territories occupied at least one spring.

TABLE 3.8. Spearman Rank Correlations of Group and Population Characteristics Using Mean Annual Data

	Breeding Population Size (10)	Percent Territories Occupied (11)	Mean Group Size (11)	Percent Nonbreeders in Pop. (11)	Breeders per Group (11)	Nonbreeders per Group (11)	Sex Ratio Breeders (11)	Sex Ratio Nonbreeders (11)	Percent Groups with 2 Breeding ♀♀ (7)
GROUP AND POPULATION CHARACTERISTICS									
Percent territories occupied (11)	.65*								
Mean group size (11)	.90***	.35							
Percent nonbreeders in population (11)	.66*	.35	.59						
Breeders per group (11)	.72*	.36	.74**	.13					
Nonbreeders per group (11)	.74*	.25	.84***	.82**	.35				
Sex ratio of breeders (11)	−.27	−.76**	−.19	−.24	.03	−.25			
Sex ratio of nonbreeders (11)	.56	.62*	.30	.53	.24	.45	−.33		
Percent groups with 2 breeding ♀♀ (7)	.99***	.77*	.94***	.79*	.78*	.79*	−.46	.92**	
Reproductive success (fledglings group^{-1}) (11)	.04	.45	−.17	−.08	.15	−.25	−.25	.05	.09
Reproductive success prior year (10)	.60	.28	.64*	.50	.61	.47	.19	−.19	.63

NOTE: Values in parentheses are the number of years of data for that variable. Sample size is the lesser of the two numbers for the variables being tested.

* $p < .05$; ** $p < .01$; *** $p < .001$.

Also shown in Table 3.7 are data on territory occupancy at Water Canyon, New Mexico, kindly provided by Peter Stacey. The Water Canyon population had a significantly lower proportion of permanently occupied territories than Hastings after both three years (Stacey, 1979a) and eleven years. This difference suggests that habitat saturation is greater at Hastings, a conclusion supported by the smaller group size and greater degree of juvenile dispersal at Water Canyon (Stacey, 1979a).

2. Total population size varied by a factor of 3.3 during the study, and breeding population size by a factor of 2.4. The coefficient of variation of population size was 25%, generally within the range of the species reviewed by Ricklefs (1973) with the exception of two populations of European tits, the willow ptarmigan, and the white stork, whose coefficients of variation for population size were between 40% and 50%.

Population size of acorn woodpeckers within the study area is a function of mean group size and the proportion of territories occupied. Using data from each breeding season, there is no significant correlation between these two components of population size ($r_s = 0.35$, $N = 11$, ns; Table 3.8). Both, however, are significantly correlated with total population size (mean group size: $r_s = 0.90$, $N = 10$, $p < 0.001$; proportion of territories occupied: $r_s = 0.65$, $N = 10$, $p < 0.05$; Figure 3.12). Controlling for the number of territories occupied, the partial correlation between breeding population size and mean group size (partial $r = 0.96$, $N = 7$, $p < 0.001$) is greater than that between breeding population size and number of territories occupied controlling for mean group size (partial $r = 0.86$, $N = 7$, $p = 0.003$). This suggests that mean group size explains a slightly larger proportion of variance in population size than does territory occupancy.

Other correlations between group and population characteristics are also given in Table 3.8. These clarify some of the correlated changes occurring in population size and group composition. For example, population size during the breeding

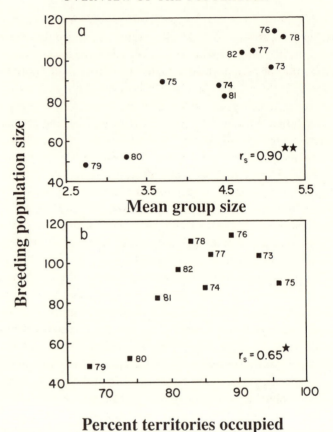

FIGURE 3.12. Total breeding population size plotted against (a) mean group size and (b) the proportion of territories occupied in the study area.

season correlates significantly and positively with the proportion of nonbreeders as well as with both the number of breeders (including the frequency of joint-nesting) and nonbreeders per group. There is no overall correlation between population size and reproductive success during either the same year or the prior year (but see section 5.3).

3. All components of acorn woodpecker group composition vary considerably among years (interyear C.V.'s range from

37 to 180%), much more so than does total population size (C.V. = 25%). Variability in the number of nonbreeders per group is especially great (C.V. = 176 to 180%), whereas the number of breeders per group varies much less (C.V. = 37 to 54%).

4. The sex ratio of the adult population is consistently biased toward males by a three-to-two margin. Considering the breeding season data only, breeding males outnumbered breeding females in all 11 years (binomial test, $p < 0.001$), significantly so in 3 of these years. Overall, 404 male breeder-years were recorded during the study compared to only 274 female breeder-years (59.6% males: binomial test, $p < 0.001$). The sex ratio among nonbreeders is also male biased, but is not as pronounced. The number of nonbreeding males outnumbered females in 9 of 11 years (binomial test, $p = 0.065$); there were no significant individual-year differences. Overall, there were 183 male nonbreeder-years compared to 124 female nonbreeder-years (59.6% males, $p < 0.001$); this measure thus yields a sex ratio identical to that of breeders.

5. Slightly less than half (45%) of breeding groups contain nonbreeders. Groups may contain up to 10 nonbreeders, but we did not observe more than 5 of the same sex within a given group. The mean number of nonbreeders per group is 1.2.

Our goal in this chapter has been to provide a basic introduction to the size and composition of the Hastings population. These data form an important basis for virtually all of the analyses in subsequent chapters. Thus, we will frequently refer to many of the details covered here and will summarize our ideas concerning the evolutionary significance of group size and composition in Chapters 12 to 14.

3.5. SUMMARY

Acorn woodpecker groups in California almost always contain at least one granary in which birds store acorns during the autumn. Granaries provide a focus for group activities and define individual territories that may be followed for a considerable length of time.

Although the Hastings population neither increased nor decreased between the start and end of the 11-year period considered here, overall population size varied by a factor of 2.4, largely in apparent response to the autumn acorn crop. Demographic responses to poor acorn crops included poor reproductive success the subsequent spring, as well as temporary or permanent abandonment of territories during the winter. The vast majority of birds are territorial residents, but floaters arise in at least four ways. The fraction of the population made up of floaters is unknown. It is, however, almost certainly small, except in winters following acorn crop failure. During such winters many birds are forced to abandon their territories and wander in search of food.

Acorn woodpecker groups at Hastings consist of a variable number of breeders and nonbreeders of both sexes. Maximum total group size at Hastings is 15 birds and mean group size during the breeding season is 4.4 ± 2.4; nonetheless, modal group size consists of a monogamous breeding pair with no nonbreeders. With very few exceptions, groups always contain individuals of both sexes. Annual fluctuations in size are relatively great in all segments of the population including breeders and adults of both sexes, first-year birds, and especially nonbreeders.

Groups may contain up to 4 breeding males (only rarely more), 3 breeding females, and 10 nonbreeders, but few groups ($<2\%$) contain more than 4 male breeders, 2 female breeders, or 6 nonbreeders. The sex composition of both breeders and nonbreeders is about 60% males. The number of male and female breeders in groups is closely approximated by a Poisson distribution. However, groups with no nonbreeders, and those with 3 or more nonbreeders of each sex, are more frequent than expected. This may result from greater success of acorn woodpeckers that disperse in coalitions of same-sexed siblings.

Slightly over 50% of all territories were occupied all eleven springs during the study. Assuming that territory occupancy is indicative of quality, this suggests that a relatively high proportion of territories in the population are of high quality. Unfortunately, lack of comparable data on other species makes it difficult to evaluate these data vis-à-vis the evolution of group living in the acorn woodpecker. Nonetheless, the pattern of territory occupancy suggests a relatively high degree of "habitat saturation," an ecological factor considered to be important to the evolution of cooperative breeding.

CHAPTER FOUR

Ecological Consequences of
Acorn Storage

*The [acorn] woodpecker is our native aristocrat. He is unruffled
by the operations of the human plebs in whatever disguise. Digger
Indians, Don Joses, or Doctors of Philosophy are all the same to
him. Wigwams, haciendas, or university halls, what matter such
frivolities, if only one may go calmly on with the main business
of life, which is indubitably the hoarding of acorns.*

W. L. Dawson, 1923

4.1. HISTORICAL PERSPECTIVE

Although the social behavior of acorn woodpeckers has been the
primary focus of recent interest in this species, their spectacular
acorn-storing habits (Figure 4.1) were recognized far earlier.
One early record comes from the Spanish explorer Herrera,
visiting Chiapas, Mexico, near the end of the sixteenth century.
He wrote:

> There are some thrushes called carpenter birds, black
> with a little red in the head and breast. They eat but
> acorns and a few dig into the bark of great pines and
> place an acorn in each hole, so tight that they cannot
> be removed by hand, and one sees many pines cobbled
> with acorns from top to bottom, one above others by
> admirable order; and they sustain themselves on the
> acorns with this prevention, and eat these acorns fasten-
> ing themselves onto the pine bark, and hammering with
> the beak. (Herrera, 1601–1615; original in Spanish)

Thus, written accounts of this aspect of acorn woodpecker
biology predate the first mention of its group-living habits

70

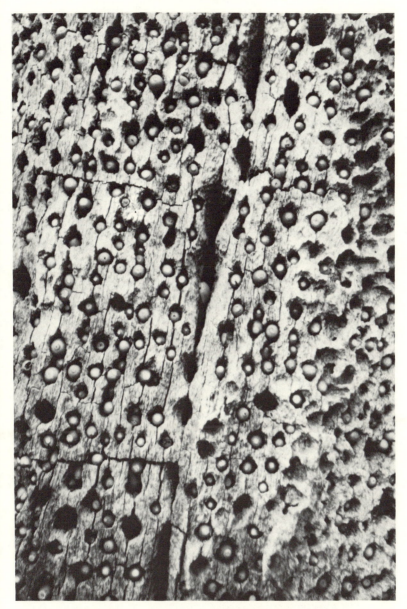

FIGURE 4.1. Detail of a granary filled with acorns.

(Henshaw, 1876) by over 250 years. Later explorers of California also frequently noted the acorn-storing habits of these birds. For example, Pedro Font, traveling through central coastal California in the latter part of the eighteenth century, mentions acorn storage by the "carpenter" birds (Bolton, 1933; quoted in Rossi, 1979).

Discussion of acorn storage took several interesting twists in the latter half of the nineteenth and the early twentieth centuries. Numerous workers, beginning with Jackson (1866), believed that the birds ate the insect larvae that often parasitized the acorns rather than the acorns themselves. The history of this idea, known fondly to acorn woodpecker biologists as the "grub theory," is entertainingly reviewed by MacRoberts (1974). Despite being discredited by Bendire (1895) and especially by Beal's (1911) dietary analyses, it continues to surface to this day.

A second concern about acorn storage that arose during this time involved its ecological significance. It was well known that acorn woodpeckers might store tens of thousands of acorns in a single tree, and that acorns make up a considerable portion of its diet. Nonetheless, it was generally considered unlikely that they could make use of them all (e.g., Dawson, 1923). Dawson's eloquent prose exemplifies the disbelief that some early naturalists had concerning the value of these communal caches:

> But, again, why does the bird hoard treasure on this lavish, irrational scale? For exercise? Perhaps. To be doing something—for the same reason that a high school girl chews gum or a callow youth sucks cigarettes, a matron does embroidery, or a middle-aged gentleman of increasing girth trots after a twinkling white ball—to kill time. Possibly, also, from force of habit. Following the blind urge of a provident instinct, the bird over-shoots the mark. Having no accurate criterion of judgment, or inhibitive power, it just goes on forever, *working*.

It is not impossible, of course, that some ancestral experience of drought and famine has fastened this lesson of
providence deep in the[ir] race; but it is much more probable that the species is hipped, and that it applies no more
reason to its life than does an old miser who goes on hoarding gold. Be this as it may, the explanation what you will,
the obsession of the California Woodpecker is undoubtedly one of the most pathetic things in nature. (Dawson,
1923:1029)

As it turns out, Dawson's conjecture was premature: acorns
appear to be important to virtually everything about acorn
woodpeckers. Of course, it is now accepted that ecological factors, including food supply, have a major influence on social
organization in general (e.g., Crook, 1965; Lack 1968; Emlen
and Oring, 1977).

Acorn woodpeckers provide a unique opportunity to examine the influence of food on a cooperative breeder since their
method of acorn storage permits the direct quantification of
their major food reserve. Thus, our goals in this chapter are: (1)
to briefly describe the pattern of acorn storage in the Hastings
population, (2) to discuss several important aspects of stored
acorns, and (3) to assess the energetic importance of stored
acorns. We will defer a discussion of the relation of these findings
to the social behavior of this species until Chapter 14, when we
will summarize evidence from our analyses bearing on the evolution of group living and cooperative breeding. Further details
of acorn storage can be found in MacRoberts and MacRoberts
(1976).

4.2. FOOD OF ACORN WOODPECKERS

At Hastings, acorns are stored in the autumn, usually between
September and December, as they mature on the trees. Stored
acorns are eaten until either the new acorn crop begins to

mature late the following summer or they run out. The relative importance of acorns as well as alternative food resources during the annual cycle is presented in Figure 4.2.

Several points are worth noting in Figure 4.2: First, immature green acorns are often not stored, but eaten directly from the trees starting in late summer until no additional acorns are present. Second, insects are captured (primarily by flycatching but also by bark gleaning) whenever weather is suitable for insect activity; they are a major source of food during the breeding season, especially for nestlings. Sap, buds, and catkins become available in January and may be used extensively into the early spring. Sapsucking drops off at this time, but again becomes prevalent in midsummer, especially once the stores of a group have been exhausted. Stored acorns are eaten at all times of the year except in late summer and early autumn, when acorns are

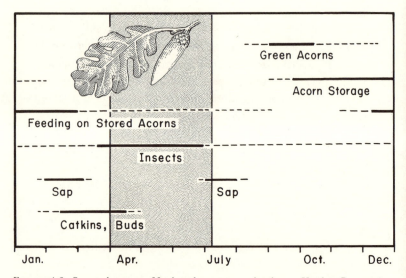

FIGURE 4.2. Seasonal pattern of food use by acorn woodpeckers at Hastings Reservation, modified from MacRoberts and MacRoberts (1976). A dotted line indicates that use of a resource is irregular depending on the year and the availability of alternative food. The main breeding season (April through early July) is stippled; during good acorn years second nests may extend into late July and autumn nests may begin in mid-August.

74

usually eaten directly off trees. However, as we discuss below, the number of stored acorns varies considerably from year to year, and stores in some groups may be exhausted well before the breeding season begins in April.

4.3. DYNAMICS OF ACORN STORAGE

Methods.

In order to quantify acorn storage, we counted stores at approximately bimonthly intervals using a hand counter and binoculars and annually sampled the stored acorns from granaries (Table 4.1). Sampled acorns were identified to species, assessed for insect damage, dried, and weighed. These samples were taken for only the last six of the seven years for which data are presented here.

Because some stored acorns are hidden in crevices, this procedure almost certainly underestimates their true numbers. However, careful examination of several limbs and trees suggests a possible error of 10 to 20%. This degree of underestimation is not sufficient to alter our conclusions.

Results

Figure 4.3 presents the total number of acorns stored in the granaries surveyed during seven years beginning in September

TABLE 4.1. Summary of Procedures Used to Sample Stored Acorns

1. Count stored acorns directly every 2 to 3 months
2. Remove a random sample of 25 to 100 acorns in December or January from each group
3. Freeze the acorns (when this was done varied among years)
4. Separate acorns by species; dry in drying oven
5. Shell the acorns; identify and separate those that are insect damaged
6. Weigh each intact acorn separately; weigh insect-damaged acorns as a group by species

(the beginning of the "acorn fiscal year"). There were 24 potentially active groups included in these surveys; the actual number that were used in any one year varied from 22 in 1974–75 to 15 in 1978–79.

The curves shown in Figure 4.3 reveal several patterns. First, there appears to be an upper limit to the number of acorns stored by the population at about 37,000 (the actual maximum counts were 36,846 in December 1975 and 36,670 in December 1977). This maximum was reached (totals between 34,000 and 37,000) in four consecutive years (1974–75 to 1977–78). Totals were somewhat lower in two of the years (1979–80 maximum = 25,800; 1980–81 maximum = 29,400) and were markedly lower only during the poor acorn year of 1978–79 (maximum = 9,950).

The second interesting pattern in Figure 4.3 is that the rates at which stores diminish varied considerably even during years

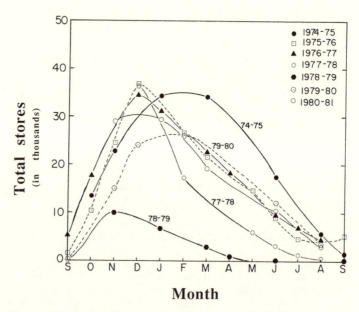

FIGURE 4.3. Total maximum number of acorns stored by the Hastings population during seven years.

when the maximum number of acorns stored was similar. For example, in 1974–75, the number of stored acorns peaked in January and only began to decline about March, very close to the beginning of the breeding season, which begins in April (see section 5.1). The more common pattern, exemplified by 1975–76, yields a peak in December, following which numbers drop off through spring and summer. In 1977–78, however, stores declined far faster than average and were nearly exhausted by spring, despite a peak as high as it had ever been during the survey period. Thus, there is considerable variability in the rate stored acorns are used independent of the maximum number of acorns stored.

Further understanding of the year-to-year pattern of acorn storage can be gleaned from Figure 4.4, which shows the average number of acorns stored per bird during each of the seven years. Note that the maximum number of acorns stored per bird is

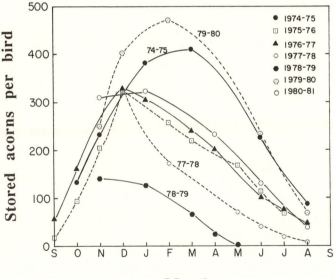

FIGURE 4.4. Number of stored acorns per bird during seven years based on the population size in January.

77

virtually identical (between 322 and 327) for four of the seven years. The maximum number of acorns stored per bird was 470 in 1979–80, even though this was one of the "medium" years in terms of total number of acorns stored. Acorns stored per bird was also higher than normal in 1974–75. The only year this measure was below the 320 range was in 1978–79, a poor crop year when several of the groups in the study area abandoned their territories.

4.4. WHAT LIMITS ACORN STORAGE?

Presumably it is to a group's advantage to store as many acorns as possible. What, then, limits acorn storage? There are at least four possibilities: the acorn crop, the time or ability of acorn woodpeckers to find and store acorns, interspecific competition for acorns, and storage facilities.

Despite its intuitive appeal, the first of these is the least likely. As discussed above, the maximum number of acorns stored is about 37,000. The area included in this analysis is 300 ha, about half (150 ha) consisting of oak canopy. Thus, a maximum of 120 acorns ha^{-1} (248 acorns ha^{-1} below oak canopy) are stored by the Hastings population.

This quantity is almost certainly small. Consider the productivity of four valley oaks on the top of School Hill (see Figure 3.1) sampled each year with 0.25 m^2 traps set by J. R. Griffin. During the 16 years between 1970 and 1985, the acorn fall below these four trees averaged 42 acorns m^{-2}. Multiplying the acorns caught under each of the four trees by their estimated canopy area yields an estimated median annual productivity of 16,700 acorns—nearly half the number stored by the entire Hastings population—for these four trees alone. In only three of the years (19%) was the estimated acorn fall below these four trees under 1,000 acorns. It is therefore probable that the study area as a whole, encompassing an estimated 55,000 oaks (based on our vegetation sampling; see section 4.9), pro-

78

duces several orders of magnitude more acorns than are stored by the woodpecker population in all but very poor years.

In good acorn years the difference between acorn storage and production is striking. In 1979, for example, the four valley oaks on School Hill produced an estimated 190,000 acorns, over five times the number stored by the entire population. Considering that none of the above estimates includes acorns removed by woodpeckers or other animals prior to maturity, it is probable that only in the worst years do acorn woodpeckers store more than a small fraction of the total acorn crop.

Additional evidence against the acorn limitation hypothesis comes from the weak, insignificant correlation between the size of the acorn crop and reproductive success, variables that should be strongly correlated if storage is limited by crop size. However, significant correlations exist between reproductive success and lateness of the crop and quality of the stored acorns (see section 4.6).

It is similarly unlikely that the number of acorns stored is limited by the length of time birds have to harvest and store them. The overall mean ($\pm$SD) rate of acorn storage in the autumn is 2.48 acorns hour^{-1} for breeders and 1.51 acorns hour^{-1} for nonbreeders (Mumme and de Queiroz, 1985). At these rates, an average-sized group of three breeders and two nonbreeders storing acorns 10 hours a day during the autumn could fill a large granary of 4,000 holes in 39 days, and even a pair of birds could fill an average granary of 2,000 holes in 41 days. This period of time is less than the normal duration of acorn availability, which generally starts in September and runs at least through November, and in some years much later.

We do not know to what extent this rate of acorn storage taxes the daily time or energy budgets of the birds. However, evidence from individual groups suggests that acorn storage is generally unrelated to group size: of sixteen groups for which we have data from five or more years, group size was positively correlated with the maximum number of acorns stored in nine, negatively correlated in six, and completely unrelated in one.

79

In only two of these groups was the correlation between group size and acorns stored significant, one positively and one negatively.

The third factor potentially constraining acorn storage is competition from other species. Since acorn woodpeckers harvest acorns from the canopy, competitors include squirrels (primarily California ground squirrels, *Spermophilus beecheyi*) and several species of corvids (yellow-billed magpies, scrub jays, and common crows). All these species, but particularly jays and crows, may take significant numbers of acorns from some territories in some years. In years of low acorn production, competition with these species may be significant. However, in fair to good years, the relatively large number of acorns being produced makes it unlikely that competition from these species represents a significant constraint on acorn storage for acorn woodpeckers in our population.

The remaining factor that might limit the number of acorns stored is the available storage space. Besides the conclusion that only a small proportion of the acorn crop is stored by the woodpeckers, several lines of evidence—direct, experimental, and anecdotal—support this hypothesis:

1. Upper limit to stored acorns. There is apparently an upper limit to the number of acorns stored by the population, as suggested by Figure 4.3. This limit is probably set by the number of holes available in which to store acorns, as granaries are usually filled each year, regardless of the size of the highly variable acorn crop. Alternatively, the number could reflect a relatively constant acorn crop (despite knowledge to the contrary), a constant amount of time available for storing acorns, or both. We consider these alternatives unlikely, but cannot disprove them.

2. Mass acorn storage. Some acorn woodpeckers are known to have huge storage facilities. Dawson (1923), for example, reported trees containing 20,000 to 50,000 acorns—as large or larger than the entire storage facilities of the Hastings study area combined. Unfortunately, a group with such a large granary has never been studied, and it is not known whether this many acorns can be stored in a single season.

However, there exist numerous anecdotal reports of acorn woodpeckers storing vast numbers of acorns, often in man-made structures where the acorns subsequently became unretrievable. Ritter (1938), for example, records a total of 62,264 acorns stored in the door and window casings of an unused house in the Sierra foothills. A recent example, reported in *Time* magazine (1984), concerns a group of acorn woodpeckers that, apparently in a single season, stored about 220 kg of acorns in a wooden water tank near Flagstaff, Arizona. Such examples show that sometimes acorn woodpeckers are able to find and store many more acorns than available storage facilities can accommodate. The use of odd storage sites, such as the Coulter pine (*Pinus coulteri*) cones reported by Jehl (1979), further supports the hypothesis that storage space, rather than the acorns themselves, may be limiting.

3. Experimental removal of stored acorns. Several groups at Hastings sometimes use old roost holes for acorn storage. Between 21 October and 13 November 1982 we took advantage of this habit at group Plaque by removing, at frequent intervals, the stored acorns cached in an old roost hole.

The results are shown in Table 4.2. During the 22 days of the experiment 2,534 acorns were stored in the hole, excluding those present initially. Overall storage rate was 115 group^{-1} day^{-1}, or 19 bird^{-1} day^{-1}. Meanwhile, the number of stored acorns counted in the remaining storage facilities increased from 1,266 to 1,657; the combined storage facilities (including the old roost hole) of this group were estimated to hold 2,300 acorns. Thus, the number of acorns stored in the granary was limited by storage facilities.

If granaries limit acorn storage, then as long as the population is relatively constant and storage facilities change slowly, it follows that the same number of acorns per bird should be stored each year. This was the case in four of seven years (Figure 4.4). Only 1978–79 stands out as a year when the birds

TABLE 4.2. Record of Stored Acorns Taken from an Old Roost Hole at Group Plaque, 21 October to 13 November 1982

| | Acorns Removed | | | | Storage Rate per | | |
Date	agrifolia	lobata	douglasii	Total	Day	Bird-day	Mass (kg, wet)
21 October	13	304	0	317	—	—	—
24 October	13	287	0	300	100	17	1.87
27 October	77	173	0	250	83	14	1.41
29 October	84	210	3	297	149	25	1.80
1 November	95	175	40	310	155	26	1.77
2 November	154	144	21	319	319	53	1.76
3 November	247	73	20	340	340	57	1.64
10 November	143	160	31	334	48	8	1.77
12 November	43	10	8	61	30	5	0.31
13 November	210	88	25	323	323	54	1.81
Total	1079	1624	148	2851	115	19	14.14

NOTE: Plaque was composed of 4 breeders and 2 first-year nonbreeders.

apparently could not obtain enough acorns to fill their granaries. The high per capita values for 1979–80 appear to be a result of the small population that year, which was in turn due to territory abandonment and poor reproductive success following the poor 1978–79 crop (Chapter 3). Similarly, the high per capita values of 1974–75 may have been largely due to the below average population that winter (Figure 3.3). Thus, there appears to be a density-dependent relationship between population size and acorns stored per bird limited by the granary facilities.

At some level acorn storage must be limited by the availability of acorns. However, the preceding evidence points to the importance of granaries, independent of the acorns stored in them, as a critical resource to acorn woodpeckers.

The granary-limitation hypothesis also raises the following question: if storage facilities limit acorn storage and stored acorns are important to survivorship and reproductive success, why do acorn woodpeckers fail to drill more holes? The answer is that they do drill more holes. However, storage holes eventually rot and crumble to the point of being unusable. Furthermore, holes are frequently placed in dead limbs or snags, which are particularly susceptible to loss or damage. It is not at all unusual for limbs containing many hundreds or even a thousand or more storage holes to fall during storms; between 1975 and 1982, we estimated that at least 12,000 holes were lost to natural causes from a total of eleven different territories, including the entire storage facilities of three territories (MacRoberts, Upper Haystack, and Big Tree). This represents a loss of about 1,700 holes year^{-1}, or nearly 5% of the total storage facilities of the population each year.

Unfortunately, it is extremely difficult to quantify the rate at which new holes are drilled. We suspect, however, that the woodpeckers are creating holes as fast as possible given the time necessary to drill one (probably about an hour), and perhaps more important, the presence of suitable substrate in which to put them. Thus, it is likely that available storage facilities are set by a balance between the rate that new holes can be created and that old holes are lost or otherwise rendered unusable.

4.5. ENERGETIC VALUE OF ACORNS AND ACORN STORES

Species Composition of Stored Acorns

We sampled acorns from granaries during six years in order to estimate the species composition and energetic content of stored acorns (see Table 4.1). Figure 4.5 presents the estimated species

All samples

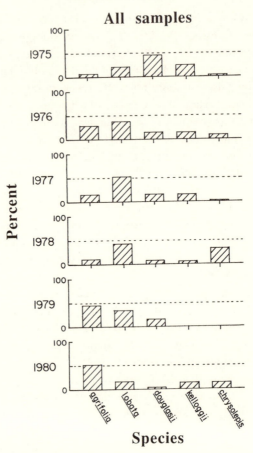

FIGURE 4.5. Estimated species composition of acorns stored during six years, 1975–76 through 1980–81, based on samples taken from granaries. Interyear differences are highly significant by ANOVA ($F_{5,20} = 4.12, p < 0.01$).

breakdown of acorn stores for each year. Specific composition varies from year to year, as indicated by a two-way ANOVA in which the effect of the year is significant ($F_{5,20} = 4.1, p < 0.01$), but that of the species is not ($F_{4,20} = 0.9$, ns). Note that three species alternated being the most abundant, and that each species except *Q. lobata* was nearly absent in at least one year.

Size and Composition of Acorns

We did not attempt to determine annual or intraspecific variation in nutritional composition. We did, however, obtain at least a rough index of interspecific variation by analyzing three acorns for each of the five species found on the study site. These analyses, performed by B. Oskroba of Colorado State University, measured crude protein by the micro-Kjeldahl method, nonstructural carbohydrates by digestion in 0.2 m NH_2SO_4 (Heinze and Murneck, 1940; Smith et al., 1964), lipids by extraction in SkellySolve F, and soluble tannins by the methods described in the Assoc. of Offic. Agric. Chemists (1965).

The results are summarized in Table 4.3. Protein and available carbohydrates vary little among the species. However,

TABLE 4.3. Composition and Energetic Content of Acorns Used by Acorn Woodpeckers at Hastings Reservation

Subgenus Species	Percent Soluble Tannins	Percent Lipids	Percent Protein	Percent Total Available Carbohydrates	Energetic Content (kJ/gm)
Quercus					
Valley oak					
(*Q. lobata*)	0.373	5.6	5.5	13.6	5.53
Blue oak					
(*Q. douglasii*)	0.384	8.3	6.6	17.4	7.49
Protobalanus					
Canyon live oak					
(*Q. chrysolepis*)	0.398	16.7	3.9	12.6	11.39
Erythrobalanus					
Coast live oak					
(*Q. agrifolia*)	0.460	24.3	7.1	13.2	13.23
Black oak					
(*Q. kelloggii*)	0.444	26.5	5.3	14.9	14.07

NOTE: Energetic content calculated using 39.8 kJ/gm for lipids, 18.8 kJ/gm for protein, and 16.7 kJ/gm for carbohydrates. Values are based on three acorns from each species.

lipid content varies considerably among the three subgenera, being low in *Quercus* (the "white" oaks), high in *Erythrobalanus* (the "red" oaks), and intermediate in *Protobalanus*. Thus, the energetic value of an acorn from either of the two red oak species (*Q. agrifolia* and *Q. kelloggii*) is two to three times greater per gram than that of the white oak species.

There are two reasons to suspect that acorns are a relatively poor food resource. First, their protein content is very low (3.9 to 7.1% of dry weight; Table 4.3). These values are markedly lower than the protein content of insects, which average 54.0% of dry weight (calculated from O'Conner, 1984, and Morton, 1973), and comparable to the average protein content of tropical fruits (6.8% of dry weight). The latter are sufficiently poor in protein to constrain the growth rates of nestlings dependent on them (Morton, 1973; Ricklefs, 1974; Foster, 1978).

Second, the tannins present in acorns (Table 4.3) presumably bind what little protein is available, making them unavailable to seed predators (Bate-Smith, 1973; Tempel, 1981; Martin and Martin, 1982). It is well known, for example, that tannins significantly reduce the nutritional value of food for chickens (Yapar and Clandinin, 1972; Marquardt and Ward, 1979), and are even associated with growth abnormalities (e.g. Elkin et al., 1978). Few experiments, however, have been conducted to determine the degree to which wild birds are affected by this problem. However, Perrins (1976) showed that blue tit nestlings fed diets containing oak tannins grew more slowly than those fed diets without tannins.

Besides nutritional composition, a second highly variable attribute of acorns is size. The mass of shelled acorns varies highly significantly ($p < 0.001$) both among years and among species in a two-way ANOVA (based on dry mass of shelled acorns of four species [all except *Q. chrysolepis*] collected over the four years 1976–79). However, acorns of all five species stored in granaries had a mean mass of about 1.2 to 1.4 g, which bore no consistent relationship to the mass of acorns available based on either traps placed below trees or random sampling (W. D. Koenig and R. L. Mumme, unpublished data). Thus, acorns of this size may have been stored preferentially.

Insect Damage and Tannin Content

The proportion of all acorns sampled from granaries that showed signs of insect damage varied considerably among years, reaching as high as 53% (Figure 4.6). Damage also varied significantly among species, with the red oaks experiencing the least insect damage. Although we do not know the causes of the annual variation in insect parasitism of acorns, the interspecific variation appears to be correlated with differences in soluble tannin concentration (Figure 4.7).

Tannins are a set of secondary plant metabolites common in oaks (Feeny, 1970; Zucker, 1983) and generally accepted as

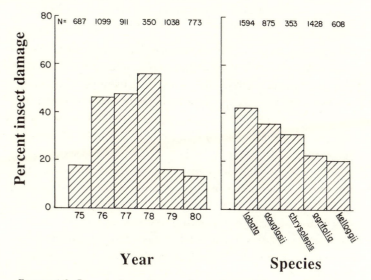

FIGURE 4.6. Percent of stored acorns insect damaged by year and species. Interyear and interspecific variation are both significant ($p < 0.001$) by ANOVA.

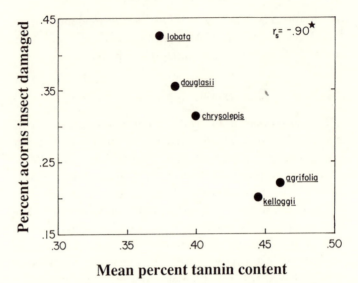

FIGURE 4.7. Percent of all acorns of five species of oaks collected from granaries that were damaged by insects, plotted against the mean percent soluble tannin. The correlation is significant by a Spearman rank test.

defense compounds against herbivory (Bate-Smith, 1973; Tempel, 1981) because of their ability to precipitate proteins (Martin and Martin, 1982). The proportion of insect damage to acorns of the five oak species was inversely correlated with the mean concentration of tannins (Figure 4.7). This relationship, like that between tannin concentration and herbivore damage in bracken ferns (Tempel, 1981), supports, at least weakly, the generally accepted assumption that herbivory has been an important selective force in the evolution of secondary metabolites as defensive substances. There are, however, at least two reasons this conclusion must be received with caution. First, the assay method may not correlate with the degree the tannins present form insoluble complexes with proteins (Martin and Martin, 1982). Second, tannin concentration may vary considerably from tree to tree (Sork et al., 1983). Thus, the small samples we analyzed are not necessarily representative of the acorns stored by the birds.

Energetic Value of Stores

We estimated the total energetic content of the acorn stores of the Hastings population for each year by considering their number, species composition, mass, insect damage, and energetic content. The maximum energetic content for each year is presented in Table 4.4 along with the estimated degree of insect damage, the energetic value of the stores per bird, and the estimated number of bird-days of energy stored in the granaries. On average, 6.7% of the total estimated energetic content of stores are lost to insects: it is likely that only a small fraction of this amount is recouped by the birds eating the larvae themselves. On a yearly basis, the percent of energy lost to insects ranged from 2.7 to 13.3%.

Dividing the total estimated energetic value of stored acorns by the population size yields the average number of kJ per bird

TABLE 4.4. Insect Damage to and Total Energetic Content of Acorn Stores at Their Winter Maximum

Year	kJ Stored	kJ Lost to Insects	Percent Insect Damage	Population Size	kJ Stored per Bird	Days of Energy Stored per Bird	Total Potential Fraction of DEE From Acorns
1975–76	450,070	19,511	4.2	123	3,659	13.8	0.065
1976–77	444,302	53,359	10.7	116	3,830	14.4	0.068
1977–78	295,075	38,499	11.5	126	2,342	8.8	0.041
1978–79	104,420	16,057	13.3	60	1,740	6.6	0.031
1979–80	362,846	11,784	3.2	57	6,366	24.0	0.113
1980–81	438,835	11,959	2.7	91	4,822	18.2	0.085
Total	2,095,562	151,173	6.7	573	3,657	13.8	0.065

NOTE: Based on an estimated daily energy expenditure (DEE) of 186 kJ bird^{-1} day^{-1} (see section 4.5) and digestive efficiency of 70%. Population size is for January. The total fraction of DEE potentially contributed from acorns is based on the period from 1 December to 1 June (213 days).

present in the granaries when stores are at their maximum (Table 4.4). For perspective, we estimated the daily energetic needs of acorn woodpeckers from the equation for total daily energy expenditure of species that do not forage in flight as a function of body mass given by Walsberg (1983):

$$\ln(TDEE) = \ln(12.84) + 0.610 \ln(MB),$$

where MB is body mass and $TDEE$ is total daily energy expenditure. Using MB equal to 80 g (a rough average for adults and first-year birds; Koenig, 1980a), yields a $TDEE$ of 186 kJ. (Preliminary work using doubly labeled water indicates that this approximation is very good: the mean daily energy expenditure of four adults captured during the breeding season in 1985 was 184.8 kJ [W. W. Weathers and W. D. Koenig, unpublished data].) Assuming 70% digestive efficiency (a reasonable estimate for insects [Kale, 1965; Krebs and Avery, 1984], but possibly high for acorns), the total daily energetic needs of an acorn woodpecker are thus about 266 kJ. Dividing the total stores per bird by this value yields the estimated number of days of energy stored per bird (Table 4.4). This value ranges from a high of 24 days in 1979–80, the years of low population size, to a low of 6.6 days in 1978–79, the year of the poor acorn crop. On average, when the granaries are full they hold about 14 days of energy per bird in the population. This represents about 6.5% of the total energetic expenditures for birds between 1 December, by which time acorn storage is usually complete, and 1 June, by which time most young are fledged (Table 4.4).

The estimated energetic value of stores present in June is given in Table 4.5. By this time the energetic content of stores is lower; on average, slightly over four days of energy reserves remain for each individual in the population. Thus, between December and June each bird uses an average of only ten days worth of acorns, amounting to about 4.6% of total energetic needs. Stored acorns cannot be the only, or even the major, source of food during this period; indeed, the birds apparently

90

TABLE 4.5. Total Energetic Content of Acorn Stores in June

Year	Total Estimated kJ Stored	Population Size	kJ Stored per Bird	Days of Energy Stored per Bird
1975–76	126,932	113	1,123	4.2
1976–77	112,783	104	1,084	4.1
1977–78	26,518	106	250	0.9
1978–79	1,813	48	38	0.1
1979–80	152,579	52	2,934	11.0
1980–81	133,295	82	1,626	6.1
Total	554,185	505	1,097	4:1

NOTE: Daily energy expenditures estimated as in Table 4.4. Population size is for the breeding season (May) excluding nestlings and fledglings.

try to "stretch out" the use of their stored acorns for as long a period as possible.

Comparison with Other Species

That acorn woodpeckers store only a small amount of their total energetic needs in their granaries contrasts sharply with data from other food-caching birds, particularly corvids. Some of the most detailed information is for Clark's nutcrackers (Vander Wall and Balda, 1977; see also Tomback, 1978). In northern Arizona this species stores piñon pine (*Pinus edulis*) nuts in small caches in the ground. Vander Wall and Balda (1977) estimated that a population of about 150 birds cached a total of 3.3 to 5×10^6 nuts year^{-1}, or between 22,000 and 33,333 nuts bird^{-1} year^{-1}. These values are two orders of magnitude more than the 325 or so acorns stored per woodpecker in an average year. Furthermore, these authors estimated that each bird stored between 109,000 and 163,400 kJ in nuts each year, 100 to 150 times the average energetic value of acorns stored by an acorn woodpecker at Hastings. They estimated the energetic content of the nuts stored by the nutcrackers to be

91

2.2 to 3.3 times the energy needed to survive the six months of winter, when the birds are dependent on stored food. Similarly massive storage has been reported by Darley-Hill and Johnson (1981), who estimated that 54% of the acorn crop of *Q. palustris* in Virginia was stored by an undetermined number of blue jays.

This striking difference reflects in part the more efficient method of storage used by acorn woodpeckers, whose acorns, once they are cached in a granary, are relatively safe from other seed predators, yet placed where virtually all can be retrieved. Nonetheless, the overall extent of food storage by acorn woodpeckers is far smaller than that by nutcrackers. It is thus all the more impressive that acorns are so important to the reproductive success of acorn woodpeckers (see section 4.6).

4.6. ACORN STORES AND BREEDING SUCCESS

In opposition to Dawson's (1923) query, the data presented above suggest that acorn stores are not a food supply on a "lavish, irrational scale," but rather a modest food cache providing insurance for occasional winter food shortages and for additional energetic needs during breeding. Furthermore, analysis of virtually any feature of the breeding success of acorn woodpeckers at Hastings leads to the conclusion that stored acorns are important to that success. Here we present several lines of evidence supporting this conclusion; some additional data are discussed in Chapter 5.

Autumn and Second Nesting

First, good acorn crops provide sufficient resources for autumn nesting, as well as second broods the following summer. The only two years that autumn nesting was recorded during the study were both excellent crop years, as indicated by the acorn fall data collected at School Hill; furthermore, the two years with the greatest incidence of second nests were the summers following these high crops (see Table 5.2).

TABLE 4.6. Reasons Groups with and without Stored Acorns Fail to Nest Successfully

	Stores Remaining	No Stores Remaining	All
No attempt observed	6 (32%)	20 (63%)	26 (51%)
Failed as eggs	3 (16)	2 (6)	5 (10)
Failed near hatching	3 (16)	6 (19)	9 (18)
Nestlings starved	4 (21)	4 (13)	8 (16)
Predation on adults	3 (16)	0 (0)	3 (6)
Total (% of all)	19 (37)	32 (63)	51 (100)

Nesting Failure

During the study, 27 groups made no nesting attempt during the spring. The most frequent correlate of such failure to attempt breeding was a lack of stored acorns: the granaries of 15 (55.6%) of the 27 groups were empty (see section 5.4).

Table 4.6 lists the causes of nesting failure depending on whether groups had stored acorns. Of the 51 groups failing, 63% had no stored. acorns. Groups without stores usually fail to attempt nesting, but otherwise tend to fail shortly after hatching, often apparently due to nestling starvation. In contrast, nests of groups with stored acorns failed more or less uniformly as eggs, near hatching, as nestlings, or due to adult predation. These patterns suggest that the energy balance of groups is simply not sufficient to raise young in the absence of stored acorns.

Comparison of Groups with and without Stored Acorns

Groups with stored acorns remaining in the spring experience considerably enhanced reproduction compared to groups lacking stores (Table 4.7). Groups with stored acorns nest significantly earlier, have larger clutch sizes, a higher proportion of eggs that hatch, fledge a higher proportion of hatched eggs,

93

TABLE 4.7. Reproductive Success of Groups with and without Stored Acorns Remaining in May

	Stores Present	Stores Exhausted	Statistic[a]	p-value
$\bar{x}$ first egg date	7 May $\pm$ 23.7 (157)	30 May $\pm$ 36.4 (20)	4.0	<0.001
Clutch size (single females)	4.55 $\pm$ 1.03 (93)	3.57 $\pm$ 0.65 (14)	3.5	<0.001
Percent eggs hatching	69.0 (578)	54.2 (59)	4.7	<0.05
Percent eggs fledging	47.6 (578)	25.4 (59)	9.2	<0.01
Percent hatched eggs fledging	68.9 (399)	46.9 (32)	5.6	<0.05
Young fledged	2.89 $\pm$ 1.87 (137)	0.51 $\pm$ 1.11 (43)	7.0	<0.001
Young alive in February	1.62 $\pm$ 1.53 (104)	0.34 $\pm$ 0.79 (32)	4.7	<0.001
Percent groups successful	82.7 (139)	23.3 (43)	51.3	<0.001
$\bar{x}$ group size	4.57 $\pm$ 2.24 (139)	3.44 $\pm$ 2.03 (43)	3.7	<0.001

NOTE: Data are $\bar{x} \pm$ SD (N groups).

[a] Clutch size, young fledged, young alive in February, and group size tested by Mann-Whitney U test (z-value given). Others by χ^2 contingency test (χ^2 value given).

fledge more young, and have more young survive through their first winter. More than four-fifths (82.7%) of groups with stored acorns fledge young, whereas less than one-fourth (23.3%) of groups without stored acorns do so. Groups with acorns are also significantly larger (Table 4.7), but the mean difference of one bird can only explain a small part of the observed reproductive differences. Thus, reproduction appears to be strongly dependent on the availability of stored acorns.

94

Size of the Acorn Crop

The strong dependence of acorn woodpeckers on acorns suggests that there might be a correlation between crop size and reproductive success. To test this prediction, we measured the acorn crop in two ways. Unfortunately, both measures have severe limitations. The first is based on the acorn trap data from School Hill discussed above, and thus does not provide a good estimate of the acorn crop over the entire study area or for anything but the four valley oaks actually sampled. Using these limited data, there are no significant correlations between crop size and either clutch size or young fledged (Table 4.8). As a slightly more robust test, we correlated the acorn fall with the number of young fledged by group School Hill, located at the sampling site; again, the correlation is not significant (Table 4.8).

The second measure involved sampling 250 individual trees each year in the autumn when the crop was at its peak. Two observers visited the trees, which included all five species of oaks spread throughout the study area. The same individual trees were sampled each year. Two estimates of the crop were made. First, the two observers both counted as many acorns as possible in an arbitrary section of the tree during a fifteen second period. The two counts were then added. Second, the observers subjectively evaluated the crop on a scale of 0 (no acorns) to 4 (bumper crop). These procedures provide an index of the acorn crop superior to the School Hill traps, but were not initiated until 1980. Thus, as of 1985 we have only five years of data available to test for correlations with acorn woodpecker reproduction. The correlations between the two crop indices and reproductive parameters are positive, but neither striking nor significant (Table 4.8).

These results must be interpreted cautiously. Nevertheless, we currently have no evidence that reproductive success of the population is correlated with the size of the acorn crop the preceding autumn, except when the crop is particularly poor.

TABLE 4.8. Spearman Rank Correlations of Acorn Crop and Storage Variables with Young Fledged and Mean Clutch Size of Single Females

| | Mean Clutch Size of Single Females | Young Fledged per Group | | Young Fledged per Female |
		Entire Population	School Hill only	
Acorn fall below 4 Q. lobata[a]	0.43 (10)	0.41 (14)	0.48 (11)	0.40 (14)
Mean crop score of 250 oak trees[b]	0.20 (5)	0.30 (5)	—	0.30 (5)
Mean number of acorns counted in 250 oak trees[b]	0.30 (5)	0.80 (5)	—	0.80 (5)
Lateness of acorn fall below 4 Q. lobata[a]	0.71 (7)	0.85*** (11)	0.63* (11)	0.89*** (11)
Maximum acorns stored	−0.09 (6)	−0.04 (7)	—	−0.11 (7)
Maximum kJ stored	0.43 (6)	0.49 (6)	—	0.49 (6)
Maximum acorns stored per bird	0.83* (6)	0.93** (7)	—	0.96*** (7)
Maximum kJ stored per bird	1.00** (6)	0.94* (6)	—	0.94* (6)
Mean kJ per stored acorn	0.94* (6)	0.89* (6)	—	0.89* (6)

NOTE: Sample size (number of years) in parentheses.
[a] School Hill acorn traps; see section 4.4.
[b] Acorn sampling procedure described in section 4.6.

The above result is expected if acorn storage is generally limited by storage facilities rather than crop size. Indeed, it suggests that features of acorn production other than crop size per se may be important to acorn woodpeckers. We will discuss two such features: timing of the crop and acorn quality.

Timing of the Acorn Crop

When acorns are on the trees late in the season, the birds are dependent on their stores a shorter period of time prior to the breeding season, making it more likely that they will still have stores available for breeding. As we saw in Figure 4.3, there is considerable variability in how fast acorn stores are depleted; this variability could partly be a function of how late acorns can be taken off the trees independent of the size of the crop.

Our measure of acorn crop lateness is again taken from the valley oak acorn fall data from School Hill. Using the timing of acorns caught in traps and assuming a Gaussian distribution of falling acorns, we estimated the last date that an arbitrarily low number of acorns would fall, and used this date as an estimate of the lateness of the crop. The correlation between the lateness index and the number of acorns falling from the sampled trees is not significant ($r_s = 0.51$, $N = 11$, ns); thus, crop size and crop lateness do not appear to be directly related. However, the correlations of the lateness index with reproductive success both of group School Hill and of the population as a whole are significant (Table 4.8).

Our observations indicate that the availability of acorns, especially of live oaks, late in the winter and even into spring is highly variable. During two years, we observed considerable numbers of live oak acorns present in some areas as late as March, and in 1985, after the period covered in detail here, acorns were still present from the bumper crop of autumn 1984 through June over a considerable part of the study area. Thus, in some years and in some territories, birds need not resort to using their stored acorns until the breeding season has actually begun, and in rare cases acorns are available on the trees throughout the spring breeding period.

Quality of Acorns

A second feature of the acorn crop that might influence woodpecker demography is the quality of the acorns. We have already noted several features that influence acorn quality, including species, size, and insect damage. Taking these variables into consideration, we estimated the mean energetic value per stored acorn in each of six years. The results indicate a significant correlation between the quality of stored acorns and reproductive success (Table 4.8).

Quantity of Stores and Reproductive Success

There are strong correlations between both clutch size and reproductive success of the population and the number and energetic value of acorns stored per capita the prior winter (Table 4.8; a representative graph of the correlation between the energetic content of stores per bird at the winter maximum and reproductive success the following spring is given in Figure 4.8). In contrast, there is a lack of significant correlations between reproductive success and the total number of acorns stored or their total energetic content (Table 4.8). These results

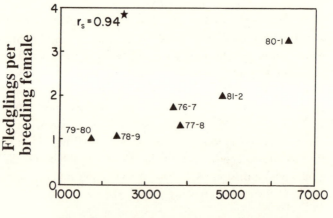

FIGURE 4.8. Reproductive success of the population (fledglings per breeding female) plotted against the total maximum kJ per individual stored in granaries during the prior winter.

indicate that success is dependent on the ratio of stored food reserves to population size.

Stored Acorns and Ashmole's Hypothesis

Ashmole's hypothesis states that variations in fecundity are the result of differences in the ratio of breeding season resources to population density (Ashmole, 1961, 1963; Ricklefs, 1980). Support for this hypothesis has been found in several studies of geographic patterns of clutch size variation in birds (e.g., Ricklefs, 1980; Koenig, 1984).

Our findings indicate that annual variation in both clutch size and reproductive success is correlated with relative resource availability, as indicated by acorn stores per capita (Table 4.8). These results thus support Ashmole's hypothesis as an explanation for a significant amount of year-to-year variation in fecundity patterns in acorn woodpeckers.

4.7. A PARADOX

Overall Energetic Importance of Stored Acorns

The finding that the success of the population is strongly dependent on acorn stores is paradoxical for two reasons. First, as discussed above, the total energetic content of stores is relatively small, and thus can supply only a small fraction of the total energy needs of the population. (Acorns as a whole still contribute substantially to the overall energy budget of acorn woodpeckers as a result of their strong day-to-day dependence on "green" acorns taken directly from the trees in late summer and early autumn, or as long as they are available; see Figure 4.2.) Furthermore, there is no evidence that acorns contain some essential nutrient that is important to reproduction; even the small amount of protein present is probably bound to tannins and indigestible (Bate-Smith, 1973; Tempel, 1981; Martin and Martin, 1982; see also section 4.5).

Thus, it is unlikely that acorns provide anything more important to the birds than energy. We know of no comparable example in which reproductive success is so strongly dependent on a food supply that is not only separated temporally from

TABLE 4.9. Food Items Brought to Nestlings

Taxon	N Items	Percent Samples	Families
Acorn fragments	27	39	—
Arthropoda			
Arachnida	1	4	—
Insecta	—	100	—
Orthoptera	8	11	Gryllacrididae, Tettigoniidae
Dermaptera	16	18	Forficulidae
Plecoptera	3	11	Perlodidae
Hemiptera	9	25	Pentatomidae, Coreidae, Scutelleridae, Miridae, Corizidae
Homoptera	8	25	Cicadellidae, Membracidae, Cicadidae
Neuroptera	4	14	Raphidiidae, Myrmeleontidae
Coleoptera	56	50	Scarabaeidae, Coccinellidae, Cerambycidae, Buprestidae, Cuculionidae, Scolytidae, Dytiscidae, Staphylinidae, Malachiidae, Hydrophilidae, Elateridae, Meloidae, Melandryidae, Chrysomelidae, Carabidae
Lepidoptera	11	21	Lasiocampidae, unidentified larvae
Diptera	14	14	Tipulidae
Hymenoptera	40	68	Formicidae, Apidae, Vespidae, Sphecidae, Halictidae, Ichneumonidae, Andrenidae, Pompilidae

NOTE: Based on 28 samples from pipe-cleaner-collared young. Families of insects are listed in the order of decreasing frequency in the samples.

the breeding season by at least six months, but also apparently supplies only a small fraction of the overall diet.

Diet of Nestlings

The second reason this striking dependence on stored acorns is paradoxical is that acorns, although present in the diet of older nestlings, are not even the major food item fed to nestlings. Table 4.9 summarizes food items collected from nestlings collared with pipe cleaners. Only 39% of samples contained acorn fragments, whereas all samples contained insects. Both Coleoptera and Hymenoptera are represented more frequently than acorns.

Thus, the importance of stored acorns to reproductive success is unlikely to be the result of their food value to nestlings. Rather, we propose that acorns are used primarily by the adults, and that their availability both allows a group to surpass the necessary energetic threshold for breeding and to feed their nestlings enough insects to permit a high probability of success. Apparently this energetic threshold is fairly tight and probably exacerbated by the highly unpredictable availability of flying insects. Thus, the few days of stored energy available in an average granary during the spring is enough to make the difference between successful breeding and failure.

4.8. ACORN STORAGE AND SOCIALITY

Can we also detect any influence of stored acorns on survivorship, or various group or population characteristics? We test for such effects in Table 4.10 by reporting Spearman rank correlations between maximum stored acorns and stored energy, both total and per bird, between adult winter survival, various group characteristics during the following spring, and two population characteristics.

TABLE 4.10. Spearman Rank Correlations of Yearly Storage Variables with Adult Survival, Group, and Population Characteristics

	Maximum Stored Acorns	Maximum Acorns Stored per Bird	Maximum kJ Stored	kJ Stored per Bird
Adult winter survival	−0.39	0.79*	0.37	0.94**
Mean group size in spring	0.93**	−0.14	0.43	−0.14
Percent adult nonbreeders	0.89**	0.00	0.60	−0.03
Breeders per group	0.76*	−0.20	0.49	0.09
Nonbreeders per group	0.86*	−0.11	0.37	−0.09
Percent groups with 2 ♀ breeders	0.99***	0.23	0.75	−0.03
Breeding population size	1.00***	0.04	0.66	−0.09
Percent territories occupied	0.75	0.36	0.83*	0.03

NOTE: $N = 7$ years for stored acorns; 6 years for kJ stored.

Adult survivorship is significantly greater in years when more acorns and energy are stored per bird (Table 4.10). This correlation may, however, be an indirect result of density-dependent adult survivorship, as relatively more acorns are stored in years of low population size (section 4.4). Most likely, both these factors contribute to variation in adult survival. In any case, in years when more acorns are stored per individual, winter survival is higher and reproductive success the following spring is greater.

None of the other variables correlates significantly with per capita acorns or energy stored. However, all correlate significantly with either the maximum number of stored acorns or the maximum total number of kJ of stored acorns. Hence, in years when the total quantity of acorns stored by the population is high, the population is large. This is both because a higher proportion of territories are occupied and because groups themselves are large. Groups contain significantly more breeders as well as nonbreeders, and more frequently contain joint-nesting females. Furthermore, a significantly higher proportion of adults are nonbreeders (Table 4.10). Thus, not only does population reproduction appear to be influenced by acorn storage, but so too are survivorship, group size, and group composition.

4.9. OAKS, ACORNS, AND TERRITORY QUALITY

We have thus far analyzed our data on a year-to-year basis. A parallel approach is to perform analyses territory-by-territory. This not only allows us to test the generality of some of the above patterns but also to examine the importance of vegetational features such as oak diversity and abundance.

Procedures

To determine reproductive success and characteristics of groups inhabiting specific territories, mean values for all years the territory was occupied were used; the number of years varied from territory to territory according to data given in Chapter 3. Vegetation characteristics were determined in the following way. A 60-m grid was superimposed on aerial photographs of the study area. Using the photographs, we went to the center of each grid point and used it for the center of a 0.04-ha circular plot (James and Shugart, 1970). Within the plot, the species and diameter at breast height of all trees were recorded; estimates of canopy, shrub, and ground cover were also made. In addition, species of trees observed within the 60-m grid area but not within the 0.04-ha plot were noted. Each sample thus covered slightly over 10% of the total area of the 60-m grid. In all, 631 vegetation plots were sampled.

Territorial boundaries of groups were assigned based on our experience observing individuals in groups over the past ten years. Using the points within each territory, we calculated the following: total number of oak species present within the territory, mean density of oaks in the territory (number of trees divided by area), and total estimated basal area of oaks present in the territory.

In addition, we calculated several variables related to the acorn stores and storage facilities of groups. We estimated the total number of storage holes present on territories based on the maximum number of stored acorns ever counted at that territory. The mean number of species of acorns stored was calculated from our granary samples; the total number of species stored was the total number of species of acorns identified in our samples during the six years samples were taken.

Results

The results of Spearman rank correlations are presented in Table 4.11. Neither territory size nor any of the three vegetation characters correlate significantly with any of the variables measuring occupancy, reproductive success, or average composition of the groups living on particular territories. However, the estimated number of storage holes on a territory has strong correlations with all these factors; territories with large storage facilities are occupied more often, have higher reproductive success, and contain more breeders and nonbreeders. Additionally, territories in which more species of acorns are stored, either in total or on average, are occupied more frequently and contain more birds. This relationship between species of acorns stored and territory occupancy remains significant or nearly significant after controlling, in partial correlation analyses, for: (a) the total number of acorns sampled from each territory ($p < 0.05$), (b) the mean number of acorns sampled per year ($p < 0.05$), and (c) the number of years acorns were sampled from each territory ($p = 0.09$). These data thus suggest that oak

104

TABLE 4.11. Correlations between Territory Characteristics and History of Groups Inhabiting the Territory

| | Percent Years Occupied | Mean Young Fledged | | Mean Group Size | Breeders per Group | Nonbreeders per Group | Percent Years with Nonbreeders |
		Occupied years	All years				
Territory size (25)	0.10	0.24	0.27	0.09	0.22	0.27	0.37
N oak species in territory (29)	−0.08	0.15	0.11	0.24	0.18	0.15	0.00
Oak density (22)	0.01	0.18	0.19	0.12	−0.03	0.23	0.16
Total oak basal area (22)	0.18	0.33	0.41	0.03	0.07	0.28	0.31
Estimated N storage holes (29)	0.72***	0.41*	0.64***	0.76***	0.70***	0.64***	0.58***
Mean N species stored (23)	0.45*	0.31	0.44*	0.43*	0.37	0.32	0.24
Total species stored (23)	0.46*	0.36	0.50**	0.61**	0.45*	0.43*	0.29

NOTE: Data are for all years that a territory was watched (see Chapter 3). Sample size (number of territories) is in parentheses.

species diversity may influence acorn woodpecker demography, a possibility first proposed by Bock and Bock (1974).

One way in which oak species number might influence territory occupancy would be if a greater oak species diversity lowers the probability of a complete crop failure (Bock and Bock, 1974). This might be the case if different ecological factors influenced the crops of different species. However, even if the same ecological conditions influence acorn production by all species, the differing maturation times for their acorns (either one or two years; see Chapter 2) would result in their crops being asynchronous.

If increased oak species diversity lowers the probability of total crop failures, we would expect the number of species of oaks to correlate with the success of birds inhabiting particular territories, but it does not (Table 4.11). One possible resolution is that the number of species of acorns stored by groups, which does correlate with success, provides a better indicator of effective diversity of acorns available to the birds than our vegetation samples. Note, however, that despite relatively high oak species numbers at Hastings—three to five species are common throughout the area—acorn crop failures and subsequent woodpecker population crashes are still relatively common (Hannon et al., 1987; see also Chapter 11).

In conclusion, the single most important feature determining average group size, composition, occupancy, and success on any particular territory is the size of storage facilities. (Although group and granary size are correlated, granaries change much more slowly than the composition of groups using them, and causality is unlikely to go the other way: large groups cannot "cause" large granaries except on a long-term basis.) There is no discernible effect of territory size, number of oak species, density of oaks, total basal area of oaks within territories, or any other vegetational variable considered. However, there is some indirect evidence that territories containing more species of productive oaks are larger and more successful than those with fewer species.

4.10. COMPARISON WITH PREVIOUS WORK

Interpopulation Studies

The strong relationship between acorns and acorn wood-pecker population ecology has been identified by several workers employing a variety of approaches. Bock and Bock (1974), using data from Audubon Society Christmas bird counts (see Bock and Smith, 1971) taken over the entire United States range of the acorn woodpecker, found densities of this species on the Pacific coast to vary linearly with oak abundance. However, there appeared to be a threshold effect of oak species number: in areas with five or more species of oaks, woodpeckers were common, while in areas with four or fewer oaks, woodpeckers were uncommon or scarce. The advantage of high oak species diversity proposed by these authors is a lower probability of total acorn crop failure resulting from asynchronous acorn production by sympatric species. Virtually no association of acorn woodpecker abundance with oak density or species number was found for populations in the Southwest. Bock and Bock (1974) suggested that this was due to the generally marginal habitat found in the Southwest, where acorn crop failures were common.

In contrast, Roberts (1979a), employing transect studies, found no correlation of woodpecker densities with oak species diversity in California, whereas in the Southwest he found a positive correlation between these two variables. Nevertheless, despite their differences, both studies are at least consistent with the interpretation that oak species diversity and abundance have important effects on acorn woodpecker populations.

Trail (1980) compared populations in southeastern Arizona with that at Hastings as reported by MacRoberts and Mac-Roberts (1976). He found no significant differences in group composition despite markedly larger storage facilities at Hastings; however, the number of adults per group was larger at Hastings ($p = 0.055$), in keeping with our finding that mean group size and the number of breeders per group are significantly related to storage facilities.

Additional interpopulation work on acorn woodpeckers has been carried out by Stacey and Bock (1978). Their discovery of a migratory and essentially nonsocial population of acorn woodpeckers at the Research Ranch in southeastern Arizona provides the only study of a population living under conditions of poor acorn availability and clearly indicates that local ecological conditions can critically influence acorn woodpecker sociality and behavior.

Intrapopulation Studies

Intrapopulation studies of cooperative breeders involving vegetational analysis have been performed on several species. For Hall's babbler, Brown and Balda (1977) found a significant relationship between flock size and vegetational indices of home-range quality among six groups followed for eight days in early spring. Similarly, Zack and Ligon (1985b) found group size and reproductive success in the grey-backed fiscal shrike to correlate with shrub cover, a vegetational feature that in turn correlated with greater insect abundance (Zack and Ligon, 1985a). Finally, Brown et al. (1983) found several vegetational correlates with group size in the grey-crowned babbler. Thus, in contrast to the results reported here for acorn woodpeckers, these studies indicate that vegetation may influence group size and other characteristics in some species of cooperative breeders. Presumably, such interspecific differences in the vegetation characters correlating with various group parameters reflect in some way the particular resources critical to each population, as we discuss at greater length in section 14.1.

Returning to acorn woodpeckers, Trail (1980) studied the effects of vegetation, acorn crop, and storage facilities on group size and composition in ten groups over two years in southeastern Arizona. Interestingly, the two years of his study varied considerably in acorn abundance. In the year of poor acorn availability, home ranges were significantly larger as birds apparently ranged widely in search of food. Groups followed were also smaller in the poor acorn year; however, the density of groups did not change. Thus, the population fluctuations associated with acorn crop variability appeared to be buffered, at least to some extent, by the woodpecker social system. Unfortunately, different groups were sampled in the two years of Trail's study; thus, his results may have been due to differences between the two populations rather than indicative of real effects of habitat or resources on acorn woodpecker demography.

If we substitute "stored acorns" for "acorn crop," we found similar correlations between food resources and group size: in years with many stored acorns, groups as well as the population as a whole were larger, but there is no significant correlation between stored acorns and the proportion of territories occupied.

For the Hastings population, MacRoberts and MacRoberts (1976) found a significant correlation between maximum group size during the winter and both storage facilities and territory size. Our results, based on the same population over a more extensive period, confirm the former correlation but indicate that, averaged over many breeding seasons, mean group size and composition are not significantly correlated with territory size (Table 4.11). Storage facilities appear to be a far more important influence on group size and composition than either territory size or vegetation.

4.11. CONCLUSION

The analyses discussed in this chapter all investigate the ecological correlates of group size and social organization. They differ considerably in their approaches as well as in their conclusions. However, one theme stands out for acorn woodpeckers: both the acorn crop and acorn storage have important effects on group size, productivity, and sociality. We now turn to an investigation of demographic processes in the Hastings population.

4.12. SUMMARY

Acorn storage is a major activity of acorn woodpecker societies. The acorn crop and various characteristics of acorns vary markedly from year to year; yet the factor limiting acorn storage in all but the worst crop years appears to be storage facilities. Consistent with this hypothesis, features other than size of the crop per se, including quality of the acorns and length of time acorns are available on the trees, correlate significantly with reproductive success. Generally, both clutch size and reproductive success are most strongly correlated with number and energy content of acorns stored per bird.

The apparent influence of acorn storage does not stop with reproductive success. On both a year-to-year and territory-by-territory basis, group size and composition are positively correlated with acorn storage, both in terms of maximum number of acorns and energy stored (year by year) or number of storage holes (territory by territory). Also, adult winter survivorship is greater in years when many acorns are stored per bird, although this may in part be an indirect result of density-dependent survival. Thus, group living is more prevalent in territories with more storage holes and in years when many acorns are stored by the population. There are no correlations between the success of groups, composition of groups, or the occupancy of territories with territory size, oak diversity (measured directly), oak density, or total basal area. However, territories with greater oak diversity as measured indirectly by the number of species of acorns stored have a higher frequency of occupancy and support larger groups, suggesting that oak species diversity may influence the quality of acorn woodpecker territories.

These findings point to the critical dependence of acorn woodpeckers on both acorns and the granaries in which they are stored. Both of these interrelated factors are likely to be important considerations in the evolution of cooperative breeding in this species.

CHAPTER FIVE

Reproductive Success, I:
General Patterns

As described in Chapter 3, acorn woodpeckers live in groups of between two and fifteen individuals consisting of a mix of birds of different ages and of both breeders and nonbreeding helpers. Thus, it should not be surprising that reproductive success is influenced by group composition.

Ecological factors also vary considerably, both from year to year and from territory to territory, and may affect reproductive success. In particular, the acorn crop is quite variable, and acorns and granaries have strong effects on reproduction and other demographic features of acorn woodpecker populations. Flying insects, another major food resource of acorn woodpeckers, also vary greatly from year to year, as well as locally even within a small site such as Hastings. Thus, given the social and ecological variables potentially influencing reproductive success, the problem of isolating the effects of any individual variable becomes unusually challenging.

The factors potentially influencing reproductive success in our population are shown in Figure 5.1. Yearly variation in ecological conditions, territory quality, prior breeding experience, and group composition are all likely to influence reproductive success both directly and indirectly. In this chapter our analysis focuses on temporal patterns and ecological influences affecting acorn woodpecker breeding success, saving the influence of differing group composition for Chapter 6.

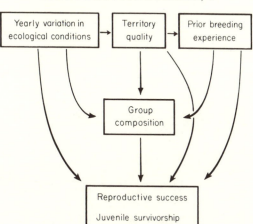

FIGURE 5.1. Path diagram of the variables affecting reproductive success and survivorship in acorn woodpeckers. Territory quality influences prior breeding experience by affecting the survivorship of adults and thus turnover rate. Both territory quality and prior breeding experience affect group composition, while all three of these factors may independently influence reproductive success. Yearly variation in ecological conditions (food availability, predation, etc.) affects territory quality, group composition, and reproductive success.

5.1. CHRONOLOGY OF BREEDING

Timing

The main breeding season for acorn woodpeckers at Hastings is from April through June, but autumn nesting in August and September also occurs. The estimated first egg date (FED) of the earliest recorded nest at Hastings was 2 April, and the latest autumn nest was begun about 23 September. The weekly distribution of FED's for all nests, separating first, second, and re-nesting attempts, is given in Figure 5.2.

The majority of nests are begun between 15 April and 15 May; 60% of all first nests are started by 5 May and 92% by 26 May. Autumn nesting is separated from the spring peak by a distinct gap during the summer; no nests were initiated during

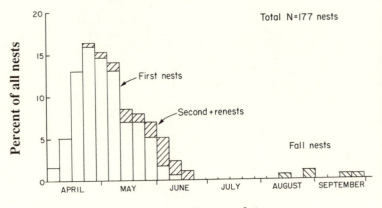

Estimated first egg date

FIGURE 5.2. The percent of nests initiated each week for all years combined between 1976 and 1982. First nests, second and renesting attempts, and autumn nests are indicated.

the 58-day period between 16 June and 13 August. (A nest initiated 9 July 1985 finally provided an exception.) Autumn nesting appears to be dependent on the same factor (acorns) as breeding the following spring (see section 5.2). Thus, the range in first egg dates for a woodpecker fiscal year can be considered to be from about 14 August to 15 June the subsequent year, with a 192-day hiatus during the winter when no nests are begun.

Autumn nests were rare: only five, involving 2.2% of group-years, were recorded during the study, two in 1974 and three in 1979. Four of these (80%) were successful. The latest autumn nest, initiated about 23 September, failed, but one nest initiated about 15 September fledged two young on 1 November.

Groups attempted second nests following successful first nests only 11 times (4.9% of all group-years of observation); 10 of these (91%) were successful. Second nesting attempts usually followed relatively early first nests: the mean estimated FED for first nests followed by an additional attempt was 14 April (SD = 7.4 days; dates ranged from 4 to 26 April). The mean time separating FED's of first and second nesting attempts was 48.6 ±

114

9.2 days ($N = 10$); since a complete nesting cycle takes about 45 days to complete (4 for egg laying, 11 for incubation, and 30–32 for the nestling period), such nests usually did not overlap. However, they did so on three occasions. The minimum time between a successful first and second nest was about 33 days, yielding an overlap of about 12 days (MacRoberts and MacRoberts 1976).

Renesting attempts following nest failure occurred 17 times; 11 of these (65%) were successful. One group made 2 renesting attempts and another made 3 renesting attempts before successfully fledging young. Thus, the majority of nests (148 of 177 [84%]) were first spring nests. Only 4% were second nests, 10% were renesting attempts, and 3% autumn nests. No significant differences in clutch size, hatchability, young fledged, or success rate were detected between first nests and second, autumn, or renests; thus all were combined for subsequent analyses.

Synchrony

Reproductive synchrony is not pronounced. Between 1975 and 1982 the total span of FED's for first nests was 66 days, ranging between 29 and 57 days (Table 5.1). Starts of second and renests extended only one week beyond the latest first nest (Figure 5.2). Standardizing by the earliest nest found in each year, 69% of all first nests were initiated within the first four weeks of each season.

Mean FED of first nests varies significantly between years, ranging from 23 April to 18 May (Table 5.1). This difference is significant, even after removing annual differences in when breeding began by standardizing by the earliest nest in each year (Kruskal-Wallis one-way ANOVA, $\chi^2 = 16.8$, df $= 7$, $p < 0.05$). Thus, timing of nests varies considerably with annual ecological conditions. As we discussed in Chapter 4, the ecological feature most likely to be causing this variation is the autumn acorn crop (see Table 4.7).

TABLE 5.1. Yearly Variation in First Egg Dates (First Nests Only)

Year	Mean ± SD (days)	Earliest	Latest	Range (days)	N Nests
1975	4 May ± 10	18 April	27 May	39	17
1976	6 May ± 9	22 April	21 May	29	17
1977	4 May ± 16	11 April	1 June	51	19
1978	7 May ± 14	8 April	25 May	47	14
1979	18 May ± 15	24 April	9 June	46	15
1980	23 April ± 17	4 April	31 May	57	21
1981	29 April ± 10	16 April	26 May	40	19
1982	29 April ± 12	13 April	27 May	44	26
All years	3 May ± 15	4 April	9 June	66	148

NOTE: Differences among years in first egg dates significant by a Kruskal-Wallis one-way ANOVA, $\chi^2 = 33.3$, df = 7, $p < 0.001$.

5.2. ANNUAL VARIABILITY

Here we focus on both yearly variation in reproductive parameters and some of the ecological factors that may cause the variation. The reproductive variables that we will consider include:

1. *Clutch size.* The number of eggs laid in a nest in groups containing only a single breeding female.

2. *Set size.* The number of eggs laid in all nests, including groups with more than one breeding female.

3. *Hatchability.* The estimated proportion of eggs surviving the 11-day incubation period that successfully hatch. Partial egg loss during incubation was rare, and eggs usually remain in the nest for several days, up to two weeks. Thus, we assume that eggs unaccounted for between incubation and hatching (in nests known to have survived to the expected hatching date) hatched, but that the young subsequently died.

116

4. Percent of eggs hatching. The percent of all eggs laid that were known or believed to hatch. As with hatchability, this measure assumes that eggs unaccounted for between incubation and hatching did in fact hatch.

5. Proportion of normal eggs fledging. The proportion of normal eggs (i.e., excluding "runt" eggs; see below and Koenig [1980b]) producing young that survive to 25 days of age (when we band them; see Chapter 2). Young surviving to this age are assumed to fledge, which occurs at 30–32 days.

6. Proportion of hatched eggs fledged. The proportion of hatched eggs producing fledglings.

7. Breeding success. A group is successful if it raised any young to fledging age during a particular acorn year (e.g., September through the following August).

8. Number of young fledged. Includes all young fledged during the acorn fiscal year.

9. Number of young fledged per bird. May be either per group member, per breeding male, or per breeding female.

10. Number of young surviving to February. Young begin to disperse in March following fledging. Thus, a February cutoff date allows a good estimate of juvenile winter survivorship.

11. Number of young surviving to February per bird. Again, may vary relative to different fractions of the group.

Breeding Success

Productivity of the Hastings population is shown in Table 5.2. Fledging success varied from 0.90 to 3.86 young per group over the 11 years (a 4.3-fold range) and the percent of groups successfully fledging young ranged from 27.8% in 1973 to 100% in 1980.

Table 5.2 also gives the percentage of groups with successful second and autumn nests. There is a strong correlation between years in which fledging success is high overall and in which second nests were produced ($r_s = 0.81$, $N = 10$, $p < 0.05$). Thus, successful second nests tend to occur in years of good

117

TABLE 5.2. Productivity of the Study Area by Years

Year	Fledglings per Group ($\bar{x} \pm SD$)	Fledglings per Breeding Female	Fledglings per Bird	Percent Groups with Second Nests	Percent Groups with Fall Nests	Percent Groups Breeding Successfully	N Groups
1972	1.56 ± 1.67	1.67	0.26	0.0	0.0	55.6	9
1973	0.94 ± 1.86	0.74	0.18	0.0	0.0	27.8	18
1974	1.22 ± 1.40	1.02	0.27	5.6	10.5	52.6	19
1975	3.86 ± 1.96	2.68	0.95	9.5	0.0	95.5	22
1976	2.25 ± 2.05	1.73	0.43	0.0	0.0	61.9	21
1977	1.86 ± 1.98	1.34	0.37	0.0	0.0	52.4	21
1978	1.29 ± 1.87	1.08	0.25	0.0	0.0	38.1	21
1979	0.90 ± 1.22	1.00	0.35	0.0	14.3	47.6	21
1980	3.58 ± 1.80	3.26	1.08	15.8	0.0	100.0	19
1981	2.35 ± 1.72	2.00	0.52	8.7	0.0	78.3	23
1982	2.97 ± 1.92	2.19	0.65	6.9	0.0	79.3	29
Mean	2.12 ± 2.01	1.75	0.48	4.5	2.2	64.1	223
C.V.	49.9%	46.0%	61.4%	—	—	—	—

NOTE: Fledglings per group vary significantly among years (Kruskal-Wallis test, $\chi^2 = 51.1$, df = 10, $p < 0.001$). Coefficient of variation is for yearly means.

reproduction. Autumn nests also occur in good (fiscal) years: both 1974 and 1979, when autumn nesting occurred, preceded spring breeding seasons of excellent productivity. This forms the basis for our decision to consider annual productivity based on the acorn fiscal year rather than on the calendar year.

Set Size and Fledging Success

The distribution of set size for groups containing only a single breeding female, two breeding females, and all nests is presented in Figure 5.3. Yearly variation in clutch size, set size, eggs surviving to hatching, and fledging success are given in Table 5.3.

118

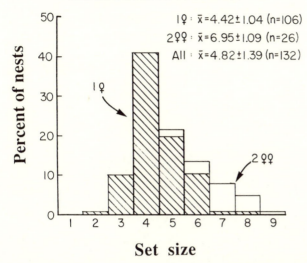

FIGURE 5.3. The distribution of the number of eggs incubated by groups, 1976 to 1982. Sets that were the product of joint-nesting females are added to the top of the histograms representing those produced by single females.

Both measures of clutch size varied significantly from year to year: mean clutch size of single females varied from 3.7 to 4.9 eggs, whereas set size varied from 3.7 to 5.3 eggs. Hatchability and the proportion of eggs producing nestlings also varied significantly among years; mean hatchability was 85.9%, whereas 67.3% of all eggs laid hatched. About 45% of all eggs and 67% of all hatched eggs eventually produce a fledgling.

The percent of all eggs that produce fledglings is similar to the mean egg success rate found by Ricklefs (1969) for sixteen hole-nesting species, but is considerably lower than the 77% egg success rate in two species of European woodpeckers reported by Pynnönen (1939) as well as the 67% mean success rate of eggs found by Lack (1954) for twenty-four studies of hole-nesting species breeding in nest boxes.

Nilsson (1986) has recently discussed relative breeding success in hole-nesting passerines using productivity (estimated young fledged per nest) as a proportion of clutch size. Data from acorn

TABLE 5.3. Yearly Variation in Reproductive Parameters Other than Fledging Success

Year	Clutch Size	Set Size	Percent Hatch-ability	Percent Eggs Hatching	Percent of Eggs Fledging	Percent of Hatched Eggs Fledging	N Eggs[a]
1976	4.10 ± 0.57 (10)	4.69 ± 1.25 (13)	94.6	86.9	54.1	62.3	61 (56)
1977	4.18 ± 0.98 (11)	4.60 ± 1.40 (15)	91.1	73.9	49.3	66.7	69 (56)
1978	4.00 ± 0.76 (8)	4.77 ± 1.36 (13)	91.9	54.8	41.9	76.5	62 (37)
1979	3.71 ± 0.61 (14)	3.71 ± 0.61 (14)	78.4	55.8	38.5	69.0	52 (37)
1980	4.92 ± 1.29 (25)	5.11 ± 1.42 (27)	84.2	60.2	39.8	66.3	133 (95)
1981	4.68 ± 1.16 (19)	4.86 ± 1.28 (22)	77.2	67.0	42.5	63.4	106 (92)
1982	4.52 ± 0.77 (19)	5.25 ± 1.58 (28)	88.0	72.7	49.0	68.0	143 (117)
Mean	4.42 ± 1.04 (106)	4.82 ± 1.39 (132)	85.9	67.3	44.9	66.8	626 (490)
χ^2	15.8	14.7	14.1	24.6	6.3	2.4	
p-value	<0.05	<0.05	<0.05	<0.001	ns	ns	

NOTE: Data are $\bar{x} \pm$ SD (N nests). Statistical tests are for differences among years by Kruskal-Wallis one-way ANOVA: all df = 6. Includes runt eggs, but not eggs destroyed by joint-nesting females.

[a] First number is the number of eggs followed; the number in parentheses is the number of eggs surviving to hatching (used for calculation of hatchability).

woodpeckers, obtained by dividing the mean number of young fledged per group (2.12; Table 5.2) by mean set size (4.82; Table 5.3), yields 44.0%, lower than either the means (±SD) for six species nesting in natural cavities (53.6 ± 3.5%) or for ten open-nesting species (56.0 ± 10.3%; Nilsson, 1986). Thus, at least by this measure of success, acorn woodpeckers are somewhat less successful than Nilsson's sample of either cavity- or open-nesting passerines.

5.3. ECOLOGICAL INFLUENCES ON REPRODUCTIVE SUCCESS

What ecological factors influence reproductive success? Here we will consider timing, weather, population density, and food. We defer discussion of the influence of group composition until Chapter 6.

Timing

Date of clutch initiation correlates strongly with several aspects of reproductive success (Table 5.4). Early nests have significantly larger clutches, fledge a higher proportion of hatchlings, and fledge more total young. There is no correlation between fledging date and hatchability, fledging success of eggs, or winter survival of young (Table 5.4).

Thus, there is an overall advantage to breeding early. Similar advantages have been documented in other birds, including the Florida scrub jay (Woolfenden and Fitzpatrick, 1984) and

TABLE 5.4. Correlations of Date of Nest Initiation (FED) with Reproductive Parameters

	FED		Standardized FED		
	r_s	p-value	r_s	p-value	N
Clutch size (single females)	−0.39	<0.001	−0.35	<0.001	86
Percent eggs hatching	0.04	ns	0.05	ns	148
Percent eggs fledging	−0.15	ns	−0.11	ns	148
Percent hatched eggs fledging	−0.43	<0.001	−0.43	<0.001	92
Young fledged	−0.34	<0.001	−0.36	<0.001	148
Percent young fledged surviving to February	−0.06	ns	0.04	ns	97

NOTE: First spring nests only. Standardized FED's are all corrected such that the first egg date for the year is 2 April.

various migratory species (e.g., the northern oriole [Williams, 1982]).

Weather

Table 5.5 presents correlations of weather variables with mean clutch size and annual reproductive success. Variables considered include winter and spring mean temperatures, rainfall, and sunshine.

TABLE 5.5. Spearman Rank Correlations of Weather and Flying Insect Abundance with Yearly Mean Reproductive Success

	Mean Clutch Size	Young Fledged per Group	Young Fledged per Breeding Female
WEATHER VARIABLES			
Winter rainfall	−0.16 (10)	−0.11 (14)	−0.13 (14)
Winter rainfall prior year	−0.29 (10)	−0.26 (14)	−0.19 (14)
Winter rainfall 2 years earlier	0.79* (10)	0.82** (14)	0.78** (14)
Winter sunshine	0.29 (7)	0.38 (11)	0.42 (11)
Mean minimum winter temperature	0.36 (7)	0.08 (11)	0.09 (11)
Mean winter temperature	0.32 (7)	0.37 (11)	0.41 (11)
Mean minimum spring temperature	0.00 (7)	−0.50 (11)	−0.52 (11)
Mean spring temperature	−0.40 (7)	−0.67* (11)	−0.69* (11)
FLYING INSECT ABUNDANCE			
Total catch yellow pan traps	−0.71 (7)	−0.71 (7)	−0.71 (7)
Total catch Malaise traps	−0.54 (7)	−0.29 (7)	−0.29 (7)
Hymenoptera in yellow pan traps	−0.46 (7)	−0.54 (7)	−0.54 (7)
Tipulidae in yellow pan traps	−0.39 (7)	−0.39 (7)	−0.39 (7)
Tipulidae in Malaise traps	0.11 (7)	0.07 (7)	0.07 (7)

NOTE: Temperatures and rainfall recorded at reservation headquarters; sunshine measured with a Campbell Stokes–type sunshine recorder. Winter (1 December to 31 March) precedes the breeding year; spring (1 April to 30 June) is concurrent with it. Sample sizes (N years) in parentheses.
* $p < 0.05$;   ** $p < 0.01$.

Only two of the variables yield significant correlations: mean spring temperature and winter rainfall two years earlier. Mean spring temperature was inversely correlated with reproductive success. Insects are a staple food for nestlings, and higher spring temperatures possibly have a detrimental effect on reproduction by drying out the vegetation, leading to a faster decline in the spring insect flush. Alternatively, higher temperatures might lead to increased nest predation by permitting greater snake activity. However, insects are not only more active during warm weather but tend to be slightly, although not significantly, more abundant during warmer springs (correlations between mean spring temperature and the total catch of the yellow pan traps, the Hymenoptera caught in the yellow pan traps, and the tipulids caught in both the yellow pan and Malaise traps were positive [all $r_s > 0.30$]; only that between mean spring temperature and the total catch in the Malaise trap was negative [$r_s = -0.13$]). Thus, the inverse correlation between mean spring temperature and reproductive success either is not meaningful or results from some unknown relationship between temperature and the breeding biology of acorn woodpeckers.

Nor is the importance of the strong correlation between reproduction and winter rainfall two years earlier obvious. Possibly it reflects a relationship between rainfall and the autumn acorn crop two years later, which in turn might influence reproduction the following spring. Our preliminary crop data, however, do not suggest a positive relationship between rainfall and crop size: the correlations between the acorn crop censuses and winter rainfall two years earlier are not significant, while those for the prior winter's rainfall are negative (r_s between winter rainfall in the prior season and the mean acorn counts = -0.94, $N = 6$, $p < 0.05$; for winter rainfall and the mean crop index $r_s = -0.77$, $p < 0.10$). It is still possible that winter rainfall influences the lateness of the acorn crop or acorn quality two years later—those aspects of the crop that we found to be important to reproductive success in Table 4.8. Additional data will be necessary to test such possibilities, however.

123

The conclusion we can draw from Table 5.5 is that the reproductive success of woodpeckers at Hastings appears to be independent of any directly interpretable effect of the weather. Rainfall may indirectly influence success through some effect on the acorn crop, but the inverse correlation with spring temperature remains unexplained.

Population Density

The correlation of overall reproductive success with population size, measured in April at the start of the breeding season, is graphed in Figure 5.4. Because of the special interest of this relationship we have included data through 1985 (11 years total). The resulting correlation is barely negative ($r_s = -0.09$) and not significant. Thus, there is no apparent density dependence of reproductive success.

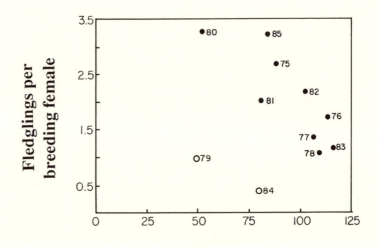

Population size

FIGURE 5.4. Correlation between reproductive success and population size measured in April for eleven years, 1975 to 1985. Correlation for all years is $r_s = -0.09$ (ns). Excluding the two crop failure years of 1979 and 1984, $r_s = -0.82$ ($N = 9$, $p < 0.05$).

However, examination of the scatter in Figure 5.4 suggests that this conclusion may be premature: excluding 1979 and 1984, which were years of acorn crop failure, there is a significant inverse correlation ($r_s = -0.82$; $N = 9$; $p < 0.05$). In fact, this relationship is expected given the density-dependent effect of stored acorns per bird on reproductive success (Table 4.8) coupled with the limited storage facilities, a relationship permitting smaller groups to store more acorns per bird (see section 4.4). Thus, except during years of catastrophic crop failure, there appears to be strong density dependence of reproductive success, similar to that documented in other hole-nesting species (e.g., the great tit; Kluijver, 1951; Lack, 1966). In years of crop failure this relationship breaks down: not only does the population crash but subsequent reproductive success is poor as well.

Food

A high proportion of the diet of nestling and adult acorn woodpeckers is composed of flying insects (Table 4.9; Figure 4.2). Thus, it is reasonable to expect that flying insect populations, which exhibit great annual variation in abundance and composition at Hastings (W. D. Koenig and R. L. Mumme, unpublished data), might correlate with reproductive success.

We sampled flying insects in two ways. First, we used a large, 3-m Malaise trap set in a field about 0.3 km from reservation headquarters. Second, we used a series of nine yellow pan traps distributed over an area 0.5 km within reservation headquarters. Both sets of traps were emptied weekly, and insects were sorted to order, dried, and weighed. The catch from April to June inclusive—the three primary breeding months—was totaled for all insects as well as for crane flies (family Tipulidae) and bees and wasps (order Hymenoptera), two favored food types (see Table 4.9).

Correlations of clutch size, young fledged per group, and young fledged per breeding female with the measures of the annual spring insect catch are shown in the bottom section of Table 5.5. There are no significant correlations between clutch size or number of young fledged with either total abundance of flying insects, abundance of tipulids, or abundance of hymenoptera using either of the two sampling schemes. In fact, twelve of the fifteen (80%) correlations shown in Table 5.5 are negative. Again, our sampling procedures are limited and warrant caution. However, these results at least suggest the surprising conclusion that reproductive success is independent of the major food resource being fed to young. Thus, breeding success is either independent of weather and insects or, at best, their effects are swamped by the much greater influence of acorns and acorn storage.

Territory Quality

Another major set of variables potentially influencing reproductive success concerns the quality of territories; indeed, territory quality has long been recognized as one of the chief confounding factors in determining the influence of group size and composition on reproductive success (Lack, 1968; Zahavi, 1974, 1976; Brown and Balda, 1977; Koenig, 1981b; Woolfenden and Fitzpatrick, 1984). We assessed the importance of territory quality on reproductive success in several ways. First, we divided territories into two categories: "permanent" territories, those occupied every spring during the study (type 1 territories; see Chapter 3), and "marginal" territories, those unoccupied at least one spring (type 2 and 3 territories). A comparison of clutch size and success of groups by territory category is given in Table 5.6.

Clutch size of single females living on permanent territories is slightly larger than that of females in marginal territories. However, no differences between groups inhabiting the two types of territories were observed in reproductive success or in the number of young surviving to February.

126

TABLE 5.6. Reproductive Success of Groups Living in Territories That
Were Occupied Every Spring (Permanent Groups) versus Territories That
Were Unoccupied at Least One Spring (Marginal Groups)

	Permanent Groups	Marginal Groups	Statistic[a]	p-value
$\bar{x}$ first egg date	10 May ± 29.3 (133)	8 May ± 14.4 (44)	0.7	ns
Clutch size (single females)	4.42 ± 1.14 (81)	3.85 ± 0.83 (26)	2.2	<0.05
Young fledged	2.28 ± 2.09 (151)	1.78 ± 1.80 (45)	1.7	ns
Young alive February	1.26 ± 1.47 (127)	0.80 ± 1.21 (45)	1.8	ns
Percent groups successful	66.2 (154)	59.7 (67)	0.6	ns

[a] Statistical tests as in Table 4.7.

A second approach to determining the factors influencing
territory quality is to correlate characteristics of territories
with the occupancy record and reproductive success of birds
inhabiting those territories. Analyses involving territory size,
vegetation characters, and variables related to acorns and
granaries were presented in Table 4.11.

We examine relative flying insect abundance estimated from
a series of four small (2-m) Malaise traps run for three years
on different parts of the study area in Table 5.7. The total catch
and number of tipulids caught during the spring (1 April to 30
June) were averaged for the three years and these values assigned to territories in the vicinity of the trap and in physically
similar areas nearby. (The four subsets of groups consisted of
those in the valley near reservation headquarters, those on
either Haystack or School Hill, those in a riparian area along
the main watercourse, and those on the Arnold; see Figure 3.1.)

127

TABLE 5.7. Spearman Rank Correlations of Flying Insect Abundance with the Occupancy and Mean Reproductive Success of the Groups Inhabiting Those Territories

	Percent Years Territory Occupied	Young Fledged	Young Surviving to February
Malaise trap catch (23)	0.17	0.29	0.44*
Number of tipulids (23)	−0.15	0.35	0.23

NOTE: Numbers in parentheses are the number of territories.
* $p < 0.05$.

None of the correlations between territory quality and the variables related to territory characteristics are significant (Table 4.11). Those involving flying insect abundance are also generally weak, although that between the Malaise trap catch and young surviving to February is marginally significant (Table 5.7).

Again, this situation changes when considering variables related to acorn storage, three of which were considered in the bottom of Table 4.11. All three acorn storage measures correlate significantly with the percentage of years territories are occupied, and the estimated number of storage holes is also correlated with the mean number of young fledged during years the territory was occupied. Thus, territories with larger storage facilities are more likely to be occupied and to produce more young per year than those with fewer storage holes. For additional discussion of these correlations, see section 4.9.

In summary, the occupancy record of territories and the success of birds residing on individual territories correlates both with aspects of the acorn crop itself and storage facilities, whereas there are generally weak correlations with other food resources (flying insects) as well as vegetation characteristics of territories. These results thus support our earlier conclusion that acorns and storage facilities appear to be of paramount importance to the breeding success of acorn woodpeckers.

5.4. CAUSES OF NESTLING MORTALITY

If reproductive success is primarily determined by food, we can predict that most nesting failure should result from brood starvation or abandonment rather than nest predation. The overall occurrence of nest failure and of failures to attempt breeding entirely are presented in Table 5.8. During the eight years in which nests were regularly followed, 26 of 175 groups (15%) did not attempt to breed at all (a small number may have begun to nest but failed early in incubation). The remaining 149 groups made 183 nesting attempts, 133 (73%) of which were successful. The estimated daily mortality rate on nests using these data is 0.71%, low compared to open-nesting species but similar to the average daily mortality rate of seven hole-nesting species (all passerines, generally breeding in nest boxes) reviewed by Ricklefs (1969). Additional discussion of daily mortality rates is presented later in this section.

TABLE 5.8. Failure to Attempt Nesting, Incidence of Nest Failure, and Daily Mortality Rate on Nests

Year	N Groups	N (%) Groups Not Attempting to Breed	Total Nesting Attempts	N (%) Nest Failures	Percent Daily Mortality[a] Rate on Nests
1975	22	1 (4.5)	24	2 (8.3)	0.193
1976	21	4 (19.0)	18	5 (27.8)	0.723
1977	19	3 (15.8)	19	8 (42.1)	1.094
1978	21	8 (38.1)	15	7 (46.7)	1.397
1979	21	6 (28.6)	19	9 (47.4)	1.426
1980	19	0 (0.0)	33	9 (27.3)	0.708
1981	23	2 (8.7)	24	4 (16.7)	0.405
1982	29	3 (10.3)	31	6 (19.4)	0.479
Total	175	27 (15.4)	183	50 (27.3)	0.709

NOTE: Autumn nests included in following year's totals.

[a] Calculated using the formula $-(\ln P)/t$ (Ricklefs, 1969), where P is the proportion of successful nests and t is the length of the nesting period (45 days).

Failure to Attempt Nesting

Of 27 groups that made no nesting attempt, 4 (14.8%) contained no breeding female, 1 (3.7%) contained no breeding male, and 15 (55.6%) had no stored acorns remaining during the spring (see section 4.6). Of the remaining 8 groups, one contained 2 unrelated breeding males; the failure of this group to attempt nesting may have resulted from an unusual degree of competition between these 2 males. This leaves only 6 groups (22.2% of all nonnesting groups) whose failure to nest was unrelated to any obvious deficiency. Thus, failure to attempt nesting is most frequently correlated with a lack of stored acorns.

Nesting Failure

Table 5.9 presents a detailed analysis of mortality of nests and eggs; sample sizes vary slightly from earlier analyses (e.g., Table 5.3) because only nests on which complete information was available on the fates of both eggs and young are included.

Of 180 nests, 20 (11.1%) failed to yield any hatched young. The major causes of failure were desertion and false starts. Causes of desertion were usually unknown, although particular females seemed to have a predilection for abandoning. One female, for example, deserted three successive nests immediately after clutch completion before successfully hatching two young in a fourth nest. False starts occurred in groups containing two joint-nesting females, and generally involved the laying of one or two eggs, which were immediately abanboned without being incubated. We do not know why this happens, but suspect that a lack of reproductive synchronization is involved. (If only one female is ready to lay, the breeding males may continue to mate-guard the second female rather than commence incubation. Possibly under such conditions the first female abandons her solo attempt after laying only one or two eggs.) Two nests were apparently lost to nest-hole

Table 5.9. Causes of Nest and Egg Mortality

	Nests	Percent of Individual Nests	Percent of Losses	Eggs	Percent of Individuals	Percent of Losses
NESTS STARTED OR EGGS LAID	180	—	—	693	—	—
Losses due to:						
Hatching failure	—	—	—	72	10.4	32.4
Runt eggs[a]	—	—	—	13 (28)	1.9 (4.0)	5.9 (12.6)
Egg destruction[b]	—	—	—	58	8.4	26.1
Nest-site competition	2	1.1	10.0	10	1.4	4.5
Adult death	1	0.6	5.0	—	—	—
Desertion, false starts	11	6.1	55.0	42	6.1	18.9
Unknown	6	3.3	30.0	27	3.9	12.2
Total losses	20	11.1	100.0	222	32.0	100.0
YOUNG HATCHED	160	88.9	—	471	68.0	—
Losses due to:						
Adult death	1	0.6	3.2	1	0.1	0.6
Desertion	3	1.7	9.7	9	1.3	5.1
Weather	3	1.7	9.7	5	0.7	2.8
Starvation	9	5.0	29.0	111	16.0	62.4
Predation	3	1.7	9.7	14	2.0	7.9
Other	2	1.1	6.5	11	1.6	6.2
Unknown	10	5.6	32.3	27	3.9	15.2
Total losses	31	17.2	100.0	178	25.7	100.0
YOUNG FLEDGED	129	71.7	—	293	42.3	—

[a] Numbers in parentheses include runt eggs lost to other sources of mortality.

[b] Includes an estimated 3.5 eggs per nest destroyed by groups containing joint-nesting females prior to 1980; beginning with 1980 the actual number of eggs destroyed in such nests is used (see section 5.4.)

competitors (one to ash-throated flycatchers and the other to European starlings), and the loss of one nest early in incubation was attributable to the loss of the breeding female to a Cooper's hawk. This leaves six nests (3.3%) that were lost for unknown reasons. Eggs in these nests simply vanished; thus, these nests may have been lost to predators. No instances of egg predation were observed during the time period covered here, but an entire clutch was lost to a gopher snake (*Pituophis melanoleucus*) in spring 1986.

Nests containing young failed for a variety of reasons. Nine nests (5.0%) were apparently lost to starvation of all young: all young were found dead in the nest. Only three nests were known or strongly suspected to have been lost to predators. One was lost to a nocturnal predator who not only took the nestlings but the nocturnally incubating male as well; this kind of predation is similar to that occurring regularly in the joint-nesting groove-billed ani (Vehrencamp, 1977). Local predators that could have been responsible for this loss include northern pygmy owls and long-tailed weasels (*Mustela frenata*).

In another case, a gopher snake pellet was found in the nest. The only case of nest predation that we witnessed during the study involved a gopher snake that ate two of three nestlings; the third was left and fledged successfully. Another ten nests were lost for unknown reasons. Young in these nests again simply vanished, and thus may have been depredated. However, eight of these groups had previously run out of stored acorns and may have failed because of food shortage.

Thus, of all total nest losses, a minimum of three (1.7%) were lost to predation. Assuming all nests lost to unknown causes were depredated, the maximum amount of nest predation is 10.6% of all nests or 37.3% of all failed nests. (Excluding the eight groups whose nests failed that had also run out of acorns, the estimated rate of nest predation is 6.1% of all nests, or 21.6% of all failed nests.) Thus, on average, predation is probably about as important as desertion and false starts (7.8% of all nests, 27.5% of all failures) and, at most, slightly more important than food shortage as an apparent cause of nesting failure. (Food shortage accounts for the loss of at least 5.0% of all nests and 17.6% of all failures; including the eight nests lost for unknown reasons but in which the group had run out of stored acorns, the percent of nests failing due to food shortage may be as high as 9.4%, or 33.3% of all failed nests.) In at least some years (e.g., 1975 and 1980), over 95% of nests are successful (Table 5.2), and thus predation is certainly low.

Few comparative data are available. Nilsson (1984), however, reported predation rates ranging from 6.2 to 32.1%, with an overall average of 15.3% ($N = 412$ nests) for six species of passerines in natural cavities. Thus, the predation rates observed in our population may be low compared to other cavity nesters.

Causes of egg mortality are also presented in Table 5.9. We followed 635 eggs, including 58 eggs that are estimated to have been laid in communal nests but destroyed by joint-nesting females prior to the initiation of incubation. We did not discover egg destruction among joint-nesting females until 1980; thus, this number is the sum of 23 eggs we observed being destroyed from 9 nests watched between 1980 and 1982 and 35 eggs we estimate were destroyed from 10 additional nests containing 2 females at which we did not look for egg destruction. This latter estimate is derived from data indicating that an average of 3.5 eggs per nest are destroyed in communal nests (52 eggs from the 15 nests reported in Mumme et al., 1983b). Such destruction occurs invariably when more than one female nests jointly and is apparently the consequence of reproductive competition among the females. For additional details concerning this phenomenon, see Mumme et al. (1983b).

The single greatest cause of egg mortality is hatching failure, accounting for the loss of 10.4% of all eggs, followed by egg destruction, which accounts for the demise of an estimated 8.4% of all eggs. Desertion and false starts are the only other major sources of egg mortality; only 3.9% of eggs were lost to unknown causes.

133

A small proportion of eggs (1.9%) fail to hatch because they are abnormally small "runt" eggs (Koenig, 1980b). However, the 13 such eggs listed in Table 5.9 represents a far higher proportion than reported for any other bird species, wild or domestic (Koenig, 1980b). Moreover, the true proportion of runt eggs laid by acorn woodpeckers is even higher than this figure for two reasons. First, three additional runt eggs were observed that were lost to other sources of egg mortality and are not listed in the "runt egg" row in Table 5.9. Second, and more important, a high proportion—about 21%—of eggs destroyed by joint-nesting females are runts (Mumme et al., 1983b). Including these eggs results in an overall frequency of runts of 4.0%. Of 134 sets of eggs, 15 (11.2%) contained one or more runt eggs.

The greatest cause of nestling loss is starvation, accounting for 16.0% of all individuals and a majority (62.4%) of nestling losses. (Starvation is assumed to be the cause of death when brood reduction occurs or when dead, uneaten nestlings are found in the nest.) Brood reduction is common, and few nests, except in very good years, fledge all young that hatch. Several miscellaneous causes account for the remaining lost nestlings. Predation is known or strongly suspected to have eliminated 14 (2.0%) nestlings, and an additional 27 nestlings (3.9%) died of unknown causes.

Thus, between 2.0 and 9.8% of all eggs are ultimately depredated, accounting for between 3.5 and 17.0% of all losses. (The upper estimates include all eggs and nestlings lost to unknown causes; we do not count egg destruction by joint-nesting females as predation.) However, the majority of nestlings lost to unknown causes were in groups that had run out of acorns and may have been victims of food shortage rather than predation. Excluding the 27 nestlings falling into this category, the remaining losses possibly attributable to predation include 5.6% of all eggs laid, or 9.8% of all losses. However, even assuming that all unknown losses are attributable to predation, the relative magnitude of nest predation we observed is only about one-fifth that in the Florida scrub jay, in which nearly 55% of all eggs laid, and 75% of all eggs lost, are depredated (Woolfenden and Fitzpatrick, 1984).

Daily Mortality Rates

The values in Table 5.9 are used to calculate daily mortality rates in Table 5.10 according to the methods presented by Ricklefs (1969). The top line of this table presents the values for the average daily mortality rate of eggs (m) during the egg stage (m_e) and nestling stage (m_n), the rate during the entire nesting cycle, and the difference between the mortality rate of eggs during the two stages ($m_e - m_n$). The second row presents similar data for nests (M), and the final row lists values for "partial losses" ($m - M$) during the various stages of the nesting cycle.

Compared to the hole- and niche-nesting species considered by Ricklefs (1969), the daily mortality rate of eggs is high (1.91% compared to a mean of 1.02% [maximum 1.53%] for

TABLE 5.10. Percent Daily Mortality Rates on Eggs and Young

	Egg Stage ($._e$)	Nestling Stage ($._n$)	Total	Difference[a] ($._e - ._n$)
Daily mortality rate of eggs (m)	2.57	1.58	1.91	0.99
Daily mortality rate of nests (M)	0.78	0.72	0.74	0.07
Partial loss ($m - M$)	1.97	0.86	1.17	—

NOTE: Calculations of mortality factors based on Ricklefs's (1969) equation $m = -(\ln P)/t$, where P is the proportion of eggs or young surviving and t is the length of the egg (15 days), nestling (30 days), or total nesting cycle (45 days) and the data in Table 5.9. Subscripts refer to egg (e) and nesting stage (n).

[a] This is the "vertical component" of Ricklefs (1969).

7 species). The apparent reason for this high value is the very high rate of partial loss of 1.17% per day (1.97% during the egg stage, 0.86% during the nestling stage). This is as high as any of the values found in Ricklefs's analysis of over 50 species and reflects the large number of individuals lost to starvation, egg destruction, and hatching failure. The daily mortality rates of nests are almost identical during the egg and nestling stages $(M_e - M_n)$. The higher rate of egg loss during the egg stage $(m_e - m_n)$ is attributable to eggs destroyed by joint-nesting females, runt eggs, and hatching failure.

These data thus suggest that, compared to starvation, predation is a relatively unimportant cause of mortality of eggs and young in the acorn woodpecker. The major source of nest and egg losses is food shortage, a result expected given the apparent strong dependence of reproductive success in this population on stored acorns.

The second most important source of egg loss is hatching failure. As discussed in detail elsewhere (Koenig, 1982; Koenig et al., 1983), hatchability varies with group composition and is significantly lower in groups with two joint-nesting females (78.9% of 90 eggs) than in groups with only a single breeding female (88.2% of 287 eggs). Hatchability is also lower in groups with two or more breeders of either sex (82.7% of 231 eggs) than groups with only a single breeding member of both sexes (91.4% of 116 eggs; both $\chi^2 = 4.1, p < 0.05$). Thus, decreased hatchability is one cost of mate-sharing and joint-nesting in this population (Koenig, 1982).

The third largest source of mortality shown in Table 5.9 is egg destruction by joint-nesting females. (As described above, however, the total amount of such losses is partially estimated.) A similarly high incidence of egg destruction has been found in another joint-nesting species, the groove-billed ani (Vehrencamp, 1977). In both species, egg destruction is the result of sexual competition uniquely related to their social breeding habits and is another cost of group living.

5.5. CONCLUSION

In this chapter we have considered the effects of a variety of ecological factors on reproductive success and examined the specific causes of breeding failure. The conclusions are in some ways paradoxical: on the one hand, the major causes of mortality of both eggs and nestlings are apparently related to food (nestling starvation) and to sociality (egg destruction by joint-nesting females and the increased amount of hatching failure in groups containing mate-sharing males or joint-nesting females). Reproductive success appears nonetheless to be uncorrelated with yearly variation in the abundance of the primary food resource being fed to nestlings (flying insects). Moreover, a substantial proportion of the population lives in groups containing cobreeding males, females, or both.

The likely resolution of the first paradox—the predominance of nestling starvation and the independence of nestling survival from the availability of their primary food resource—can be found in the dependence of reproductive success on stored acorns and, ultimately, the acorn crop itself (see Chapter 4). Although young are fed primarily insects, the energy budget of adults is apparently tight and access to stored acorns for their own use and for emergency food for nestlings is critical to reproductive success. Other ecological factors influence reproductive success: in particular, success declines seasonally and reproductive success is strongly density dependent in years other than when the acorn crop fails. However, acorns and acorn storage appear to be paramount to acorn woodpecker reproductive success.

The resolution of the second paradox—the high degree of sociality in the population despite the high costs in terms of egg loss—is found in the compensating benefits accruing to individuals that nest jointly or share mates. Detailed investigation of such benefits constitutes a major theme of Chapters 6, 7, and 13.

5.6. SUMMARY

The breeding season of acorn woodpeckers at Hastings Reservation is governed by the acorn fiscal year. Autumn nesting, beginning in mid-August, is uncommon but occurs in a small proportion of groups in years of excellent acorn crops. After a winter hiatus, the main breeding season begins in early April. Second broods are also unusual, but may occur when many stored acorns remain from the prior autumn's acorn crop. Breeding is not particularly synchronous among years, and first egg dates vary significantly from year to year.

Annual variability in reproductive parameters is high. The proportion of groups successfully raising young varied between 27.8 to 100%, and the number of young fledged per group varied between 0.90 to 3.86. Clutch size also varies significantly from year to year.

There is a strong seasonal influence on reproductive success: earlier nesting attempts have larger clutches and fledge more young than later nests. There is strong density dependence in annual reproduction except in years of acorn failure. However, we can find no evidence that weather, either during the spring or the prior winter, correlates with reproductive success. (A strong correlation of reproductive success with winter rainfall two years earlier may indicate some as yet unconfirmed influence of rainfall on the acorn crop.) Similarly, reproductive success appears to be generally independent of the abundance of flying insects, which are the major food for nestlings. Differences in territory size or vegetation also appear to have little overall effect on territory occupancy or reproductive success.

Groups unsuccessful at nesting failed for numerous reasons; 15% failed to attempt nesting at all. These groups sometimes lacked a breeding female but frequently lacked stored acorns. Similarly, nest failure was often correlated with an absence of acorn stores.

Predation was not the major source of nest failure. Assuming that all eggs and nestlings lost to unknown causes were depredated, predation accounted for at most the loss of 9.8% of all eggs laid, and for 17% of all losses. The largest cause of egg mortality was hatching failure, followed by egg destruction by joint-nesting females; the largest source of nestling loss was apparently starvation. Daily mortality rates are especially high for partial nest loss, reflecting the large losses due to hatching failure, egg destruction, and starvation.

CHAPTER SIX

Reproductive Success, II:
Group Composition

In this chapter we discuss the interactions between group composition and reproductive success. This aspect of cooperative breeding has traditionally received considerable attention, as the effects of helpers on the reproductive success of the breeders they are aiding can influence the evolution of both group living and apparently altruistic traits.

Our previous analyses of group composition and reproductive success (Koenig, 1981b; Koenig et al., 1983) have stressed the relationship between group size and mean number of young fledged per group and per capita, with preliminary analyses of the effects of differing numbers of breeders on reproductive success. We will briefly present updated results of similar analyses here. However, because acorn woodpecker groups consist of variable numbers of breeders and nonbreeders, there is little question that analyses based on group size alone are simplistic (see Woolfenden and Fitzpatrick, 1984), even if heuristically useful (Koenig and Pitelka, 1981). We are now confident of our ability to categorize the breeding status of individuals (see Koenig et al., 1984), and are able to perform analyses of the effects of group composition on reproductive success in a more finely tuned fashion. The goal of this chapter is therefore to isolate, for groups and for the individuals concerned, the effects of the number of breeding males, breeding females, and nonbreeders on reproduction, and discuss the possible reasons for the observed patterns.

6.1. GROUP SIZE AND TURNOVERS

There is a significant overall effect of group size on reproductive success of acorn woodpecker groups at Hastings (Table 6.1; Figure 6.1a). In terms of number of young fledged, groups of six to nine (with the exception of the poor showing of groups of seven) tend to have a clear edge over other groups; similarly, groups of size six and eight have the most young surviving to February, before emigration of young may occur (see Chapter 8). These differences are the result of larger sets of eggs (Table 6.1) and thus larger broods in successful nests (Figure 6.1b). There is no significant difference in the probability that groups of different size will successfully fledge young (Figure 6.1c), and the frequency of second nests is too low to cause major differences in productivity (see Table 5.2).

Reproductive success per capita yields a very different picture: pairs fledge the most young per individual (Figure 6.1a) and have more young surviving to February per bird than any other group size with the exception of groups of size six (Table 6.1); none of these differences is significant, however.

An earlier analysis of reproductive success in this population (Koenig, 1981b) indicated that, besides yearly variation, the single most important factor influencing reproductive success is whether or not the group has undergone a turnover in breeders from the prior year. Data on the effect of turnovers are presented in Table 6.2. There is a strong, highly significant difference between groups experiencing a turnover compared to those that have not. Differences are significant for both the probability of success and the size of broods in successful nests. Differences between groups in which the breeding males, breeding females, or breeders of both sexes changed from the prior year are not significant.

Helpers rarely remain in their natal group as nonbreeders following the replacement of breeders of either sex (see section 3.2). Consequently, a higher proportion of groups not experiencing a turnover contain nonbreeders (63% of 127 group-

141

TABLE 6.1. Clutch Size and Reproductive Success as a Function of Group Size

Group Size	Clutch Size	Set Size	Young Fledged	Young Surviving to February	Young Fledged per Capita	Young Surviving to February per Capita
2	4.00 ± 0.76 (25)	4.00 ± 0.76 (25)	1.33 ± 1.35 (51)	0.70 ± 0.89 (46)	0.67 ± 0.68	0.35 ± 0.45
3	4.29 ± 0.85 (21)	4.78 ± 1.34 (27)	1.78 ± 1.85 (50)	0.91 ± 1.33 (32)	0.59 ± 0.62	0.30 ± 0.44
4	4.47 ± 1.13 (15)	5.18 ± 1.56 (22)	2.18 ± 1.84 (40)	1.06 ± 1.27 (32)	0.54 ± 0.46	0.27 ± 0.32
5	4.50 ± 0.85 (10)	5.23 ± 1.64 (13)	2.79 ± 2.35 (28)	1.64 ± 1.73 (22)	0.56 ± 0.47	0.33 ± 0.35
6	4.50 ± 1.00 (4)	4.80 ± 1.10 (5)	3.63 ± 3.11 (8)	2.83 ± 3.06 (6)	0.60 ± 0.52	0.47 ± 0.51
7	4.25 ± 0.50 (4)	4.25 ± 0.50 (14)	2.29 ± 2.20 (11)	1.55 ± 1.57 (11)	0.33 ± 0.31	0.22 ± 0.22
8	4.67 ± 0.58 (3)	5.50 ± 2.07 (6)	4.14 ± 0.69 (7)	2.50 ± 1.22 (6)	0.52 ± 0.09	0.31 ± 0.15
9	5.67 ± 2.08 (3)	5.80 ± 1.48 (5)	3.13 ± 2.80 (8)	1.14 ± 1.68 (7)	0.35 ± 0.31	0.13 ± 0.19
10	4.00 ± — (1)	5.67 ± 1.53 (3)	2.67 ± 1.94 (3)	1.29 ± 1.25 (5)	0.27 ± 0.19	0.13 ± 0.13

11	—	4.00 ± — (1)	1.00 ± 1.41 (2)	0.00 ± — (1)	0.09 ± 0.13	0.00 ± —
12	—	—	0.00 ± — (2)	0.00 ± — (2)	0.00 ± —	0.00 ± —
Mean	4.33 ± 0.94 (86)	4.82 ± 1.41 (111)	2.11 ± 2.01 (219)	1.15 ± 1.45 (172)	0.48	0.26
χ^2 (df)	6.9 (7)	16.1 (7)	25.2 (8)	15.7 (8)	5.8 (8)	10.2 (8)
p-value	ns	<0.05	<0.001	<0.05	ns	ns

Note: Data are $\bar{x}$ SD (N). Statistical tests are by Kruskal-Wallis one-way ANOVA; groups of size 10 to 12 are combined for testing (8 to 12 for clutch size).

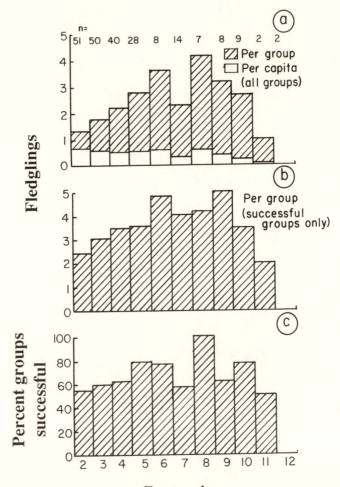

FIGURE 6.1. (a) Fledglings per group and per capita versus total group size. Differences among groups are significant for fledglings per group, but not fledglings per capita (see Table 6.1). (b) Fledglings per group by group size, excluding groups failing to fledge any young. Differences among groups are significant (combining groups with ten or more birds, $\chi^2 = 30.3$, df = 8, $p < 0.001$). (c) Mean percent of groups successfully fledging young in a year by group size. Differences among groups not significant (combining groups with eight or more birds, $\chi^2 = 6.8$, df = 6, ns).

TABLE 6.2. Influence of Turnovers in Breeders on Reproductive Success

	Male Turnover Only	Female Turnover Only	Turnover of Both Sexes	All Turnovers	No Turnover	Statistic[a]
Young fledged	1.10 ± 1.29 (10)	0.97 ± 1.76 (29)	1.08 ± 1.48 (37)	1.04 ± 1.57 (76)	2.69 ± 2.03 (134)	5.8***
Young fledged, successful nests only	2.20 ± 0.84 (5)	2.80 ± 1.99 (10)	2.67 ± 1.05 (15)	2.63 ± 1.38 (30)	3.61 ± 1.48 (100)	3.4****
Young alive in February	0.60 ± 0.97 (10)	0.52 ± 1.29 (25)	0.50 ± 0.80 (22)	0.53 ± 1.06 (57)	1.48 ± 1.56 (107)	4.2***
Percent success	50.0 (10)	37.9 (29)	40.5 (37)	40.8 (76)	75.2 (134)	23.3***

NOTE: Data are $\bar{x} \pm$ SD (N).

[a] Comparison is for all turnovers versus no turnover. Young fledged and young alive in February tested by Mann-Whitney U test (z-value given); percent success tested by χ^2 contingency test. All comparisons of male turnover versus female turnover, male turnover versus turnover of both sexes, and female turnover versus turnover of both sexes not significant.

*** $p < 0.001$.

years) than those in which a turnover had taken place in the prior year (16% of 74 group-years); this difference is highly significant ($\chi^2 = 39.4$, $p < 0.001$). Because of this interdependence, groups without nonbreeders can be expected to do worse than those with nonbreeders independent of the effects of helpers on reproduction per se. We will therefore carefully attempt to untangle the effects of turnovers, particularly when we examine the effects of nonbreeders on reproductive success.

6.2. EFFECTS OF MATE-SHARING

Males

The number of breeding males in a group ranges from one to four (Figure 3.6). Total group reproductive success increases significantly with the number of breeding males (Table 6.3; Figure 6.2). Success per male, however, is greatest for singletons, although not significantly so. These results corroborate our earlier findings (Koenig et al., 1983).

Females

Acorn woodpecker groups may contain as many as three breeding females (Mumme et al., 1983b; Hannon et al., 1985; see Figure 3.6). However, three females are rare (we know of only one case during the period covered here) and are not considered further.

Groups containing two joint-nesting females incubate significantly larger sets of eggs than do groups with a single female (Figure 5.3; Table 6.4), despite egg destruction, which accounts for the loss of an estimated 33% of all eggs laid by such females (Mumme et al., 1983b). Also, successful nests of groups with two females fledge significantly more young than those with only one female. Other comparisons are not significant, although success is generally the same or slightly greater for groups with

TABLE 6.3. Effect of Breeding Males on Reproductive Success

| | Number of Breeding Males | | | | χ^2 | p |
	1	2	3	4		
$\bar{x}$ first egg date	4 May ± 13 (59)	3 May ± 16 (53)	1 May ± 14 (25)	25 April ± 13 (4)	1.5	ns
Clutch size	4.60 ± 1.22 (52)	5.00 ± 1.48 (52)	5.21 ± 1.53 (24)	3.67 ± 1.15 (3)	6.6	ns
Young fledged	1.73 ± 1.64 (93)	2.25 ± 2.02 (71)	3.31 ± 2.48 (35)	3.50 ± 1.30 (4)	14.3	<0.01
Young fledged, successful nests only	2.93 ± 1.07 (55)	3.48 ± 1.44 (46)	4.30 ± 1.94 (27)	3.50 ± 1.30 (4)	12.2	<0.01
Young alive in February	0.84 ± 1.16 (69)	1.20 ± 1.35 (56)	2.03 ± 1.97 (30)	2.33 ± 2.34 (3)	11.4	= 0.01
Young fledged per male	1.73 ± 1.64	1.13 ± 1.01	1.10 ± 0.83	0.88 ± 0.32	5.2	ns
Young alive in February per male	0.84 ± 1.16	0.60 ± 0.68	0.68 ± 0.66	0.58 ± 0.59	0.7	ns
Percent success	59.1 (93)	65.3 (72)	78.4 (37)	100.0 (4)	6.5	ns

NOTE: Statistical tests by Kruskal-Wallis one-way ANOVA except for percent success, which is by a χ^2 contingency test; df = 3. Data are $\bar{x} \pm$ SD (N); N of groups identical for per group and per male analyses. SD of mean first egg date in days.

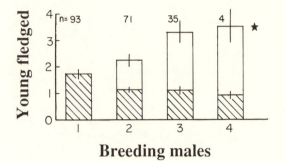

Breeding males

FIGURE 6.2. Number of young fledged per group (total height of bars) and per breeding male (hatched area only) by number of breeding males. Data are $\bar{x} \pm$ SE. Differences in young fledged per group are significant (Kruskal-Wallis one-way ANOVA, $p < 0.05$).

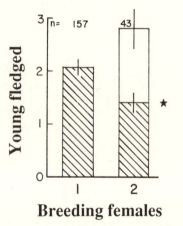

Breeding females

FIGURE 6.3. Number of young fledged per group (total height of bars) and per breeding female (hatched area only) by number of breeding females. Data plotted are $\bar{x} \pm$ SE; difference in young fledged per female is significant.

two females. On a per-female basis, singletons are significantly more successful than those nesting jointly.

Thus, groups with joint-nesting females produce more total young due to larger clutches and broods in successful nests, but success per female is greater among singletons. These conclusions are similar to those for breeding males. Differences between

148

TABLE 6.4. Effect of Breeding Females on Reproductive Success

| | Number of Breeding Females | | | |
	1	2	Statistic	p
$\bar{x}$ first egg date	3 May ± 15 (111)	3 May ± 14 (29)	0.0	ns
Eggs incubated	4.33 ± 0.94 (86)	6.77 ± 1.31 (22)	6.3	<0.001
Young fledged	2.08 ± 1.88 (157)	2.81 ± 2.49 (43)	1.9	=0.06
Young fledged from successful nests only	3.17 ± 1.37 (103)	4.28 ± 1.62 (29)	3.6	<0.001
Young surviving to February	1.13 ± 1.23 (126)	1.65 ± 2.06 (31)	0.6	ns
Young fledged per female	2.08 ± 1.88	1.41 ± 1.25	2.1	<0.05
Young surviving to February per female	1.13 ± 1.23	0.83 ± 1.03	1.0	ns
Percent success	66.3 (160)	65.1 (43)	0.0	ns

NOTE: Statistical tests for number of young fledged, eggs incubated, and young surviving to February by Mann-Whitney U (z-value given); for percent success by χ^2 contingency test. Data are $\bar{x} \pm$ SD (N). SD for mean first egg date in days.

singleton and joint-nesting females are not as strong as those for males, however.

6.3. EFFECT OF NONBREEDERS

The number of nonbreeders in groups ranges from zero to ten; about half of all groups contain no nonbreeding helpers (see Chapter 3). The effects of different numbers of nonbreeders on reproductive output are considered in Table 6.5. Overall, no significant differences exist among groups containing varying numbers of nonbreeders in any of the measures of success either

TABLE 6.5. Effect of Nonbreeders on Reproductive Success.
I. Number of Nonbreeders Present

Number of non-breeders	All Groups			Groups with 2 Breeders		
	Young fledged	Young alive in February	Percent success	Young fledged	Young alive in February	Percent success
0	1.65 ± 1.27 (112)	0.88 ± 1.35 (93)	55.8 (113)	1.36 ± 1.35 (50)	0.71 ± 0.90 (45)	56.0 (50)
1	2.44 ± 1.92 (36)	1.26 ± 1.34 (23)	72.2 (36)	1.73 ± 1.67 (15)	0.50 ± 0.76 (8)	60.0 (15)
2	3.53 ± 2.27 (19)	2.33 ± 1.98 (15)	85.7 (21)	2.33 ± 1.94 (9)	1.00 ± 1.26 (6)	66.7 (9)
3	2.00 ± 1.86 (9)	2.14 ± 1.35 (7)	66.7 (9)	2.38 ± 2.13 (8)	1.17 ± 1.33 (6)	62.5 (8)
4	3.36 ± 2.02 (14)	1.18 ± 1.16 (11)	78.6 (14)	—	—	—
5+	3.07 ± 2.17 (15)	1.73 ± 1.49 (11)	73.3 (15)	—	—	—
χ^2	7.0	5.5	1.9	1.0	1.0	0.1
df	4	4	4	2	2	2
p-value	ns	ns	ns	ns	ns	ns

NOTE: Statistical test is Kruskal-Wallis one-way ANOVA (for young fledged and alive in February) and χ^2 contingency test (for percent success). Tests compare groups with differing numbers of nonbreeders (excluding those with no nonbreeders).

for all groups or groups containing breeding pairs (plus nonbreeders) only. There is a slight tendency for groups with more than one nonbreeder to fledge more young and have more surviving to February, but tests comparing the success of groups with a single nonbreeder with those containing two or more nonbreeders are not significant.

Further, there are no significant differences in success according to the sex of nonbreeders in groups containing only a single nonbreeder (Table 6.6; a similar result was reported for Florida scrub jays by Woolfenden and Fitzpatrick, 1984). Thus, for the

TABLE 6.6. Effect of Nonbreeders on Reproductive Success.
II. Sex of Nonbreeder

| | Sex of Nonbreeder | | | |
	♂	♀	Statistic	p
ALL GROUPS				
Young fledged	2.25 ± 2.00 (16)	2.60 ± 1.88 (20)	0.4	ns
Young alive in February	0.92 ± 1.25 (12)	1.64 ± 1.43 (11)	1.5	ns
Percent success	62.5 (16)	80.0 (20)	0.6	ns
TRIOS ONLY				
Young fledged	1.70 ± 1.89 (10)	1.80 ± 1.30 (5)	0.1	ns
Young alive in February	0.43 ± 0.79 (7)	1.00 ± — (1)	—	ns
Percent success	50.0 (10)	80.0 (5)	0.3	ns

NOTE: Only groups containing a single nonbreeder of the appropriate sex are used. Top part includes all groups regardless of number of breeders; second includes only groups containing a breeding pair plus the nonbreeder. Comparisons by Mann-Whitney U and χ^2 contingency tests.

remainder of our analyses, we will combine all groups containing nonbreeders without regard to number or sex. In so doing, we acknowledge that we are overlooking an important feature of nonbreeding helpers; in fact, to the extent that additional nonbreeders beyond the first may fail to increase the fitness of group members, the existence of groups with several nonbreeders—common in our population—becomes considerably more problematical than explaining the existence of the first nonbreeder. We will, however, return to this issue in Chapter 12 when we explicitly consider the fitness effects of nonbreeders.

Comparing groups with and without nonbreeders, we find strong, significant effects of nonbreeders on all aspects of group reproductive success (Table 6.7). In this table we have also compared the effects of groups with and without nonbreeders

151

TABLE 6.7. Effect of Nonbreeders on Reproductive Success. III. Presence or Absence of Nonbreeders

					$\bar{x}$ Number Nonbreeders	
	No Nonbreeders	Nonbreeders Present	Statistic	p	All groups	Those with Nonbreeders
$\bar{x}$ first egg date	5 May ± 16 (82)	30 April ± 12 (66)	1.5	ns	—	—
Young fledged	1.65 ± 1.80 (112)	2.86 ± 2.03 (93)	4.5	<0.001	1.21 ± 1.82	2.64 ± 1.86
Bad years (4)	0.76 ± 1.21 (45)	1.92 ± 2.05 (25)	2.5	<0.01	1.02 ± 1.39	2.00 ± 1.35
Fair years (4)	1.21 ± 1.52 (34)	3.00 ± 1.69 (34)	4.0	<0.001	1.49 ± 1.91	2.94 ± 1.70
Good years (3)	3.33 ± 1.78 (33)	3.41 ± 2.16 (34)	0.6	ns	1.10 ± 2.06	3.12 ± 2.42
Young fledged from successful nests only	2.98 ± 1.45 (62)	3.80 ± 1.43 (70)	3.7	<0.001	—	—
Young alive in February	0.88 ± 1.35 (93)	1.66 ± 1.56 (67)	3.6	<0.001	—	—
Percent success	56.6 (113)	75.8 (95)	7.5	<0.01	—	—

NOTE: Tests by Mann-Whitney U (z-value given) and χ^2 contingency tests. SD of mean first egg date in days. In the breakdown involving bad, fair, and good years the number of years in each category is in parentheses.

during "good" years (mean fledglings per group >2.50), "fair" years (mean fledglings per group >1.50 and <2.50), and "bad" years (mean fledglings per group <1.50; these data are presented in Table 5.2). The influence of nonbreeders is significant and positive in both bad and fair years, but not in good years.

We analyzed this effect in more detail by examining the effect of nonbreeders during each year separately (Table 6.8). The influence of nonbreeders varies markedly from year to year.

TABLE 6.8. Effect of Nonbreeders on Reproductive Success.
IV. Yearly Variation in Young Fledged

Year	No Nonbreeders	Nonbreeders Present	Increment Due to Nonbreeders	Ratio No Nonbreeders/ Nonbreeders Present
1972	0.67 ± 1.15 (3)	3.00 ± 1.41 (2)	2.33	0.22
1973	0.00 ± 0.00 (6)	2.33 ± 2.73 (6)	2.33	0.00
1974	1.10 ± 1.20 (10)	1.83 ± 1.72 (6)	0.73	0.60
1975	4.00 ± 2.14 (11)	3.70 ± 1.83 (10)	−0.30	1.08
1976	1.25 ± 1.91 (8)	2.92 ± 1.93 (12)	1.67	0.43
1977	1.17 ± 1.75 (12)	2.78 ± 1.99 (9)	1.61	0.42
1978	0.70 ± 1.34 (10)	1.82 ± 2.18 (11)	1.12	0.39
1979	0.84 ± 1.26 (19)	1.50 ± 0.71 (2)	0.66	0.56
1980	3.15 ± 1.57 (13)	4.50 ± 2.07 (6)	1.35	0.70
1981	1.36 ± 1.21 (11)	3.27 ± 1.73 (11)	1.91	0.42
1982	2.78 ± 1.39 (9)	2.89 ± 2.25 (18)	0.11	0.96
Total	1.65 ± 1.80	2.86 ± 2.03	1.21	0.58

In 1973, for example, groups without nonbreeders produced no young whatsoever, whereas in 1975 groups without nonbreeders fledged slightly more young than groups containing nonbreeders.

In Figure 6.4 the relative reproductive success of groups with and without nonbreeders is plotted against the mean number of young fledged per group. For both number of young fledged and number of young surviving to February there is a positive,

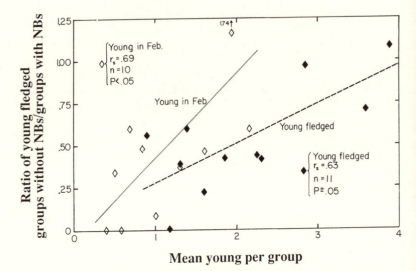

FIGURE 6.4. The ratio of the mean number of young fledged and young surviving to February by groups without nonbreeders to groups with nonbreeders plotted against the mean number of young fledged for all groups during each year.

significant relationship between these variables. Thus, as suggested by the analysis in Table 6.7, the apparent influence of nonbreeders is strongest in bad years when the overall reproductive output is poor. During such years the success of groups with nonbreeders may be four or more times that of groups without nonbreeders. During good years, however, when the overall number of young fledged per group is high, there is at most only a small increase in success correlated with the presence of nonbreeders.

6.4. AN ANALYSIS OF VARIANCE

Table 6.9 compares the effects of all three components of group composition (number of breeding males, breeding females, and presence of nonbreeders) on reproductive success by means of an analysis of variance. The number of breeding males and the presence of nonbreeders significantly influence reproductive success while controlling for the other two components of group composition, whereas the number of breeding females has little effect.

As we saw in section 6.1, there is a strong correlation between the presence of nonbreeders and the absence of turnovers. The lower part of Table 6.9 controls for this interdependence by adding turnovers to the ANOVA. In this analysis, the significant effect of nonbreeders disappears entirely, leaving the number of breeding males as the only significant group composition factor.

TABLE 6.9. Analysis of Variance of the Effect of Group Composition on Reproductive Sucess

	$\bar{x}$ square	df	F-value	β	Adjusted Mean Values			
					0	1	2	3
GROUP COMPOSITION ONLY								
Breeding males	17.1	2	4.9**	0.25	—	1.79	2.26	3.12
Breeding females	0.6	1	0.2	0.03	—	2.21	2.38	—
Nonbreeders	50.3	1	14.3***	0.29	1.74	2.92	—	—
Explained ($R^2 = 0.18$)	12.3	11	3.5***	—				
Residual	3.5	145	—	—				
GROUP COMPOSITION AND TURNOVERS								
Breeding males	14.9	2	5.0**	0.24	—	1.76	2.36	3.00
Breeding females	1.9	1	0.6	0.06	—	2.18	2.48	—
Nonbreeders	9.8	1	3.3	0.14	1.98	2.58	—	—
Turnovers	67.4	1	22.5***	0.36	1.23	2.79	—	—
Explained ($R^2 = 0.31$)	16.1	14	5.4***	—				
Residual	3.0	136	—	—				

NOTE: Categories are breeding males 1, 2, 3; breeding females 1, 2; nonbreeders and turnovers 0 (none), 1 (present). Grand mean: 2.24 young fledged per group. All interactions between variables not significant.
** $p < 0.01$;   *** $p < 0.001$;   other $p > 0.05$.

There thus appears to be no significant effect of nonbreeders on reproductive success independent of the influence of turnovers. As before, the effect of breeding males is greater than that of breeding females. Overall, 31% of the variance ($R^2 = 0.31$) in reproductive success of groups can be explained by these three components of group composition in conjunction with prior group history.

6.5. DISCUSSION

Reproductive Success and Prior Breeding Experience

Prior breeding experience, as indicated by presence or absence of a turnover in the breeders during the prior year, has a strong influence on group reproductive success (Table 6.2). In contrast with the Florida scrub jay, in which pairs containing one experienced partner had similar clutch sizes and reproductive success as pairs in which both partners were experienced (Woolfenden and Fitzpatrick, 1984), groups of acorn woodpeckers had poor success whether breeders of one or both sexes were inexperienced.

This difference between the two species may reflect the different primary agents of nest mortality: predation in Florida scrub jays and starvation in acorn woodpeckers. Possibly the threat of nest predation in jays can be mitigated by a single experienced individual, whereas in order for woodpeckers to overcome starvation and hatching failure (Table 5.9), breeders of both sexes must have prior experience together. Both starvation and hatching failure would often decrease brood size, and, indeed, groups with no turnovers have both a higher probability of success and fledge more young from successful nests.

Prior breeding experience has important indirect effects as well (see Figure 5.1). Groups experiencing turnovers in breeding composition generally do not contain nonbreeding helpers the following year. (In fact, the low success of such groups means that they are relatively unlikely to have nonbreeding helpers until two years following the turnover.) Thus, the poor success of groups without nonbreeders can in part be attributed to the indirect effect of turnovers on reproductive success rather than the influence of nonbreeders per se.

We therefore have attempted to control for turnovers in analyzing the effects of group composition of reproductive success. However, as pointed out by Woolfenden and Fitzpatrick (1984:248), this interdependence of demographic variables may have important consequences to life history strategies. We realize that we may be obscuring important features of this system by treating such interrelated variables as independent. The difficulties of doing otherwise require more sophisticated demographic modeling, procedures that we consider in Chapters 12 and 13.

Reproductive Success and Group Size

Reproductive success is highest in groups of eight birds (Figure 6.1a). There are at least two group size effects that might produce this pattern:

1. Lowered nest predation. Nest predation might be reduced in larger groups as a consequence of either enhanced predator detection (Pulliam and Caraco, 1984), more effective predator mobbing (Hoogland and Sherman, 1976), or both. Indeed, reduced nest predation is a means by which larger groups experience greater reproductive success in Florida scrub jays (Woolfenden and Fitzpatrick, 1984) and bicolored wrens (Austad and Rabenold, 1985).

We have little data relevant to this hypothesis, in part because of the relative rarity of nest predation in our population (see section 5.4). However, as we discuss in section 7.5, the alarm calls given by acorn woodpeckers are consistent with the hypothesis that predator detection increases in larger groups. Although adult death is a minor source of nest loss (Table 5.9), increased predator detection may serve to increase nesting success. More important, acorn woodpeckers readily mob nest predators such as snakes, and the increased mobbing made possible by larger group size may result in a lower incidence of nest loss.

The nest predation hypothesis would be supported if the group-size-dependent probability of success matched that of the average number of young fledged. Although this is arguably true (cf. Figures 6.1a and 6.1c), differences in success are not significantly related to group size. Instead, larger groups are more successful primarily because they fledge larger broods from successful nests (Figure 6.1b).

With increasing group size, predator detection and mobbing can be expected to either increase indefinitely or to peak asymptotically. Thus, the lower overall success of groups larger than eight is not consistent with this hypothesis.

2. Increased feeding rates. Mumme (1984) found increased incubation attendance and increased feeding rate, both per nest and per nestling, in groups of up to eight acorn woodpeckers; feeding rates in groups larger than eight appear to decline. This close parallel with the size-dependent pattern of reproductive success (Figure 6.1a), combined with the significance of nestling starvation (section 5.4), suggests that increased feeding is the proximate factor leading to the increased success of larger groups. This interpretation must be made cautiously, however, as a positive correlation between group size, nestling feeding rates, and reproductive success could be spurious, due to interdependence with a fourth, unmeasured variable such as territory quality (Brown et al., 1978; Wilkinson and Brown, 1984).

Increased reproductive success parallels increased feeding rates in several other cooperative breeders (e.g., the white-fronted bee-eater [Emlen, 1981], pied kingfisher [Reyer, 1980, 1984, 1986], chestnut-bellied starling [Wilkinson, 1982; Wilkinson and Brown, 1984], purple gallinule [Hunter, 1985], and dunnock [Davies, 1986]). This is by no means a universal feature of such species, however. In the grey-crowned babbler, for example, the number of nonbreeding helpers has no effect on feeding rates, although reproductive success is positively correlated with group size (Brown et al., 1978; Brown and Brown, 1981b). A similar result has been found in the white-browed sparrow weaver (Lewis, 1981, 1982). In both species, helpers appear to allow breeders to renest more rapidly rather than increase the brood size of individual nests.

Reproductive Success and Number of Breeders

Reproductive success increases significantly with additional breeding males but not with additional breeding females, particularly after controlling for turnovers and aspects of group composition (Table 6.9). This difference between the sexes is not likely to be a consequence of dissimilar size-dependent feeding patterns: breeding females do not adjust their feeding rates to group size or the number of breeders in the group, whereas breeding males tend to *decrease* their feeding rate in larger groups and groups with more breeders (Mumme, 1984). There are at least two possible explanations.

1. Decreased nest predation. This hypothesis requires that nest defense in groups containing several breeding males should be more effective relative to groups with only a single breeding male, whereas there should be no difference in groups containing one or two breeding females.

We do not know how nest defense varies according to the number of breeders within groups. However, breeding females and breeding males defend granaries with equal vigor (Mumme and de Queiroz, 1985). This suggests that additional breeding males are unlikely to be more effective at dissuading potential predators than are additional breeding females.

2. Reproductive competition. Reproductive competition is particularly intense between joint-nesting females. Such competition leads to considerable egg destruction (Mumme et al., 1983b), as well as decreased hatching success and increased frequency of runt eggs (Koenig, 1980b; Koenig et al., 1983; see also section 5.4). Competition also occurs among mate-sharing males (Mumme et al., 1983a), but does not generally involve activities that are likely to decrease reproductive success. This disparity in the forms of intrasexual competition could lead to increased reproductive success in groups with additional breeding males, but not in groups with an additional breeding female.

Because of the more destructive female-female competition, this hypothesis predicts that set size and brood size should increase relatively little with an additional breeding female compared to with additional breeding males. Contrary to this prediction, set size increases significantly with number of breeding females but not with breeding males; number of young fledged in successful nests increases with both the number of breeding males and breeding females (Tables 6.3 and 6.4). Thus, our data do not support this hypothesis for the differing patterns of size-dependent reproductive success for breeding males and females. Additional data will be necessary to resolve this issue.

Reproductive Success and Nonbreeding Helpers

Reproductive success is significantly greater in groups containing nonbreeding helpers (Table 6.7), particularly in years when breeding conditions are poor (Figure 6.4). There are no apparent differences between the effects of nonbreeding males and nonbreeding females (Table 6.6). Furthermore, there is no enhancement of reproductive success in groups containing two or more nonbreeders over those with only a single nonbreeder (Table 6.5).

Nonbreeding helpers contribute significantly less than breeders to all aspects of nest care and territory defense (Mumme, 1984; Mumme and de Queiroz, 1985). Breeding acorn woodpeckers do not adjust their feeding rates to the presence of nonbreeding helpers (Mumme, 1984), in contrast with several other species of cooperative breeders (e.g., Gaston, 1973; Brown, 1978; Lewis, 1982). Thus, groups with nonbreeding helpers have higher incubation attendance and feeding rates than smaller groups without nonbreeders (Mumme, 1984). Again, this correlation between nonbreeders, reproductive success, and feeding rates suggests, but does not prove, a direct relationship. A more detailed understanding of the proximate factors leading to the observed effects of nonbreeders on reproductive success requires data and analyses beyond the scope of this book.

In any case, nonbreeders do not have a detrimental effect on group reproductive success, as suggested by Zahavi (1974) for the Arabian babbler (but see Brown, 1975). It is less clear, however, whether the presence of nonbreeding helpers directly increases the reproductive success of groups. Part of the reason for this ambiguity is that the strong influence of nonbreeding helpers is largely an artifact of the correlation between turn-overs in breeding composition and the presence of nonbreeding helpers: groups in which turnovers have occurred are much less successful and also rarely have nonbreeding helpers the following year. Nonbreeders have no significant overall influence on reproductive success once the effect of prior group history is statistically removed.

Similarly equivocal results have been obtained by Wool-fenden and Fitzpatrick (1984:195) on the Florida scrub jay, in which a helper effect remains, but is not significant, when controlling for experience and territory quality. Given the extensive nature of these studies, it seems possible that unequivocally significant results will not emerge even with additional years of data. Rather, the overall fitness effect of nonbreeders on reproductive success in these and many other cooperative breeders, if any, may truly be small and of questionable evolutionary significance. A similar proposal has been made by Jamieson (1986) and Jamieson and Craig (in press).

We make this suggestion despite the fact that a few apparently clear exceptions exist. One such example is the stripe-backed wren. In this tropical cooperatively breeding species, groups with two or more helpers outproduce pairs or trios (pairs plus one helper) by a factor of six, a pronounced effect unlikely to be erased by differences in territory quality or breeding experience (Rabenold, 1984, 1985; see also Chapter 14).

Another example of a pronounced increase in reproductive success as a result of helpers was reported by Reyer (1980) in a population of pied kingfishers inhabiting a poor environment; a second study population living in a more productive area showed no such enhancement. In both of these species it appears clear that helpers contribute positively to group reproductive success. Apparent reproductive enhancement as a result of helpers has been suggested for other species (e.g., the superb blue wren; Rowley, 1965), but generally the effect has not been enough to eliminate the possibility that territory quality or other variables unrelated to helpers per se caused the observed differences (see Koenig, 1981b).

A third example of a species in which helpers have been shown to have a positive effect on reproductive success is the grey-crowned babbler. Brown et al. (1982), in the only experimental study of the effects of helpers thus far, removed all but a single helper from groups following fledging of first broods; comparison with control groups of similar original size in which birds were caught but not removed indicated subsequent reproductive enhancement due to helpers by a factor of three.

However, because of the brevity of the study, their conclusions are debatable (Counsilman, 1977b; Lewis, 1981; Woolfenden and Fitzpatrick, 1986). In particular, any short-term disruption of the social behavior within groups resulting from the mass removal of a large fraction of group members (a consequence that we consider quite likely) would lead to similar results even if helpers produced no reproductive enhancement under undisturbed conditions. The extreme enhancement of reproductive success in groups containing multiple as compared to a single helper is in fact matched among cooperative breeders only by the stripe-backed wren. Hence, Brown et al.'s results cannot be representative of cooperative breeders in general.

Nevertheless, their work offers an important alternative to the often equivocal results stemming from the empirical approaches of virtually all other studies of cooperative breeders to date, including the acorn woodpecker. A combination of long-term empirical study and judicious experimental procedures would be particularly fruitful in unambiguously solving the question of whether helpers increase the reproductive success of the groups in which they live.

Despite the absence of a significant reproductive effect of nonbreeders when we combine all data and control for prior experience and group composition, nonbreeders may augment success in our population during years when ecological conditions are poor (Table 6.7; Figure 6.4). Poor years follow poor acorn crops, which occur about once every four or five years at Hastings (Hannon et al., 1987). Thus, a breeder can expect to suffer at least one such poor year during its lifetime, and a strong enhancement of reproductive success due to nonbreeders during that year may be biologically important.

Per Capita Reproduction and Cooperative Breeding

On a per capita basis, acorn woodpeckers that share mates— both breeding males and breeding females—almost invariably have lower reproductive success than individuals breeding on their own. Thus, an individual male produces more offspring breeding by himself than if he shares paternity equally with one or more other males, and a single female does better than if she shares maternity equally with another female. This pattern of reproductive success is consistent with the hypothesis that mate-sharing is forced on the individuals involved by ecological constraints, since they would apparently have higher reproductive success breeding on their own (see also Koenig, 1981b; Koenig and Pitelka, 1981).

However, other factors, including differences in survivorship and the assumption of equal parentage, must be considered before this conclusion can be accepted. (We will return to this problem in Chapter 13.) Furthermore, even if ecological constraints are important, this does not exclude other factors; recent theoretical treatments have suggested that relatedness, reproductive success, sexual conflict, and reproductive bias may all be importantly interrelated in group-living societies (Vehrencamp, 1979, 1983; Stacey, 1982; Davies, 1985; Davies and Houston, 1986).

For example, optimal group composition in acorn woodpeckers differs for the two sexes: a single male is most successful (as indicated by per capita annual reproductive success), but being shared by three or four males results in the greatest reproductive success for a female (indicated by total group success). Similarly, a single female has the greatest success by herself, but breeding male(s) have higher fitness if two joint-nesting females are present. Such conflicts are apparently important in influencing the highly variable mating system in the dunnock (Davies, 1985, 1986; Houston and Davies, 1985; Davies and Houston, 1986) and could also be significant here.

It is equally critical to consider *lifetime* reproductive success of the individuals involved rather than the annual reproductive output (Cavalli-Sforza and Bodmer, 1961; Grafen, 1982; Clutton-Brock, 1983, in press; Woolfenden and Fitzpatrick, 1984; Koenig and Albano, 1987). In order to carry out such analyses, it is necessary to consider what effects mate-sharing and helping behavior have on the second major component of fitness: survivorship. We will examine this demographic feature in the next chapter.

6.6. SUMMARY

Acorn woodpecker annual reproductive success varies significantly with group size and composition. In general, total group reproductive success increases with increasing group size as well as increasing numbers of breeding males, breeding females, or the presence of nonbreeding helpers. Per capita reproductive success, however, declines with increasing group size and number of breeders of either sex. Nonbreeders appear to have a strong positive effect on group reproductive success, particularly during years of otherwise poor reproduction.

Despite this pattern of reproductive enhancement, various analyses controlling for potentially confounding variables largely erase the apparent effects of increasing group size. When controlling for group composition and turnovers in breeding group composition, only the number of breeding males still produces a significant enhancement of group reproductive success; the effect of nonbreeders is largely eliminated. Thus, there appears to be little overall enhancement of reproductive success due to the presence of nonbreeders within groups independent of other correlated characters, except possibly during poor years.

The lack of an unequivocal effect of nonbreeders on reproduction is a result also found in many other cooperative species, including the intensively studied Florida scrub jay. With few exceptions, the effects of nonbreeders on reproductive success, independent of territory and experience effects, appear to be small in cooperatively breeding species. Evidence from the stripe-backed wren and pied kingfisher, however, suggests that at least in some species a clear enhancement to reproduction due to nonbreeding helpers may occur. Brown et al.'s (1982) experimental study also supports the existence of a positive effect of helpers in the grey-crowned babbler.

Results based on annual reproduction are not sufficient to answer questions concerning the evolution of cooperative breeding. Other variables, such as sexual conflict and reproductive bias, are also likely to be important. Moreover, it is necessary to consider the effects of helpers on lifetime reproduction rather than annual reproduction in order to determine the evolutionary consequences of this phenomenon.

CHAPTER SEVEN

Survivorship of Breeders

Fitness is best measured by lifetime reproductive success. Thus, a thorough evaluation of individual fitness must incorporate data on the potential for initial and repeated reproduction— hence, survivorship. The purpose of this chapter, then, is to explore the ecological and social variables influencing the survivorship of acorn woodpeckers.

Here we deal only with the survivorship of breeders, which are sedentary. Because it is impossible to distinguish directly between mortality and undetected emigration from the study area, estimating the survivorship of dispersal-prone first-year birds and nonbreeding adult helpers is more difficult. Survivorship of these individuals will be addressed in Chapter 8.

7.1. GENERAL PATTERNS

Annual

Once an acorn woodpecker acquires breeding status in a territory, it will usually remain there until death. Under most circumstances, established breeders that disappear can be presumed dead. Exceptions occur when established breeders abandon their territories following acorn crop failures, granary destruction, failed initial breeding attempts, or forced eviction by intruders (MacRoberts and MacRoberts, 1976; Koenig et al., 1984; Hannon et al., 1987; see also Chapter 3). We know of thirteen such cases (eight males, five females) in which the disappearance of breeders probably represented territory abandonment and subsequent emigration attributable to one or more of the above factors. These individuals are excluded from the following survivorship analyses. We also have observed five cases (three females, two males) of established breeders emigrating within the study area under conditions not attributable to any of the factors cited above. These cases suggest that a few of the breeders that we assumed dead, actually emigrated from the study area. However, considering the samples involved (443 breeder-years involving 167 individuals; see Table 7.1), these biases are likely to be small and are unlikely to affect our conclusions.

Consistent with our treatment of annual reproductive success (Chapter 5), calculation of annual survivorship was based on the acorn fiscal year—1 September to 31 August. Percent survivorship for each year was calculated by determining the number of birds alive at the end of the year compared to the number at risk during the period. Only banded breeders were used. Since September 1973, 9 years of survivorship data, including 96 individual males (273 male breeder-years), and 71 individual females (170 female breeder-years) were available for analysis.

Variation in annual survivorship of male and female breeders is shown in Table 7.1. Overall, male survivorship was 82.4% and female survivorship was 71.2%. Among years, survivorship varied from 65.7% (1977–78) to 100% (1973–74) for males and 52.2% (1977–78) to 100% (1973–74) for females. Although female survivorship is consistently lower than that of males, male and female annual survivorship is positively correlated (Figure 7.1). This suggests that some of the ecological factors affecting breeder survivorship are the same for the two sexes. A likely candidate is the energetic content of acorn stores, which we showed to be correlated with adult winter survival in Table 4.10.

Table 7.1. Variation in Annual Survivorship of Male and Female Breeders

	Males			Females		
Year	Survived	Died	% survival	Survived	Died	% survival
1973–74	13	0	100.0	8	0	100.0
1974–75	18	3	86.7	12	3	80.0
1975–76	19	4	82.6	11	8	57.9
1976–77	29	7	80.6	13	3	81.3
1977–78	23	12	65.7	12	11	52.2
1978–79	22	8	73.3	9	7	56.3
1979–80	30	3	90.9	16	3	84.2
1980–81	32	5	86.5	19	4	82.6
1981–82	39	6	86.7	21	10	67.7
Total	225	48	82.4	121	49	71.2
C.V.[a]	—	—	14.7	—	—	22.0
	$\chi^2 = 14.1$, df = 8, $p < 0.1$			$\chi^2 = 15.2$, df = 8, $p < 0.1$		

[a] Coefficient of variation based on arcsin transformed values.

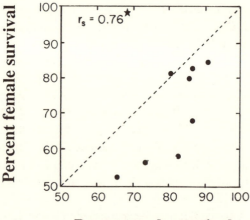

Figure 7.1. Relationship between annual survivorship of male and female breeders. Along the dotted line survivorship of males equals that of females. Data from 1973–74 are excluded because of small sample size.

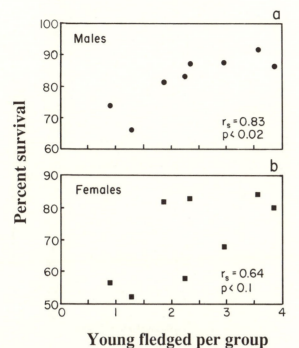

FIGURE 7.2. Relationship between young fledged per group and annual survivorship of (a) male and (b) female breeders. Data from 1973–74 excluded.

The survivorship of breeding males is positively correlated with mean fledging success per group during the same year (Figure 7.2). This indicates that good reproductive years (e.g., breeding seasons following good acorn crops that extend late into the winter; see Chapter 4) tend to be years of high survivorship as well. The correlation for female breeders is also positive, but not significant (Figure 7.2).

For both male and female breeders, annual survivorship is negatively correlated with population density at the end of the previous year (Figure 7.3). Although both correlations fall short of statistical significance, they nonetheless suggest density-dependent mortality. (A combined χ^2 test of the hypothesis using the probabilities for males and females as independent

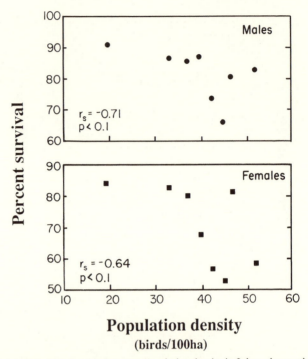

FIGURE 7.3. Relationship between population density in July and annual survivorship of male and female breeders in the following year (1 September–31 August). Data from 1973–74 excluded.

results [Winer, 1971:49] is significant; $\chi^2 = 10.4$, df = 4, $p < 0.05$.) A similar density-dependent effect was noted for reproductive success excluding years of acorn crop failure in Chapter 5. Additional discussion of factors potentially regulating population size is postponed to Chapter 11.

Seasonal

The seasonal pattern of breeder mortality is shown in Figure 7.4. Male mortality is constant through all seasons ($\chi^2 = 0.3$, df = 3, $p > 0.5$), whereas mortality of breeding females is highly

170

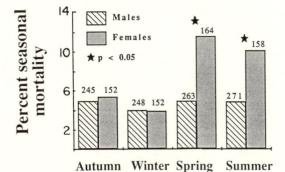

FIGURE 7.4. Mortality rates of male and female breeders during autumn (September–November), winter (December–February), spring (March–May), and summer (June–August). Sample size (number of breeder-seasons) are shown for each category. Asterisk denotes significant differences (by χ^2 test) between males and females.

seasonal: females are much more likely to die during spring and summer than during autumn and winter ($\chi^2 = 8.9$, df $= 3$, $p < 0.05$; see also Koenig et al., 1983). Female mortality during spring and summer is significantly greater than that of males during these seasons ($\chi^2 = 6.4$ and 4.5, respectively, both $p < 0.05$). During the autumn and winter, however, male and female mortality rates are identical ($\chi^2 = 0.03$ and 0.00, respectively, both $p > 0.5$; Figure 7.4).

Seasonal differences in male and female mortality would have been even more pronounced were it not for the acorn crop failure of 1978–79. Of the 14 recorded autumn-winter female deaths, 6 (42.9%) occurred in 1978–79 (Table 7.2). Female mortality during the autumn and winter of 1978–79 was 46.2% (an annual rate of 71.1%), significantly greater than the average 5.8% (11.3% annual) autumn-winter mortality rate observed during the other 8 years of the study. Male autumn-winter mortality was slightly higher during 1978–79, but not significantly so (Table 7.2). The female-biased mortality during this year was probably a consequence of male domination over females under conditions of food stress (Hannon et al., 1987).

171

TABLE 7.2. Autumn-Winter Mortality of Male and Female Breeders during the Acorn Crop Failure of 1978–79 and Other Years

	Males		Females	
	1978–79	*Other years*	1978–79	*Other years*
Survived	20	203	7	129
Died	5	17	6	8
Percent mortality	20.0	7.7	46.2	5.8
	$G = 3.1, p < 0.1$		$G = 12.4, p < 0.001$	

NOTE: G-value adjusted by Williams's correction (Sokal and Rohlf, 1981).

7.2. AGE, BREEDING EXPERIENCE, AND SURVIVORSHIP

The age-specific survivorship schedules for known-age male and female breeders are shown in Table 7.3. Because sample sizes for most age classes are quite small, particularly for females, statistical interpretation is difficult. Overall, however, survivorship of breeders does not appear to be strongly related to age. Furthermore, we can find no evidence of senescence, at least among our small sample of males older than seven years and females older than six years (Table 7.3).

Another method for looking for senescence is to compare survivorship of known- versus unknown-age breeders (Woolfenden and Fitzpatrick, 1984). Because our sample of unknown-age breeders undoubtedly includes some very old birds who were already breeders when the study began, the sample of unknown-age breeders should show lower survivorship rates if senescence were an important cause of breeder mortality. Indeed, for both males and females, survivorship is slightly higher among known-age breeders than among unknown-age breeders (83.6% versus 81.7% for males and 75.6% versus 71.4% for

TABLE 7.3. Age-Specific Survivorship of Known-Age Male and
Female Breeders

Males				Females			
Age	Survived	Died	% survival	Age	Survived	Died	% survival
1	14	2	87.5	1	4	1	80.0
2	21	9	70.0	2	10	2	83.3
3	19	3	86.4	3	8	4	66.7
4	17	2	89.5	4	5	3	62.5
5	13	2	86.7	5	4	1	80.0
6	8	1	88.9	6 +[a]	14	4	77.8
7 +[b]	33	5	86.8				

[a] Includes unknown-age birds that were at least six years old.
[b] Includes unknown-age birds that were at least seven years old.

females). However, these small differences do not approach
statistical significance (Lee-Desu survivorship statistic $D = 0.08$
for males and $D = 1.14$ for females, both $p > 0.25$; Hull and
Nie, 1981).

Thus, our data indicate no relationship between age and
survivorship. It is nonetheless probable that some form of age-
dependent mortality sets an upper limit to the potential life
span of acorn woodpeckers (see, for example, Botkin and Miller,
1974; Woolfenden and Fitzpatrick, 1984). We therefore arbitrarily set maximum longevity at 15 years in subsequent analyses. This matches the age of the oldest-known male currently
alive in the population ($\male$100, banded as an adult nonbreeding
helper in July 1972 and still breeding as of 1986, thus probably
15 years old); the oldest known females are 11 years old ($\female$193,
banded as a nestling in 1975, and $\female$260, banded as an adult in
spring 1976, both of which were still breeding as of 1986). Additional years of demographic data will be needed before we are
able to empirically demonstrate that a limit to the life span of
acorn woodpeckers exists, however.

173

Prior breeding experience also appears to have little effect on breeder survivorship. Annual survivorship of breeding males that were known or suspected to be experienced breeders was 84.2% ($N = 190$ breeder-years), not significantly greater than the 77.3% ($N = 75$ breeder-years) for inexperienced breeders ($\chi^2 = 1.7$, $p > 0.15$). For females, survivorship was actually slightly lower among experienced females (69.0%, $N = 116$ breeder-years) than among inexperienced females (75.0%, $N = 48$ breeder-years; $\chi^2 = 1.0, p > 0.20$). Limiting the analysis to just those birds with known previous breeding histories produces similar results.

Because of the apparent independence of age and breeder survivorship in our sample, we can calculate a composite survivorship curve using known- and unknown-age breeders. These data for males and females are shown in Figure 7.5. The difference in survivorship between males and females is significant

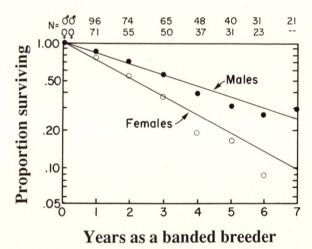

Years as a banded breeder

FIGURE 7.5. Proportion of male and female breeders surviving one to seven years after banding. The x-axis represents years as a banded breeder rather than age per se, and each point represents the surviving proportion of all birds that could have been alive x years after having become a banded breeder. The lines correspond to the expected proportions given annual survivorship rates of 82% for males and 71% for females.

(Lee-Desu survivorship statistic $D = 4.52$, $p < 0.05$; Hull and Nie, 1981). Also plotted in Figure 7.5 are the expected survivorship curves given a constant annual survivorship rate of 82% for males and 71% for females. None of the observed proportions differs significantly from the corresponding expected values.

7.3. SURVIVORSHIP AND TERRITORY QUALITY

In addition to influencing reproductive success, territory quality may have important effects on survivorship. Such effects may have significant consequences not only for the longevity of the residents but also for the probability of territorial inheritance by offspring. This latter phenomenon is discussed at length in Chapter 9. Here we discuss the effects of territory quality on the survivorship of resident breeders.

We used two indices of territory quality. The first was the number of acorn storage holes present on the territory. Male and female annual survivorship rates for territories with different numbers of storage holes are shown in Table 7.4. With the

TABLE 7.4. Annual Survivorship of Male and Female Breeders in Relation to the Number of Acorn Storage Holes on the Territory

Number of Storage Holes	Males			Females		
	Survived	Died	% survival	Survived	Died	% survival
<1,000	18	8	69.2	13	10	56.5
1,000–2,000	25	4	86.2	22	6	78.6
2,000–3,000	78	14	84.8	45	17	72.6
3,000–4,000	48	12	80.0	16	6	72.7
>4,000	58	10	85.3	33	10	76.7
Total	227	48	82.5	129	49	72.5
	$\chi^2 = 4.4$, df $= 4$, $p > 0.2$			$\chi^2 = 3.9$, df $= 4$, $p > 0.2$		

175

possible exception of the smallest granaries ($<1,000$ storage holes), survivorship of both males and females appears to be independent of this measure of territory quality.

A second measure of the effect of territory quality on survivorship involved categorizing breeders according to the frequency with which their territories were occupied (the type 1, 2, and 3 territories of Chapter 3). In this analysis we compared the survivorship of breeders living on territories that were permanently occupied (213 male and 122 female breeder-years) to that of breeders living on territories that were unoccupied one or more springs (62 male and 56 female breeder-years). This division entails a potential circularity: low breeder survivorship might lead directly to lower occupancy rates at a particular territory.

Survivorship of males was 85.0% on permanently occupied territories compared to 74.2% on territories not permanently occupied. This difference approaches statistical significance (Lee-Desu $D = 3.53$, $p < 0.07$), suggesting that males may live longer in better territories. Female annual survivorship was slightly higher in permanently occupied territories (74.2% versus 67.9%) but the difference does not approach statistical significance ($D = 0.76$, $p > 0.35$). Thus, neither analysis suggests a strong statistical effect of territory quality on breeder survival.

7.4. SURVIVORSHIP AND GROUP COMPOSITION

The relationship between group composition and survivorship is critically important to an understanding of the evolution of cooperative breeding (Brown, 1978, 1982; Vehrencamp, 1978, 1979; Ligon, 1983; Wiley and Rabenold, 1984; Woolfenden and Fitzpatrick, 1984). However, the analysis of this relationship presents special statistical difficulties.

For example, in most cooperative breeders, including acorn woodpeckers, group composition may vary considerably over the lifetime of individuals. As a result, it is usually impossible

TABLE 7.5. Annual Survivorship of Male and Female Breeders in Relation to Total Group Size

	Total Group Size				
	2–3	4–5	6–7	8+	Total
MALES					
Survived	36	73	49	69	227
Died	23	12	9	4	48
Percent survival	61.0	85.9	84.5	94.5	82.5
	$\chi^2 = 27.1$, df $= 3$, $p < 0.001$				
FEMALES					
Survived	29	47	21	32	129
Died	15	16	11	7	49
Percent survival	65.9	74.6	65.6	82.1	72.5
	$\chi^2 = 3.6$, df $= 3$, $p > 0.3$				

to measure easily the effect that a particular group composition has on the life span of an individual. Because of this problem, we (as most workers; e.g., Vehrencamp, 1978; Reyer, 1984; Woolfenden and Fitzpatrick, 1984) analyze the effect of group composition on annual survivorship rather than on life span. For surviving individuals, we use the average group composition that existed during the year in question. For individuals that die, we use the group composition parameters that existed at the time of death.

The relationship between annual survivorship of male and female breeders and total group size is shown in Table 7.5. For males, annual survivorship increases sharply with increasing group size, and this relationship holds across a wide range of group sizes. With the possible exception of very large groups (eight or more individuals), female survivorship appears to be largely unrelated to total group size.

Tables 7.6 and 7.7 show annual survivorship in relation to the number of breeding and nonbreeding individuals in the

TABLE 7.6. Annual Survivorship of Male and Female Breeders in Relation to the Number of Breeding Birds in the Group

	Males				Females			
TOTAL BREEDERS	2	3	4	5+	2	3	4	5+
Survived	41	84	63	39	42	45	30	12
Died	21	12	8	7	21	17	7	4
Percent survival	66.1	87.5	88.7	84.8	66.7	72.6	81.1	75.0
	$\chi^2 = 15.3$, df $= 3$, $p < 0.005$				$\chi^2 = 2.5$, df $= 3$, $p > 0.3$			
BREEDING MALES	1	2	3+		1	2	3+	
Survived	45	97	85		48	60	21	
Died	19	15	14		27	13	9	
Percent survival	70.3	86.6	85.9		64.0	82.2	70.0	
	$\chi^2 = 8.7$, df $= 2$, $p < 0.05$				$\chi^2 = 6.3$, df $= 2$, $p < 0.05$			
BREEDING FEMALES	1	2			1	2		
Survived	181	46			91	38		
Died	43	5			39	10		
Percent survival	80.8	90.2			70.0	79.2		
	$\chi^2 = 2.5$, $p > 0.1$				$\chi^2 = 1.5$, $p > 0.1$			

group. Breeding male survivorship is substantially higher in groups with more than two breeders, and this appears to result primarily from the presence of additional breeding males (Table 7.6). Breeding female survivorship is slightly higher in groups with more than two breeders and two or more breeding males, but the differences are small and not as pronounced as in males. For both male and female breeders, the presence of two breeding females is associated with slightly higher survivorship, but the differences are not significant.

Breeding male survivorship increases dramatically as the number of nonbreeders increases (Table 7.7). The effect of non-breeding first-year birds is especially strong, but the relationship between breeder male survivorship and number of adult non-

TABLE 7.7. Annual Survivorship of Male and Female Breeders in Relation to the Number of Nonbreeding Group Offspring Present in the Group

	Males				Females			
TOTAL NONBREEDERS	0	1–2	3–4	5 +	0	1–2	3–4	5 +
Survived	42	77	53	55	30	42	30	27
Died	24	16	4	4	14	20	9	6
Percent survival	63.6	82.8	93.0	93.2	68.2	67.7	76.9	81.8

$\chi^2 = 25.4$, df $= 3$, $p < 0.001$ $\qquad$ $\chi^2 = 2.9$, df $= 3$, $p > 0.2$

	Males			Females		
ADULT NONBREEDERS	0	1–2	3 +	0	1–2	3 +
Survived	112	68	47	67	38	24
Died	34	9	5	23	18	8
Percent survival	76.7	88.3	90.4	74.4	67.9	75.0

$\chi^2 = 7.4$, df $= 2$, $p < 0.05$ $\qquad$ $\chi^2 = 0.9$, df $= 2$, $p > 0.2$

	Males			Females		
FIRST-YEAR BIRDS	0	1–2	3 +	0	1–2	3 +
Survived	58	112	57	41	65	23
Died	31	13	4	31	14	4
Percent survival	65.2	89.6	93.4	56.9	82.3	85.2

$\chi^2 = 28.0$, df $= 2$, $p < 0.001$ $\qquad$ $\chi^2 = 14.7$, df $= 2$, $p < 0.01$

NOTE: The significant relationship between female survivorship and the number of first-year birds is probably artifactual (see section 7.4).

breeders is also significant. Furthermore, breeding male survivorship is enhanced in the presence of both male and female nonbreeders. Breeding male survivorship is 90.9% in the presence of male nonbreeders ($N = 88$) and 89.5% in the presence of female nonbreeders ($N = 105$), compared to 78.6% ($N = 187$) when male nonbreeders are absent and 78.2% ($N = 170$) when female nonbreeders are absent. Thus, annual survivorship of breeding males appears to increase when nonbreeders of any age or sex are present.

For breeding females, the presence of first-year birds also appears to have a strong effect on annual survivorship (Table 7.7).

However, this relationship is probably an artifact: in our survivorship analyses, we reclassify all first-year birds as adult nonbreeders in their first spring, when female mortality is most severe (Figure 7.4). Thus, most female mortality occurs when first-year birds are, by definition, absent, resulting in a spurious relationship between female survivorship and the presence of first-year birds.

This interpretation is supported by the absence of a relationship between female survivorship and the number of adult nonbreeders present in the group (Table 7.7). Furthermore, a within-season analysis of the relationship between female survivorship and the presence of nonbreeders suggests no correlation between these factors. This methodological bias does not, however, affect our conclusion that male survivorship is significantly higher when nonbreeders are present, as male mortality is distributed evenly throughout the year (Figure 7.4; see also section 7.1, above).

In summary, annual survivorship of breeding males increases sharply with group size, primarily due to the presence of nonbreeders, although other breeding birds (particularly other breeding males) may have an effect as well. Annual survivorship of breeding females, however, appears to be insensitive to the presence of nonbreeders and may be influenced only weakly by the presence of additional breeders.

7.5. UNTANGLING THE CORRELATES: A MULTIVARIATE APPROACH

The relationships between group composition and breeder survivorship may be confounded by at least two variables: territory quality and previous breeding experience (Brown, 1978). As shown above, two independent measures of territory quality are positively, although not significantly, correlated with survivorship of breeding males. Similarly, breeding birds in large groups (i.e., with many nonbreeders) are usually experienced breeders.

To untangle the relationships among survivorship, group composition, territory quality, and prior breeding experience, we performed a log-linear analysis of our survivorship data using a logit model. Logit models are analogous to standard multiple regressions except that they use categorical rather than continuous variables.

We determined the effects of five dichotomous independent variables on survivorship. The effect of each independent variable was assessed while simultaneously controlling for the influence of the other variables. The independent variables used in the logit model were territory quality (permanently occupied territories versus nonpermanently occupied territories), prior breeding experience (inexperienced versus experienced breeders), the number of breeding males (one breeding male versus two or more breeding males), the number of breeding females (one versus two or more), and the presence of nonbreeders (nonbreeders absent versus nonbreeders present).

Complete data were available for a total of 265 male breeder-years and 164 female breeder-years. The analysis follows the methods outlined in Knoke and Burke (1980) and Dixon (1983). Because several of the 64 cells in the contingency table were empty, 0.5 was added to all cells prior to analysis. This standard procedure is employed to ease computation and to produce a more conservative test (Knoke and Burke, 1980).

The results of this analysis are shown in Table 7.8. A logit model including the five two-way interactions between survivorship and each independent variable provided a good representation of the survivorship data for both male and female breeders ($\chi^2 = 14.3$ and 20.7, respectively; df = 26; both $p > 0.75$). Higher-order interactions were excluded, as their addition failed to provide a significantly improved fit between the data and the logit model for either males or females.

The results of the logit model confirm the main conclusions of the preceding section. Among breeding males, survivorship

TABLE 7.8. Logit Analysis of the Effects of Five Dichotomous Independent Variables on the Annual Survivorship of Breeding Males and Females

Interaction of Survivorship With	Likelihood Ratio (χ^2)	df	p-value
MALES			
Territory quality	1.69	1	ns
Breeding experience	0.20	1	ns
Number of breeding males	4.17	1	<0.05
Number of breeding females	0.14	1	ns
Number of nonbreeders	12.43	1	<0.001
Overall fit of model	$\chi^2 = 14.3$, df = 26, $p > 0.7$		
FEMALES			
Territory quality	0.75	1	ns
Breeding experience	0.42	1	ns
Number of breeding males	2.04	1	ns
Number of breeding females	0.34	1	ns
Number of nonbreeders	1.88	1	ns
Overall fit of model	$\chi^2 = 20.7$, df = 26, $p > 0.7$		

is significantly greater when nonbreeders are present than when nonbreeders are absent ($\chi^2 = 12.4$, $p < 0.001$). The presence of additional breeding males also significantly enhances survivorship of breeding males ($\chi^2 = 4.2$, $p < 0.05$). Territory quality, prior breeding experience, and the number of breeding females in the group have no significant overall influence on male survivorship (Table 7.8).

Although the fit of the logit model for breeding males was not improved by the addition of higher-order interaction terms, the interaction of breeding male survivorship, number of breeding females, and the presence of nonbreeders approached statistical significance ($\chi^2 = 3.7$, $0.1 > p > 0.05$). When nonbreeders were absent, survivorship of breeding males was only 55.8% ($N = 52$) in groups with one breeding female but 100% ($N = 11$) in groups with two breeding females ($\chi^2 = 7.7$,

$p < 0.01$). When nonbreeders were present, however, male survivorship was similar between groups with one breeding female (88.3%, $N = 162$) and groups with two breeding females (87.5%, $N = 40$).

In contrast to the results for breeding males, the logit model suggests that female survivorship is independent of territory quality, prior breeding experience, and group composition (Table 7.8). None of these independent variables had a significant influence on female survivorship.

In summary, the survivorship of breeding males appears to be significantly enhanced by the presence of additional group members, particularly nonbreeders but also additional breeding males. Additional breeding females may also improve male survivorship in groups where nonbreeders are absent. Survivorship of breeding females, in contrast, is only weakly influenced by group size and composition.

7.6. DISCUSSION

Sexual and Seasonal Differences in Breeder Survivorship

The major causes of mortality in adult acorn woodpeckers at Hastings are not well known, and our data indicate that they may differ for males and females. One major cause of mortality is undoubtedly predation by Cooper's hawks, which are common residents in the study area. Since the study began in 1971, this species has been observed to kill acorn woodpeckers at Hastings on three occasions, and may constitute the single most important agent of mortality in adult woodpeckers. This seems especially likely for breeding males, as male survivorship is constant throughout the year and not clearly associated with any seasonal food shortages or reproductive events (Figure 7.4).

Annual variation in acorn food resources may also significantly influence breeder mortality, as suggested by the relatively high autumn-winter mortality of males and females following

183

the 1978–79 acorn crop failure (Table 7.2), the positive correlation between adult winter survival and the energetic content of stored acorns (Table 4.10), and the positive correlation between annual reproductive success and breeder survivorship (Figure 7.2; see also Hannon et al., 1987).

Whether acorn-related food shortages increase starvation, stress, disease, predation, or other agents of mortality is unknown. However, annual variation in the survivorship of breeders is neither statistically significant (Table 7.1) nor as pronounced as annual variation in reproductive success (Chapter 5). The coefficient of variation in annual survivorship is 14.7% for males and 22.0% for females (Table 7.1), considerably smaller than the comparable coefficients of annual variation for young fledged per group (49.9%) and young fledged per bird (61.4%; Table 5.2).

Unlike breeder survivorship in some cooperatively breeding birds (e.g., Florida scrub jays; Woolfenden and Fitzpatrick, 1984), the survivorship of breeders in acorn woodpeckers differs significantly between the sexes, and is higher for males. This sexual difference in survivorship is entirely attributable to a sharp increase in female mortality during the spring and summer months (Figure 7.4).

The factors responsible for this increase are not clear. The six-month period of high female mortality includes virtually all activities associated with reproduction (Chapter 5) and molt (Spray and MacRoberts, 1975). However, only a few of these activities are performed differentially by males and females (Mumme, 1984; Mumme and de Queiroz, 1985). Thus, we can propose only three plausible hypotheses that could explain why females are more susceptible to physiological stress, starvation, disease, predation, or some other form of mortality during spring and summer.

1. Cost of egg production. The energetic costs of egg formation and egg laying may stress breeding females or subject them to risks greater than those experienced by males, thereby increasing female mortality. Egg production is the single reproductive activity of acorn woodpeckers that is performed only by females. Other sexual asymmetries in reproductive activities are quantitative, not qualitative.

2. Greater investment in nest care. Compared to breeding males, breeding females make significantly greater investments in the diurnal incubation of eggs, diurnal brooding of nestlings, and feeding of young (Mumme, 1984). Only for nocturnal incubation, nocturnal brooding, and nest sanitation do males contribute more than do females. However, it seems likely that the energetic expenditures of males differentially engaged in these activities do not compensate for the activities performed more often by females, for the following reasons:

a. Acorn woodpeckers nest and roost in the same types of tree cavities (MacRoberts and MacRoberts, 1976). Thus, roosting away from the nest is likely to be as dangerous as nocturnal incubation and brooding, performed mostly by males.

b. In energetic terms, diurnal incubation and brooding are probably more costly than their nocturnal equivalents. Gessaman and Findell (1979) found that incubation in American kestrels nesting in boxes could be accomplished at the level of adult resting metabolism, at least when temperatures were relatively warm. They suggested that the costs of incubation may be minimal in well-insulated natural cavities such as those generally used by acorn woodpeckers. Furthermore, a major energetic cost of incubation is the rewarming of cold eggs (Vleck, 1981). This cost is expected to be negligible during the continuous nocturnal incubation performed by males.

Diurnal incubation imposes a cost in both time and energy. Nesting female acorn woodpeckers, for example, spend roughly 32% of daylight hours (compared to 25% for breeding males) incubating eggs, reducing the amount of time they can devote to other activities (Mumme, 1984).

c. Nest sanitation, performed predominantly by males, is a relatively minor aspect of nest care routinely performed following a small fraction of feeding visits (Mumme, 1984) and seems unlikely to entail any significant energetic cost.

185

These considerations render it likely that the diurnal aspects of nest care performed differentially by females require a greater overall energetic investment than the nocturnal and sanitary activities performed more frequently by males. Thus, it is possible that this asymmetry leads directly or indirectly to the observed decrease in female survivorship during spring and summer.

3. Increased territorial defense. Because of a female-biased intruder sex ratio and sex-specific defensive behavior, breeding females play a significantly greater role in intraspecific territory defense than breeding males (Mumme and de Queiroz, 1985). Furthermore, territorial intrusions by conspecifics occur most frequently during the spring. Thus, the increased energetic costs and risks associated with territory defense are borne disproportionately by females and may render them less resistant to predation or other causes of mortality.

These three hypotheses are not mutually exclusive. We suspect that the increased energetic costs associated with all three factors (egg production, nest care, and territory defense) contribute directly or indirectly to the high rate of female mortality during the spring and summer.

Higher mortality of female breeders have been found in several other cooperatively breeding birds (Rowley, 1965, 1981; Reyer, 1980). For the splendid wren, a species in which females perform all incubation and brooding activities, Rowley (1981) attributed the lower rate of female survivorship to the females' greater role in nest care. Unlike acorn woodpeckers, however, the higher female mortality was not limited just to the breeding season.

Survivorship and Group Composition

Survivorship of breeding males is considerably enhanced by the presence of additional group members, regardless of their age, sex, or reproductive status. This effect is not an indirect consequence of differences in territory quality or breeding experience. We can suggest three plausible hypotheses that could explain this trend.

1. Increased foraging efficiency. Foraging efficiency may be higher in larger groups than in smaller groups (Bertram, 1978; Rabenold and Christensen, 1979; Pulliam and Caraco, 1984). This particular advantage of group living is unlikely to apply to acorn woodpeckers. Unlike some cooperatively breeding birds, acorn woodpeckers do not forage in groups. Also, as discussed in Chapter 4, the number of acorns that groups are able to store is ultimately set by granary size. Thus, additional group members are likely to reduce the number of stored acorns available on a per capita basis, potentially decreasing survivorship.

2. Cooperative sharing of tasks. Dangerous or energetically expensive tasks may be shared cooperatively in larger groups, thereby reducing the risk of starvation, disease, or predation experienced by individual males (Chase, 1980; Brown, 1982). Although this is an attractive hypothesis, the data on the sharing of such costly tasks are equivocal. For example, individual male contributions to acorn storage and incubation of eggs decline as group size increases, supporting this hypothesis (Mumme, 1984; Mumme and de Queiroz, 1985). However, individual contributions to other important aspects of cooperative behavior, such as granary maintenance and feeding nestlings, do not decrease significantly.

Further evidence against this hypothesis stems from the fact that breeders make significantly greater investments than do nonbreeders in nearly all forms of cooperative behavior (Mumme and de Queiroz, 1985). Thus, this hypothesis predicts that male survivorship should be higher in the presence of additional breeders (who share cooperative tasks more equitably) than in the presence of nonbreeders. However, the enhancement of male survivorship is more pronounced in the presence of nonbreeders (Tables 7.5–7.8).

3. Improved predator detection. Predator detection may be significantly enhanced in larger groups, thereby reducing predation on breeding males (Pulliam and Caraco, 1984). Although we have no data that bear directly on this hypothesis, the circumstantial evidence is suggestive. Acorn woodpeckers regularly give alarm calls in response to the approach of aerial predators, particularly Cooper's hawks (MacRoberts and MacRoberts, 1976). If all group members give alarm calls, this hypothesis predicts that the addition of any other group members, regardless of sex or reproductive status, should enhance the efficiency of predator detection and improve the survivorship of other group members. This prediction is supported by the data on breeding male survivorship.

These hypotheses are not mutually exclusive. There is considerable evidence from a variety of species that the three advantages of living in groups listed above interact (Caraco, 1979a; Pulliam and Caraco, 1984; Sullivan, 1984). Thus, any attempt to categorize the advantages associated with group living may be misleading. Nonetheless, the current evidence suggests that improved predator detection is probably the most important proximate factor explaining the positive relationship between group size and breeder male survivorship. Studies testing this hypothesis are in progress.

For breeding females, the relationship between group composition and survivorship is not clear. Female survivorship is slightly higher in larger groups, particularly in groups with two or more breeding males, but the effects are weak (Tables 7.5–7.8). In fact, in the logit analysis of female survivorship, territory quality, prior breeding experience, and three aspects of group composition all failed to explain a significant proportion of the variation in female survivorship (Table 7.8).

We conclude that the heavy spring and summer mortality impacts females more or less indiscriminately. Breeding is either risky or energetically expensive for female acorn woodpeckers, and the costs of reproduction are largely fixed; that is, reproduction carries with it dangers or energetic burdens that are not readily shareable and thereby not greatly reduced by the presence of other group members. Such fixed costs include egg formation and individual contributions to incubation of eggs and feeding of young. These latter two activities are performed primarily by breeding females, and female contributions to incubation and feeding do not decrease significantly as group size increases (Mumme, 1984).

It has long been suspected that survivorship of cooperative breeders may be enhanced by living in groups. However, relatively few studies have addressed this question empirically, and several of those that have found no evidence of higher survivorship among birds living in larger groups (e.g., green woodhoopoes [Ligon and Ligon, 1978b], groove-billed anis [Vehrencamp, 1978], stripe-backed wrens [Rabenold and Christensen, 1979], and Galápagos mockingbirds [Kinniard and Grant, 1982]).

Two recent studies, however, have demonstrated a significant positive relationship between group size and breeder survivorship. Woolfenden and Fitzpatrick (1984) found that survivorship of breeding male and female Florida scrub jays is 84.7% when nonbreeders are present versus 76.8% when nonbreeders are absent. Similarly, Reyer (1984) found that return rates of pied kingfisher breeders were higher when they were assisted by two or more nonbreeding helpers compared to one or no nonbreeders. The data, however, are statistically significant only for females. Our analyses indicate that the survivorship of breeding male acorn woodpeckers is enhanced by the presence of other group members, even when the effects of confounding variables such as territory quality and prior breeding experience have been factored out.

However, it should be kept in mind that the enhancement of breeder survivorship is not necessarily attributable to the alloparental reproductive behavior—that is, helping-at-the-nest—performed by these nonbreeders. In fact, the data suggest that this augmentation of survivorship may be more directly attributable to the benefits of living in social units in group territories rather than to alloparental behavior per se (see above). Only if the alloparental behavior performed by non-breeders *directly* increases breeder survivorship (either through reduced energy demands on breeders or through reduced predation during breeding) can higher survivorship be viewed as a benefit that accrues to breeders as a direct result of helping (Brown, 1985a) rather than as a benefit of group territoriality.

Ideally, the relationship between group composition and survivorship should be investigated experimentally. Although experimental removal of helpers has been used to investigate the effect of helpers on reproductive success (Brown et al., 1982; see Chapter 6), no such experimental investigation of survivorship has been attempted. This is not surprising, as such an experiment would be exceedingly difficult: survivorship data in long-lived organisms such as cooperatively breeding birds accumulate excruciatingly slowly. Furthermore, because survivorship is often difficult to measure as anything other than a categorical variable (percent alive or dead), large samples are required to demonstrate a statistically significant effect.

Regardless of these difficulties, data on survivorship are critical to a thorough understanding of the evolution of cooperative breeding, as very small differences in annual survivorship can have profound effects on lifetime reproductive success. This topic will be addressed in detail in Chapters 12 and 13.

7.7. SUMMARY

The annual survivorship of acorn woodpecker breeders is 82.4% for males and 71.2% for females. The significantly higher rate of female mortality is a direct consequence of high female mortality during the spring and summer. We suggest that energetic burdens and risks associated with egg formation, nest care, and territory defense (activities all performed solely or disproportionately by females) are responsible for the sex-differential survivorship. These costs may result in greater susceptibility of breeding females to predation, starvation, disease, or some other cause of mortality. Breeder survivorship of both males and females is not significantly influenced by age, prior breeding experience, or territory quality.

Female survivorship is also independent of group composition. For males, however, survivorship is positively correlated with group size, and the relationship is significant even when the effects of other variables are statistically controlled. In addition, the group-size-related enhancement of male survivorship appears to be largely independent of the sex, age, and breeding status of other group members. Thus, males appear to profit directly (at least in terms of survivorship) from living in cooperative groups, a result that has important implications for the relative costs and benefits of different social strategies.

CHAPTER EIGHT

Sex Ratio, Survivorship, and Dispersal of Young

Young acorn woodpeckers are born into an unusual milieu: socially, they find themselves in an extended family consisting of nestmates and older siblings along with an eclectic mix of parents, aunts, uncles, cousins, and in some cases grandparents and other relatives. Ecologically, they live in a situation that apparently discourages dispersal and independence, thereby rendering it likely that they will remain in their natal group for an extended period of time before becoming breeders. When they finally do disperse, many will join with other siblings of the same sex so as to compete more effectively for reproductive vacancies; some, however, whose opposite-sexed parents have died, will "inherit" their natal territory and breed there, often with one or more relatives of the same sex.

This social organization is among the most complex documented among vertebrates (see Brown and Brown, 1981a, and Frame et al., 1979, for descriptions of similarly complex social structures in Mexican jays and African wild dogs, *Lycaon pictus*) and raises numerous questions. First, what proportion of offspring disperse right away and what proportion remain at home as nonbreeders or as breeders? Do these proportions vary from year to year or between sexes, and if so, why? How long do birds defer reproduction, and how far must they go to find breeding vacancies? Does the sex ratio of offspring produced by groups vary with group composition, perhaps so as to yield larger sets of same-sexed offspring within groups or to preferentially produce helpers of the same sex?

In this chapter we will address these questions by analyzing the fates of 447 color-banded nestlings. Our goal is to quantify the patterns and variety in fates of offspring, paying special attention to demographic questions concerning parameters such as sex ratio, age at first reproduction, probability that dispersers obtain breeding opportunities, and distance of dispersal. The answers to these questions will provide us with data to which we will refer in later chapters when we calculate a synthetic life table and when we discuss in greater detail hypotheses for the evolution of cooperative breeding.

8.1. OVERALL DISAPPEARANCE OF YOUNG

The vast majority of young acorn woodpeckers remain in their natal groups until at least their first spring. During the course of the study, we recorded only four individuals (three males and one female), 1.1% of the 361 nestlings we banded through 1981, that emigrated from their natal groups prior to late February of their first year. The one female left her natal group in September, about four months postfledging; her fate is unknown but she was observed intruding at a nearby territory the following July. A second case involved a male who left his natal group in August, possibly with two other siblings, only two months after fledging and before his postjuvenal molt. This bird also emigrated to an unknown destination, but returned to his natal group the next May, following the death of his mother, and subsequently bred there along with his father. The final two individuals known to have emigrated prior to their first spring were males who joined with an older brother and moved to a nearby territory in late December, six months after fledging. Juveniles may also be forced to leave their natal group in poor acorn years (see section 11.2), but this was not observed until 1983, following the period covered here.

Other juveniles were occasionally observed to follow older individuals (known or presumed siblings) to a group into which the latter had immigrated, but none of these were known to have remained more than a few months. Juveniles also may "disperse" when their entire group abandons the territory due to an acorn crop failure; several such cases are reported by MacRoberts and MacRoberts (1976). As young do not leave their natal social unit in such cases, however, they do not affect the conclusion that the vast majority of acorn woodpeckers remain in their natal groups up to their first spring.

We can estimate the total number of birds emigrating prior to their first spring if we assume that the dispersal pattern of such young is generally the same as that of older individuals, approximately 50% of which obtain breeding space within the study area (see Table 8.6). Using this figure, we estimate that an additional four individuals beyond those discussed above emigrated out of the study area prior to their first spring, and thus that the total of such individuals comprises about 2.2% of the juvenile population. In fact, though, only half (two) of those reported above moved within the study area, and the other two moved outside the area and were detected only because they happened to return later on. Therefore, it is possible that we detected all young dispersing prior to their first spring. In any case, these data indicate that over 95% of young remain in their natal group up to their first spring, and thus we may assume with reasonable confidence that young disappearing prior to that time die.

During the 11-year period considered here we color-banded a total of 447 nestlings within the main study area. Nestlings were banded at 21 to 25 days of age, 5 to 8 days prior to fledging. The number of young banded, the percent and sex ratio of those surviving to their postjuvenal molt in the autumn (at which time they can be sexed), and the percent surviving to Feb-

TABLE 8.1. Survivorship and Sex Ratio of Young Banded as Nestlings by Year

Year	Nestlings Banded	N Surviving to Be Sexed (% of banded)	N Males (% of sexed)	N Females (% of sexed)	N Surviving to February (% of banded)
1972	13	9 (69.2)	7 (77.7)	2 (22.2)	9 (69.2)
1973	17	10 (58.8)	6 (60.0)	4 (40.0)	9 (52.9)
1974	14	8 (57.1)	4 (50.0)	4 (50.0)	7 (50.0)
1975	61	38 (62.3)	16 (42.1)	22 (57.9)	33 (54.1)
1976	44	24 (54.6)	14 (58.3)	10 (41.7)	20 (45.5)
1977	41	30 (73.2)	16 (53.3)	14 (46.7)	28 (68.3)
1978	29	7 (24.1)	4 (57.1)	3 (42.9)	5 (17.2)
1979	22	15 (68.2)	10 (66.7)	5 (33.3)	14 (63.6)
1980	66	52 (78.8)	31 (59.6)	21 (40.4)	43 (65.2)
1981	54	41 (75.9)	20 (48.8)	21 (51.2)	37 (68.5)
1982	86	58 (67.4)	27 (46.6)	31 (53.4)	49 (57.0)
Total	447	292 (65.3)	155 (53.1)	137 (46.9)	254 (56.8)
C.V.[a]	—	17.2%	—	—	19.1%
χ^2 (df = 10)	—	34.4***	—	—	29.7***

NOTE: Autumn nests included in subsequent year's totals. Statistical tests compare the yearly proportions of young banded that survived to be sexed and the yearly proportions that survived to February. None of the sex ratios differ significantly from 50:50 by the binomial test.

[a] Coefficient of variation of annual survivorship based on arcsin transformed values.

ruary are presented for each year in Table 8.1. An average of 65.3% of nestlings banded survived to the autumn postjuvenal molt, a figure that varied significantly between years from a low of 24.1% in 1978 to a high of 78.8% in 1980. Of the 292 fledglings surviving to be sexed, 254 (87.0%) survived to February; thus, the brunt of mortality on fledglings occurs prior to this time. Overall survivorship to February averaged 56.8%, a figure that also varied significantly from year to year. Coefficients of variation for first-year survivorship were nonetheless modest: for juveniles surviving to their postjuvenal molt, the

C.V. = 17.2% whereas for young surviving to February, the C.V. = 19.1% (both values calculated using the arcsin transformed percentages from each year). This level of variation is only slightly greater than the 14.1 to 15.2% C.V.'s found for annual variation in survivorship of adult breeders (Chapter 7).

We will return to the problem of discriminating survivorship from dispersal among juveniles and older nonbreeders below.

8.2. SEX RATIO

Nestling and fledgling acorn woodpeckers are not sexually dichromatic (Spray and MacRoberts, 1975). Although distinguishable from adults by their dark eyes and subtle plumage differences, juveniles of both sexes resemble adult males in crown coloration. Adult coloration, consisting of a wide black band on the crown of females separating the red nape from the white forehead, is attained at the postjuvenal molt, occurring between August and late September. Table 8.1 lists the sex ratio of first-year birds beginning at this time. There is a slight overall bias toward males, but it is not significantly different from a 50:50 ratio, nor is the sex ratio biased in any of the individual years. Assuming that juvenile mortality is not sex biased, these data imply that the sex ratio at fledging is even.

There are, however, ecological reasons, for acorn woodpeckers and for some other cooperative breeders, to suspect that the fitness of breeders could be increased by producing offspring biased toward one sex under certain circumstances. In red-cockaded woodpeckers, for example, Gowaty and Lennartz (1985) reported that "nontenured" females (those not having bred previously in the study area) appear to bias the sex ratio of their offspring toward males. Since the vast majority of helpers in this species are males, Gowaty and Lennartz interpret this result as possibly being an instance of local resource enhancement (Clark, 1978); that is, an adaptation of females breeding for the first time in the study area to produce sons that will be more helpful to them in subsequent years, thereby reducing the sons' cost relative to daughters and leading to a sex-ratio bias in favor of males (see also Trivers and Willard, 1973; Burley, 1982; Charnov, 1982).

The version of local resource enhancement suggested by Gowaty and Lennartz (1985) involves kinship asymmetries within groups. An alternative version, discussed by Emlen et al. (1986), proposes that the biased sex ratio may be thought of as male nonbreeding helpers "repaying" the breeders, thereby rendering their own production less costly than females'. This model hypothesizes that the sex-ratio imbalance should depend on the difference in the proportion of nonbreeders of each sex that act as helpers and the magnitude of the helper contribution relative to breeders.

Ligon and Ligon (MS) report a similar phenomenon for the green woodhoopoe, in which there is a bias in the sex ratio of offspring produced in the first nests of the year of females living in groups with two or less helpers; in this case, however, the bias is toward females. The Ligons suggest that the most likely explanation for this phenomenon is the smaller size of females in this species, rendering their production less expensive and therefore adaptive in small flocks and early in the season, when food is possibly scarce.

In acorn woodpeckers, male and female offspring are not dichromatic and both sexes remain as nonbreeders in approximately equal numbers (Chapter 3; see also section 8.3). There is a slight preponderance of male nonbreeders, but this bias is counteracted by the tendency for nonbreeding helper females to participate more in incubation and the feeding of nestlings than do nonbreeding helper males (Koenig et al., 1983; Mumme, 1984). Furthermore, the effects of nonbreeders on reproductive success (Chapter 6) suggest that the fitness gain to breeders with nonbreeding helpers is small or even nonexistent. Unlike the grey-crowned babblers studied by Brown et al. (1978), breeders do not decrease their nest attendance when nonbreeding helpers are present (Mumme, 1984). Thus, whether one sex should be favored by adults according to the repayment model of Emlen et al. (1986) is ambiguous.

Biasing the sex ratio of offspring could nevertheless be adaptive in acorn woodpeckers. As we describe elsewhere (Koenig et al., 1983), dispersal frequently occurs in unisexual sibling units. An important advantage of such units is that they are more competitive and are much more likely to win power struggles (Koenig, 1981a) for reproductive vacancies than are single birds (Hannon et al., 1985). Hence, we predict that, if possible, parents should bias the sex ratio of their offspring so as to produce as large a cohort of same-sexed siblings as possible (see also Ligon and Ligon, MS).

We did not laparotomize offspring, and thus have no data on sex-ratio bias at hatching or fledging. Instead, we used offspring surviving past the postjuvenal molt in early autumn, and searched for sex-ratio bias in two ways. First, we compared the sex ratio of broods of two, three, and four offspring with the expected proportions based on polynomial expansions; in no case was the observed value significantly different from those expected by chance. Thus, despite the plausibility of the "larger support group" hypothesis, we have no evidence that parents bias the sex ratio of offspring in order to produce cohorts of predominantly one sex. Ligon and Ligon (MS) also reject this hypothesis for green woodhoopoes, despite a similar dependence on allies of the same sex in that species (Ligon and Ligon, 1982, 1983).

Second, we tested for an overall sex bias among offspring produced by groups of different compositions. The results of these analyses—which included comparing groups on the basis of presence and sex ratio of nonbreeding helpers, number of breeding males and females, territory type, breeder experience, and productivity of the year (Table 8.2)—were invariably negative. Thus, we conclude that either acorn woodpeckers do not influence the sex ratio of their offspring or that the effect that they impose is minor and undetectable, at least with our sample sizes.

TABLE 8.2. Sex Ratio of Offspring Surviving to the Postjuvenal Molt Divided by Various Criteria

Criterion	Males	Females	Percent Males	χ^2	p-value
PRESENCE OF NONBREEDERS					
None	61	56	52.1		
Present	86	72	54.4	0.1	ns
SEX OF NONBREEDERS					
Males only	19	15	55.8		
Females only	21	16	56.7	0.0	ns
SEX RATIO OF NONBREEDERS					
Males > females	36	32	52.9		
Males = females	86	78	52.4	0.5	ns
Males < females	25	18	58.1	(df = 2)	
NUMBER OF FEMALE BREEDERS					
1	103	95	52.0		
2+	44	32	57.8	0.5	ns
NUMBER OF MALE BREEDERS					
1	48	36	57.1		
2	50	54	46.2	2.1	ns
3+	49	37	56.9	(df = 2)	
TERRITORY TYPE					
Always occupied	125	101	55.3		
Unoccupied ≥ 1 year	22	24	47.8	1.6	ns
TURNOVER IN BREEDERS					
No turnover	121	100	54.7		
Turnover	22	24	47.8	0.5	ns
PRODUCTIVITY OF YEAR					
Poor	24	16	60.0		
Fair	57	47	54.8	1.5	ns
Good	74	74	50.0	(df = 2)	

8.3. FATE OF OFFSPRING

Four fates are possible for offspring surviving their first winter: they can remain in their natal group as nonbreeders, inherit their natal group and breed, emigrate and become a breeder locally (i.e., elsewhere within the study area), or disappear (that is, die or emigrate out of the study area). The cumulative proportion of surviving first-year birds that undergo each of these

TABLE 8.3. Fate of Offspring by Sex and Age, 1971–78

	1st Year	2nd Year	3rd Year	4th Year	$\bar{x}$ Age (Years)
NONBREEDER IN NATAL GROUP					
Males	66.7 (40.0)	24.7 (14.8)	12.3 (7.4)	2.5 (1.5)	1.53
Females	60.3 (33.6)	23.5 (13.1)	4.4 (2.5)	1.5 (0.8)	1.41
INHERITED NATAL TERRITORY					
Males	4.9 (3.0)	12.3 (7.4)	16.0 (9.6) *	17.3 (10.4) *	2.08
Females	1.5 (0.8)	4.4 (2.5)	5.9 (3.3)	7.4 (4.1)	2.41
KNOWN (LOCAL) EMIGRANTS					
Males	12.3 (7.4) *	24.7 (14.8)	27.2 (16.3)	29.6 (17.8)	1.83
Females	1.5 (0.8)	14.7 (8.2)	23.5 (13.1)	23.5 (13.1)	2.31
DISAPPEARED OR DEAD					
Males	16.0 (49.6) **	38.3 (63.0) *	44.4 (66.7) *	50.6 (70.4)	2.05
Females	36.8 (64.8)	57.4 (76.2)	66.2 (81.1)	67.6 (82.0)	1.63
BIRDS ENTERING INTERVAL AS NONBREEDERS IN NATAL GROUP[a]					
Males	81	54	20	10	—
Females	68	41	16	3	—

NOTE: The first set of values is the cumulative percent of birds surviving to be sexed that suffered the given fate by the end of their nth year. Values in parentheses are the estimated percent of all fledglings suffering the fate, assuming an even sex ratio among fledglings disappearing prior to being sexed ($N = 108$). Comparisons, done between first set of values only, are by a χ^2 contingency test ($*p < 0.05$, $**p < 0.01$).

[a] Two males and one female (2% of original sample) still remained at the start of their fifth year.

fates by sex and age, up to the start of their fifth year, are listed in Table 8.3.

Remaining as Nonbreeders

Figure 8.1 graphs the proportion of young surviving to February that remain as nonbreeders. Approximately 60% of surviving females and 67% of surviving males live as nonbreeders in their natal territory during their first spring; the remainder have attained breeding status either by dispersal or inheritance. By the time birds are two years old, nearly a quarter are still nonbreeders. Between 4 and 12% are still nonbreeders at age three, and by year four all but a very small proportion have either died or attained breeding status. The oldest nonbreeders we recorded were two individuals, one male and one female, who remained in their natal territories through their fifth year.

There is a slight trend for males to remain as nonbreeders longer than females: again including only young surviving to

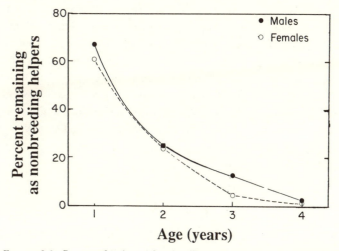

FIGURE 8.1. Percent of male and female offspring surviving to their first February that are still nonbreeders in their natal group through four years of age.

their first spring, the mean number of breeding seasons males spend as nonbreeders is 1.08, compared to 0.91 for females. As a result, males are somewhat overrepresented as nonbreeders among the older age classes (Table 8.3), but none of these differences between the sexes is significant.

Territorial Inheritance

Nonbreeding helpers may become breeders in their natal territory if all possible parents of the opposite sex (e.g., all opposite-sexed breeders alive when they were born) die and are replaced by individuals from outside the group (Koenig and Pitelka, 1979; Koenig et al., 1984). The frequency with which young inherit their natal territory and become breeders in this fashion is relatively low (Table 8.3; see also Chapter 9). Males, however, become breeders in their natal territories significantly more frequently than females: by the end of four years, 17.3% of males have inherited compared to only 7.4% of females (Table 8.3).

This difference follows directly from the observations that (1) there tend to be more male than female cobreeders in groups (1.84 versus 1.19; Chapter 3), (2) survivorship of female breeders is significantly lower than that of male breeders (Chapter 7), and (3) young inherit their natal groups following the death of all cobreeders of the opposite sex. As a result, male nonbreeders more often outlive their mother (and their mother's cobreeders), thereby inheriting their natal territory, than female nonbreeders outsurvive their father (and their father's cobreeders). We will reconsider the consequences of territorial inheritance by group offspring in Chapter 9.

Local Emigration

During their first spring, 12.3% of males emigrated locally (moved to other groups within the study area), whereas only 1.5% of females did so (Table 8.3). By the end of their fourth

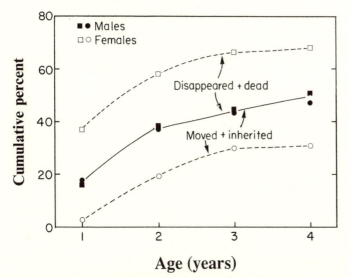

FIGURE 8.2. Cumulative percent of male and female offspring that are known to have become breeders (either through local emigration or inheritance of their natal territory) and that have disappeared (either through death or emigration outside of the study area) through four years of age. The lines for the cumulative percent of males that disappeared and that became breeders are congruent.

year, 29.6% of males and 23.5% of females are known to have emigrated.

Combining individuals who were known either to have emigrated or inherited their natal territories (that is, all those who are known to have assumed breeding status within the local population), significantly more males than females bred in their first year ($\chi^2 = 6.5$, $p < 0.05$), and by the end of four years, 46.9% of males are known to have bred locally compared to only 30.9% of females (Figure 8.2).

Disappearance and Death

Whereas significantly more males emigrate nearby or inherit their natal territory, more females disappear entirely from the study area, the difference being significant through the end of the third year (Table 8.3; Figure 8.2). By the end of four years, 50.6% of males and 67.6% of females have disappeared prior to assuming breeding status within the study area, a difference suggesting that females disperse farther (and therefore are lost to the study) more frequently than are males. These figures do not differentiate between individuals that died and those that obtained breeding positions outside the study area, an analysis we pursue below.

Annual Variation

Yearly variation in the fates of nonbreeders is considered in Table 8.4. The number of nonbreeders present in the popula-

TABLE 8.4. Yearly Variation in Fates of Nonbreeders. I. Both Sexes Combined

Year	N Nonbreeders[a]			Percent Disappeared or Left Study Area	Percent Moved Locally or Inherited	Percent Remained as Nonbreeder
	1st yr.	Older	All			
1972	18	11	29	6.9	10.3	82.8
1973	7	27	34	14.7	5.9	79.4
1974	9	27	36	30.6	38.9	30.6
1975	39	12	51	19.6	7.8	72.5
1976	22	37	59	22.0	23.7	54.2
1977	27	32	59	10.2	15.3	74.6
1978	7	45	52	69.2	23.1	7.7
1979	15	4	19	21.1	21.1	57.9
1980	42	11	53	24.5	26.4	49.1
1981	34	27	61	19.7	23.0	57.4
Total	220	233	453	24.7	20.8	54.5
χ^2 (df = 9)	—	—	—	71.4	22.1	90.2
p-value	—	—	—	<0.001	<0.01	<0.001

[a] This is the number of first-year birds fledged during the year that survived to February of the subsequent year, plus the number of older nonbreeders beginning the year in their natal groups.

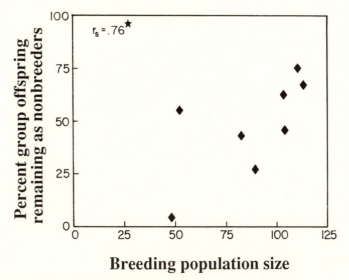

FIGURE 8.3. The percent of offspring remaining as nonbreeders in their natal territory in their first spring versus the total breeding population size.

tion during each year, including first-year birds surviving to their first spring and older nonbreeders beginning the year in their natal groups as nonbreeders, are listed. The proportion of individuals suffering each of the three fates varied highly significantly from year to year.

There is a significant positive correlation between the percent of group offspring that remain as nonbreeders in their first year and the breeding population size the subsequent spring ($r_s = 0.76$, $N = 8$, $p < 0.05$; Figure 8.3), and a significant negative correlation between the percent offspring known to be breeding (those that inherited their natal group or filled breeding vacancies in the local population) and breeding population size ($r_s = -0.82$, $N = 8$, $p < 0.05$). These correlations suggest that when the population is large and the ecological constraints against dispersal great, the proportion of young remaining as nonbreeders increases, whereas the proportion of

TABLE 8.5. Yearly Variation in Fates of Nonbreeders. II. Comparison of Males and Females

Year	N Nonbreeders		Percent Disappeared[a]		Percent Moved Locally or Inherited		Percent Remained as Nonbreeder	
	♂♂	♀♀	♂♂	♀♀	♂♂	♀♀	♂♂	♀♀
1972	16	13	0.0	15.4	18.8	0.0	81.2	84.6
1973	21	13	4.8	30.8	0.0	15.4	95.2 **	53.8
1974	24	12	20.8	50.0	45.8	25.0	33.4	25.0
1975	31	20	12.9	30.0	9.7	5.0	77.4	65.0
1976	36	23	19.4	26.1	30.6	13.0	50.0	60.9
1977	34	25	5.9	16.0	17.6	12.0	76.5	72.0
1978	30	22	73.3	63.6	26.7	18.2	0.0 *	18.2
1979	7	12	0.0	33.3	14.3	25.0	85.7	41.7
1980	31	22	16.1	36.4	38.7 *	9.1	45.2	54.5
1981	32	29	12.5	27.6	21.9	24.1	65.6	48.3
Total	262	191	19.1 **	32.5	25.2 **	14.7	55.7	52.8

NOTE: Includes both first-year and older nonbreeders. Starred comparisons are significantly different by a χ^2 contingency test or a Fisher exact test (used when the observed value for one category is <3).

[a] Includes six males (2.3%) and 13 females (6.8%) that left the study area but were seen again.

those that breed on their own decreases. We will return to this result in Chapter 12.

Finally, the fates of nonbreeders are detailed by year and sex in Table 8.5. On average, 19.1% of nonbreeder males and 32.5% of nonbreeder females disappear during any one year, whereas 25.2% of nonbreeder males and only 14.7% of nonbreeder females move within the study area or inherit their natal territories; both differences are significant. There is no significant difference in the proportion of males and females remaining as nonbreeders during a particular year.

8.4. EMIGRATION AND SURVIVORSHIP

Field studies involve finite study areas. This inevitably imposes problems when considering dispersal. Primary among these is distinguishing individuals dying from those emigrating and breeding outside the study area. As we were unable to survey surrounding areas thoroughly, the only tractable solution to this dilemma is to assume that the number of individuals successfully dispersing out of the study area equals the number that immigrated into it during the same time period. This steady state approach yields a reasonable estimate of dispersal as long as the population size is constant and the study area constitutes a representative fraction of the available habitat; that is, it is neither a dispersal sink (importing young from outside the local population) nor a dispersal fountain (exporting young to the outside population).

The first of these assumptions is met reasonably well by our population during the study period (see Chapter 3). Unfortunately, we have no objective way to evaluate whether the Hastings population acts as a dispersal sink or a dispersal fountain. Thus, we are forced to undertake the following analyses with the reservation that we do not know if the Hastings population meets the assumption of a steady state.

The first part of Table 8.6 lists the number and percent of nonbreeders for each sex that locally emigrated (line 2), inherited their natal territory (line 3), or disappeared without attaining breeding status within the local study population (line 4). Also listed are the number of immigrants known to have entered the study population (line 5) and the number of these that remained in their new groups through at least one breeding season (line 6). Given the two assumptions discussed above, these figures allow us to estimate a series of characteristics concerning both the birds disappearing from the study area and the population of nonbreeding helpers as a whole.

The results of these calculations (Table 8.6) suggest that virtually all (95.5%) nonbreeders that disappear after their first

TABLE 8.6. Estimation of the Proportion of Nonbreeders Dispersing Off the Study Area and Breeding

	♂♂	♀♀	*All*
1. N nonbreeders	116	90	206
2. N known emigrants within study area (%)	41(35.3)	22(24.4)	63(30.6)
3. N inheriting natal territory (%)	25(21.6)	6(6.7)	31(15.0)
4. N disappearing or leaving study area (%)	50(43.1)	62(68.9)	112(54.4)
5. N immigrants entering study area	45	62	107
6. N immigrants that bred[a]	34	48	82
ESTIMATES[b]			
Percent nonbreeders disappearing that get a territory (5/4)	90.0	100.0	95.5
Percent birds disappearing that breed[c](6/4)	68.0	77.4	73.2
Percent all nonbreeders obtaining a territory ([2 + 3 + 5]/1)	95.7	100.0	97.6
Percent all nonbreeders that obtain a territory and breed[a] ([2 + 3 + 6]/1)	83.6	82.2	83.0
Percent nonbreeders obtaining a territory that did so within the study area ([2 + 3]/[2 + 3 + 5])	59.5	31.1	46.8
Percent nonbreeders obtaining a territory and breeding that did so within the study area[a,c] ([2 + 3]/[2 + 3 + 6])	64.9	35.1	52.0

[a] These figures include only those individuals immigrating that persisted through a breeding season.

[b] Equations refer to numbered rows in first part of table.

[c] For these calculations, three males and two females that immigrated within the study area but did not survive through a breeding season are subtracted from the figures in line 2.

winter obtain a territory, and a high proportion (73.2%) of all birds disappearing survive at least one breeding season in their new territories. If we include nonbreeders that obtained breeding positions locally (that is, either inherited their natal territory or emigrated within the study area), we estimate that

97.6% of nonbreeders eventually attain breeding status some-where, and that 83.0% successfully survive at least one season as breeders. An estimated 46.8% of nonbreeders that obtain breeding status and 52.0% of those successfully making it through a breeding season as breeders do so within the study area.

An important corollary of these estimates is that the survi-vorship of birds as nonbreeders following their first winter is apparently quite high, as virtually all nonbreeders surviving to their first spring live to obtain a breeding situation successfully. For example, given that the average length of time birds spend as nonbreeders is about one year (calculated from Table 8.3), the mean annual survivorship of male nonbreeders based on the values in Table 8.6 are on the order of 96% for males and 100% for females. These figures are unrealistically high. Even taking into account potential discrepancies due to sampling error, however, these calculations suggest that survivorship of nonbreeders is at least as high, and probably higher, than that of breeders (generally 70 to 85%, see Chapter 7). A reasonable estimate for mean annual survivorship of nonbreeders living in their natal territory is probably on the order of 90%.

One independent check on these values can be made by performing life-table analyses (e.g., Custer and Pitelka, 1977). These are done in Chapter 10, and they indicate that non-breeding helper survivorship of 0.90 year^{-1} yields a reason-ably self-perpetuating population. A similar type of check can be made using a slight modification of Fitzpatrick and Woolfenden's (1986) equation calculating equilibrium group size from fecundity (m) and annual survivorship values for breeders (ℓ_{br}), nonbreeding helpers (ℓ_h), and first-year birds (ℓ_1). Group size equals $(x + \gamma)$, where x is the number of breeders in an average group and γ, the "standing crop" of older nonbreeders, is defined by

$$\gamma = \frac{(m)(\ell_1) - (1 - \ell_{br})}{(1 - \ell_h)}.$$

209

Mean breeding group size in our population is 4.39 and the mean number of breeders per group (x) is 3.03 (Table 3.6). Thus, γ should equal $(4.39 - 3.03) = 1.36$. Using known values for m (0.70, calculated as the mean number of fledglings group^{-1} [Table 5.2] divided by the mean number of breeders group^{-1} [Table 3.6]), ℓ_{br} (0.78, combining males and females from Table 7.1) and ℓ_1 (0.57 [Table 8.1]), the value of ℓ_h yielding γ equal to 1.36 is 0.87. Thus, our estimate of 90% annual survivorship for nonbreeding helpers is demographically plausible.

Data in Table 8.6 also confirm a sexual difference in the distance of dispersal (see also Table 8.3 and Figure 8.3). A considerably higher proportion of males than females that eventually attain a territory do so within the study area (59.5 to 64.9% of males compared to 31.1 to 35.1% of females). This suggests that, as in most other birds (Greenwood, 1980), females disperse farther, on average, than males. We will focus on this sex bias in dispersal distance in section 8.7.

8.5. AGE AT FIRST REPRODUCTION

In Table 8.7 we estimate the age at first reproduction using the data in Table 8.6 and the raw data used to produce Table 8.3. Of birds surviving through their first winter, an estimated 31.7% of males and 39.7% of females breed when one year old. Another 39.8% of males and 36.8% of females breed first when they are two years old, and a small proportion fail to breed until they are five or even six years old. The mean age of first reproduction is 2.08 years for males and 1.91 years for females, using all individuals known or estimated to have bred. Thus, as before (section 8.3), both males and females defer breeding on average until their second year, and there is a slight tendency for females to breed earlier than males. In this case, however, the sex bias is dependent on the inclusion of birds estimated to have obtained breeding positions outside of the study area.

TABLE 8.7. Estimation of the Age at First Reproduction

	N Disappeared	N Bred Obs.	N Bred Est.	N Helped	Percent First-Time Breeders Abs.	Percent First-Time Breeders Rel.	Cumulative Percent First-Time Breeders Abs.	Cumulative Percent First-Time Breeders Rel.
MALES ($N = 81$)								
1st year	13	14	25.7	54	31.7	33.4	31.7	33.4
2nd year	18	16	32.2	20	39.8	41.9	71.5	75.3
3rd year	5	5	9.5	10	11.7	12.4	83.2	87.6
4th year	5	3	7.5	2	9.3	9.8	92.5	97.4
5th year	0	1	1.0	1	1.2	1.3	93.7	98.7
6th year	0	1	1.0	0	1.2	1.3	94.9	100.0
Mean age	2.05	2.10	2.08	1.58	—	2.08	—	—
FEMALES ($N = 68$)								
1st year	25	2	27.0	41	39.7	39.7	39.7	39.7
2nd year	14	11	25.0	16	36.8	36.8	76.5	76.5
3rd year	6	7	13.0	3	19.1	19.1	95.6	95.6
4th year	1	1	2.0	1	2.9	2.9	98.5	98.5
5th year	0	0	0.0	1	0.0	0.0	98.5	98.5
6th year	1	0	1.0	0	1.5	1.5	100.0	100.0
Mean age	1.72	2.33	1.91	1.47	—	1.91	—	—

NOTE: Includes only young banded as nestlings and surviving to be sexed, 1971–78. For calculation of percent individuals breeding, 90% of the males and 100% of the females that disappeared are assumed to have bred outside the study area (see Table 8.6). Obs. = observed number breeding; Est. = estimated number breeding, including those disappearing that are assumed to have bred; Abs. = absolute; Rel. = relative to all those that eventually bred.

The only other cooperative breeder for which comparable data exist is the Florida scrub jay (Woolfenden and Fitzpatrick, 1984; see also Woolfenden and Fitzpatrick, 1986). From Table 4.1 in Woolfenden and Fitzpatrick (1984), the mean age of first reproduction (using only individuals observed to have

211

bred) is 2.96 years for males and 2.46 years for females. Thus, male Florida scrub jays remain as nonbreeders nearly a year longer than male acorn woodpeckers, whereas female scrub jays delay reproduction only slightly longer than female woodpeckers. The major factor producing this difference is the proportion of surviving juveniles that breed in their first year. For the jays, Woolfenden and Fitzpatrick (1986) found that less than 1% of both males and females breed in their first year, whereas we estimate that 31.7% of male and 39.7% of female acorn woodpeckers breed as yearlings (Table 8.7).

TABLE 8.8. Characteristics of Unisexual Unit Dispersal

	·Males	Females	All	Statistic
$\bar{x}$ number of same-sexed siblings present when ego dispersed	1.73 ± 1.21 (60)	0.95 ± 0.85 (42)	1.41 ± 1.14 (102)	3.4***
$\bar{x}$ number siblings ego dispersed with	1.10 ± 0.88	0.48 ± 0.59	0.84 ± 0.83	3.4***
Percent birds dispersing alone	31.6	57.1	42.1 ⎫	
Percent birds dispersing with at least one sibling	68.4	42.9	57.9 ⎭	5.6*
N birds dispersing out of natal group	60	42	102	
Percent siblings present that moved with ego	63.5	50.0	59.7	
Percent times sibs were present, but ego dispersed alone	18.4	37.9	25.6 ⎫	
Percent times sibs were present, at least one of which moved with ego	81.6	62.1	74.4 ⎭	2.7, ns
N times sibs were present	49	29	78	

NOTE: Data include only individuals leaving their natal group. Ego refers to a dispersing nonbreeding helper. Statistical tests compare males and females using a χ^2 contingency test or Mann-Whitney U test.

*** $p < 0.001$, * $p < 0.05$.

8.6. DISPERSAL IN UNISEXUAL SIBLING UNITS

Data concerning dispersal in unisexual sibling units is reported in Table 8.8. On average, males disperse with 1.1 siblings, whereas females disperse with only 0.5 siblings, a significant difference. Approximately 68% of males disperse with at least one other sibling, whereas only 43% of females do so.

These differences (reported earlier in part in Koenig et al., 1983) appear to be due in part to the smaller pool of nonbreeding females (see Table 3.6). This is evident from a comparison of the number of siblings dispersing together with the number present in groups at the time of dispersal (that is, the available pool of siblings with which individuals could have dispersed). Males disperse with an average of 63.5% of the male siblings present in their groups, whereas females disperse with 50.0% of their female siblings (Table 8.8). This difference is not, however, significant.

Looking only at groups in which siblings were present at the time of dispersal, males dispersed alone in 18.3% of cases, whereas females dispersed alone 37.9% of the time. This sex difference is great enough to suggest a real difference, but is not significant.

8.7. DISTANCE OF DISPERSAL

Known dispersal events (including individuals attaining breeding status in their natal groups) were categorized as to the sex of the individual involved and the distance between the individual's natal territory and that to which it emigrated (zero for birds inheriting their natal territories, otherwise taken to be the distance between the primary granaries of the two groups). We also noted whether the individual involved was dispersing for the first time ("primary" dispersal) or had already dispersed at least once. (Individuals dispersing a second time were usually birds who had returned to their natal territory following an

unsuccessful first breeding attempt [see Koenig et al., 1984].)
We also noted whether or not the individual remained through
at least one breeding season on its new territory.

We recorded 77 dispersal events by males and 52 by females.
Primary dispersal constituted 83.1% of the male and 82.7% of
the female dispersal events. Of the males, 84.4% remained
through at least one breeding season, and 82.7% of the females
did so. There were no significant differences in mean dispersal
distance between primary and secondary dispersal events or
between birds that remained through one breeding season compared to those that did not. Thus, all dispersal events were
combined in subsequent analyses.

The results are presented in Table 8.9 and graphed in Figure
8.4. The mean dispersal distance for birds detected in our popu-

TABLE 8.9. Distance of Dispersal (Kilometers)

	♂♂	♀♀	All
Mean ± SD (N) distance	0.35 ± 0.32 (77)	0.48 ± 0.44 (52)	0.40 ± 0.38 (129)
Root-mean-square dispersal distance	0.47	0.65	0.55
Corrected root-mean-square dispersal distance	0.67	0.90	0.79
Estimates of true root-mean-square dispersal distance			
a	—	—	1.42
b	—	—	2.42
c	—	—	4.88

NOTE: Difference in mean dispersal distance between the sexes not significant
(Mann-Whitney U test, $z = 1.3$, ns). "Corrected" RMS dispersal distance is
corrected for systematic bias due to the finite diameter of the study area
(Barrowclough, 1978). Estimates for the true RMS dispersal distance include
all long-distance dispersers; those going greater than the 2-km diameter of
the study area are assumed to follow the distributions a, b, and c graphed
in Figure B.1. See section 8.7 and Appendix B.

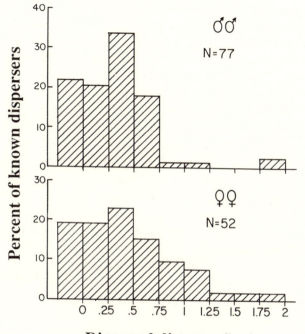

FIGURE 8.4. Distance of dispersal of males and females, divided into 0.25-km intervals. Individuals in the first interval are those inheriting and breeding in their natal territory. Mean dispersal distance is 0.35 km for males and 0.48 km for females. The longest dispersal events recorded were 1.69 km for a male and 1.75 km for a female.

lation is 0.35 km for males and 0.48 km for females. The difference between the sexes is not significant, but is in accord with other data suggesting that females tend to disperse farther than males (see section 8.3). Also in Table 8.9 are root-mean-square (RMS) dispersal distances, a value of importance in calculating effective population size and in understanding population structure (e.g., Wright, 1943, 1946; Barrowclough, 1980; see section 10.2).

One limitation of these data is the systematic bias due to the limited size of our study area. This problem can be overcome in part using Barrowclough's (1978) technique to correct for bias up to distances as great as the longest dimension of the study site, which in our case is about 2 km. The corrected RMS dispersal distances obtained by applying this method to our data are presented in Table 8.9. Corrected values are 39 to 43% greater than the observed RMS dispersal distances.

Unfortunately, the bias introduced by our inability to detect long-distance dispersers—individuals dispersing farther than the longest dimension of the study area—is not so easily removed. That this bias is considerable is suggested by two observations. First, the corrected RMS dispersal lengths excluding dispersers in the longest-distance class are only about 64% of the RMS including all data for both sexes. This indicates that the study area was not sufficiently large to adequately estimate the degree of long-distance dispersal (Barrowclough, 1978).

Second, only 59.5% of male and 31.1% of female nonbreeding dispersers obtain territories within the study area (Table 8.6); the remainder apparently disperse outside of the study area and go undetected.

It is impossible to take these "missed" dispersers fully into account in determining the true dispersal pattern of the population; we can only guess how far they go once they leave the study area. However, we can estimate the proportion of long-distance dispersers, and then incorporate a plausible dispersal distribution for these individuals in our calculation of the RMS dispersal distance.

This is done in Appendix B by subtracting the proportion of dispersers "picked up" by Barrowclough's correction from the estimated number that we failed to detect. From this analysis, we estimate that 9.9% of males, 43.8% of females, and 24.6% of all nonbreeders dispersed farther than the 2-km diameter of the study area. Dispersal distributions were plotted, and we then chose three plausible distributions for the "missing" fraction of the population from which we calculated RMS dispersal distances. Details of these procedures are presented in Appendix B.

216

The most conservative distribution, which is certainly a minimum estimate, yields a root-mean-square distance of 1.42 km. The most liberal, which assumes that a small proportion of individuals disperse as far as 100 km, results in a root-mean-square distance of 4.88 km. The intermediate RMS estimate is 2.42 km. We suspect, but cannot be certain, that these values bracket the true RMS dispersal distance (see Appendix B for additional discussion).

8.8. COMPETITION FOR SPACE

Competition and Dispersal

A corollary of the hypothesis that habitat saturation plays an important role in leading to delayed dispersal is that nonbreeders should disperse whenever possible. Thus, given that, on average, dispersal is delayed one year and that a majority of acorn woodpeckers do not assume breeding status until they are at least two years old, we can predict that there should be a shortage of breeding opportunities relative to the number of nonbreeders available to fill them.

There are several ways to estimate the severity of this shortage. One simple way is presented in Table 8.10, where we compare the number of nonbreeders with the number of breeding vacancies arising during a particular year. Nonbreeders are divided into first-year birds and older nonbreeders; we have also listed the number of birds filling vacancies as well as the number of vacancies arising. (The former is usually greater because of dispersal in sibling units.)

As expected, the number of potential recruits to the population far exceeds the number of vacancies in almost all years. On average, there were only 0.193 vacancies for every nonbreeding male and 0.339 vacancies for every nonbreeding female. (As a result of unisexual sibling dispersal, the "effective" number of

TABLE 8.10. Estimation of an Index of Breeding Competition

Year	First-Year Birds Surviving to February	N Older Nonbreeders in March	A Total Nonbreeders	B Breeding Vacancies Opening	C N Birds Filling Vacancies	B/A	C/A
MALES							
1972–73	9	7	16	5	7	0.313	0.438
1973–74	5	15	20	1	1	0.050	0.050
1974–75	4	9	13	4	5	0.308	0.385
1975–76	22	8	30	4	7	0.133	0.233
1976–77	12	12	24	3	6	0.125	0.250
1977–78	16	13	29	5	8	0.172	0.276
1978–79	4	9	13	5	9	0.385	0.692
1979–80	7	0	7	6	8	0.857	1.143
1980–81	25	4	29	4	12	0.138	0.414
1981–82	17	9	26	3	7	0.115	0.269
Total	121	86	207	40	70	0.193	0.338
$\bar{x}$ annual value	—	—	—	—	—	0.260	0.415
FEMALES							
1972–73	9	4	13	6	6	0.462	0.462
1973–74	2	9	11	2	2	0.182	0.182
1974–75	5	3	8	7	9	0.875	1.125
1975–76	17	1	18	5	8	0.278	0.444
1976–77	10	5	15	2	3	0.133	0.200
1977–78	11	11	22	7	10	0.318	0.455
1978–79	3	8	11	7	7	0.636	0.636
1979–80	8	1	9	3	3	0.333	0.333
1980–81	17	4	21	5	16	0.238	0.286
1981–82	17	8	25	8	14	0.320	0.560
Total	99	54	153	52	68	0.339	0.444
$\bar{x}$ annual value	—	—	—	—	—	0.377	0.468

NOTE: Vacancies include those opening between the prior June and May. Column C includes the total number of birds in unisexual units.

vacancies increases to 0.338 per male and 0.444 per female.) Thus, there are about three nonbreeding females for every vacancy arising in the population and about five nonbreeding males.

A second, indirect method for estimating the relative shortage of breeding opportunities, based purely on life-table parameters, has been derived by Fitzpatrick and Woolfenden (1986). In their terminology, we wish to estimate the parameter D_o, representing average annual number of breeding vacancies divided by the average number of potential recruits that could fill those vacancies:

$$D_o = \frac{(1 - \ell_{br})(1 - \ell_h)}{m\ell_1 - (1 - \ell_{br})\ell_h},$$

where m, ℓ_{br}, ℓ_h, and ℓ_1 are all defined as in section 8.4. Substituting the appropriate values for fecundity and survivorship (see section 8.4), we get

$$D_o = \frac{(0.22)(0.10)}{(0.70)(0.57) - (0.22)(0.90)} = 0.11.$$

These calculations suggest that, on average, only about one vacancy exists for every nine nonbreeding helpers. Dispersal in unisexual sibling units does not change this estimate, since calculations are in effect made in units of breeder-deaths, each of which is filled, on average, by a single nonbreeding helper, rather than vacancies per se, which are often filled by more than one nonbreeder. This value is considerably lower than that of 0.27 derived for Florida scrub jays by Fitzpatrick and Woolfenden (1986) and suggests that space competition may be even more intense in acorn woodpeckers than in Florida scrub jays.

Sex and Age-Related Asymmetries

Although competition for space is considerable for both sexes, it is apparently greater for males, at least according to the

TABLE 8.11. Estimated Probability that Nonbreeding Helpers Obtain a Breeding Position as a Function of Their Age

Age	Males (N)	Females (N)	All (N)	χ^2
First year	0.317 (81)	0.397 (68)	0.354 (149)	0.7
Second year	0.596 (54)	0.610 (41)	0.602 (95)	0.0
Third year	0.475 (20)	0.813 (16)	0.625 (36)	3.0
Fourth year	0.750 (10)	0.667 (3)	0.731 (13)	—
> Fourth year	0.667 (3)	0.500 (2)	0.600 (5)	—
> Second year	0.576 (33)	0.762 (21)	0.648 (54)	—
Overall mean	0.458 (168)	0.523 (130)	0.486 (298)	—

NOTE: Data from Table 8.7. Includes birds estimated to have obtained breeding positions. N is the number of birds entering that age class. Statistical comparisons are all $p > 0.05$.

calculations in Table 8.10. This conclusion is supported by Table 8.11, in which we present data on the probability that nonbreeders will obtain breeding positions as a function of age, based on estimated number of birds breeding obtained from Table 8.7. Females have a higher probability of breeding at all ages than do males, at least through their third year (beginning at that time sample sizes are small and differences unlikely to be important). Similarly, Fitzpatrick and Woolfenden (1986) found evidence for a lesser degree of competition for space among female Florida scrub jays.

For both sexes the probability of obtaining a breeding position increases with age. In part, this may be a consequence of the higher probability of having younger siblings with which to disperse, combined with the observation that birds with siblings are more successful at dispersal in both Florida scrub jays (Fitzpatrick and Woolfenden, 1986) and acorn woodpeckers (Hannon et al., 1985). As pointed out by Fitzpatrick and Woolfenden (1986), this means that early dispersers are at an even greater disadvantage than indicated by our calculations, thereby setting up a positive feedback system further enforcing delayed breeding. These authors also provide an excellent discussion of age- and dominance-related factors influencing competition for space.

Why Do Females Fight Harder to Attain Breeding Status?

From Table 8.10 we can conclude that competition for space is more intense for males than females. This is true even after taking into consideration the larger sibling units of males: assuming that the average dispersing unit of males consists of 2.10 individuals and that of females 1.48 individuals (from Table 8.8), there are an average of 2.47 male sibling units per male vacancy and only 1.99 female units per female vacancy. Nonetheless, females apparently disperse farther (section 8.7), suggesting that females search for vacancies over a wider area (see also Mumme and de Queiroz, 1985). Furthermore, females fight more intensely to fill vacancies than do males, as indicated by the greater number of birds and the more intense competition that occurs during power struggles to fill female vacancies (Koenig, 1981a).

We do not have a satisfactory answer to this apparent contradiction. Possibly the more intense competition between cobreeding females (e.g., Koenig et al., 1983), and the relatively small fitness benefit compared to that accruing to cobreeding males (see Chapter 13), makes it more desirable for females to strike out on their own and compete for vacancies by themselves. Males, in contrast, may be better off remaining longer in the hope of acquiring a larger sibling support group in order to compete more effectively (Hannon et al., 1985). Once a large support group of brothers was available, the probability of winning a power struggle nearby would be enhanced. Furthermore, males are more likely to inherit and breed in their natal territories (section 8.3; Chapter 9) than are females. Territorial inheritance thus provides a means by which mates can become breeders without competing for limited breeding vacancies with the population of nonbreeders.

221

Correlations with Nondispersal of Young

The habitat saturation hypothesis states that offspring are forced to remain in their natal territories, and thus predicts a correlation between vacancies arising in the population in a particular year and the proportion of offspring remaining as nonbreeders. There is, however, no correlation between these two variables for the eight years between 1975 and 1982 ($r_s = -0.10$, $\mathcal{N} = 10$, $p > 0.20$; data from Tables 8.4 and 8.10). This finding may be primarily a result of relatively little variation in the number of vacancies arising each year: in all but one year (1976–77) between nine and twelve vacancies were recorded for males and females combined.

For a similar test using data from a wider range of ecological conditions, Emlen and Vehrencamp (1983) correlated the retention of offspring with the proportion of new or newly occupied territories arising each year from both Hastings and the two populations in the Southwest (Water Canyon, New Mexico,

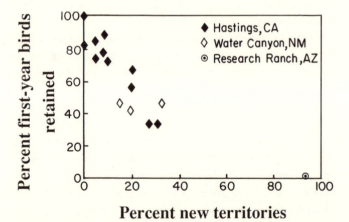

Percent new territories

FIGURE 8.5. Retention of first-year acorn woodpeckers versus the proportion of territories that become vacant during the preceding year, based on yearly data from Hasting Reservation (California), Water Canyon (New Mexico), and the Research Ranch (Arizona), compiled by Emlen and Vehrencamp (1983) with Hastings data from four additional years added. Overall $r_s = -0.87$, $\mathcal{N} = 16$, $p < 0.001$. For Hastings data only, $r_s = -0.85$, $\mathcal{N} = 13$, $p < 0.01$.

and the Research Ranch, Arizona) studied by Peter Stacey. (This differs from the prior analysis by including only new territories and excluding vacancies arising within established groups.) These data, with four additional years (1981–82 through 1984–85) from the Hastings population added, are graphed in Figure 8.5. Both the overall correlation and the correlation using only the data from Hastings are highly significant. Thus, there is a strong tendency for offspring to disperse and breed on their own as new territories become relatively more common.

Space Competition and Ecological Constraints

Intuitively, we might predict that, because the index of breeding competition is high (Table 8.10), a high proportion of birds, especially males, would never successfully obtain a breeding position. Indeed, this is generally assumed to be a corollary of the ecological constraints model for the evolution of delayed dispersal among nonbreeding helpers. However, as discussed in section 8.4, apparently almost *all* nonbreeders surviving to their first spring obtain breeding positions eventually.

This paradoxical result highlights an important feature of the ecological constraints model. In particular, space competition per se is not the critical issue (contra Brown, 1987): as we argued in Chapter 1, the number of potential recruits exceeds the number of available vacancies in many territorial species that are not cooperative breeders.

Take, for example, tit populations (genus *Parus*), which have large clutch sizes and may fledge over ten young pair^{-1} year^{-1} (see Lack, 1966). That space competition is considerable in populations of several species of parids has been demonstrated directly by removal experiments (Krebs, 1971; Ekman et al., 1981), and indirectly by the settlement patterns of yearlings (van Balen, 1980). In all these studies, the birds (often juveniles) replacing removed birds had previously either lived in marginal habitats or as floaters. (The fact that the number of new recruits ultimately exactly equals the number of vacancies opening up in a stable population is not relevant; the competition of interest is in how many birds are available to fill those vacancies and what proportion of them eventually die for lack of acquiring one.)

Recent work by Smith (1984) on the winter social organization of black-capped chickadees provides data directly relevant to the relationship, or lack thereof, between space competition and cooperative breeding. In this noncooperative breeder, young birds commonly follow one of two strategies during the winter, either joining and staying in a winter flock ("flock regulars") or switching between flocks. For birds pursuing either of these strategies, Smith (1984) calculated an index of the number of vacancies divided by the number of birds competing for them, a value analogous to the index of breeding space competition (D_o) defined by Fitzpatrick and Woolfenden (1986) and calculated above for acorn woodpeckers. In Smith's study, the index of breeding space competition was 0.24 for flock regulars and 0.28 for flock switchers.

These values indicate that space competition, measured in similar ways, is as great in this species as it is in many cooperative breeders, including the acorn woodpecker (the index of breeding space competition, as calculated directly from Table 8.10, is 0.19 for males and 0.34 for females), the groove-billed ani (0.59 for males and 0.75 for females; Koford et al., 1986; Bowen et al., MS), and the Florida scrub jay ($D_o = 0.27$; Fitzpatrick and Woolfenden, 1986). Although additional data, particularly from noncooperative breeders, are desirable, these data indicate that space competition, although perhaps necessary, is not sufficient to explain the evolution of cooperative breeding.

Additional evidence for this contention is provided by adult floaters in a wide variety of bird populations (e.g., Stewart and Aldrich, 1951; Hensley and Cope, 1951; Knapton and Krebs, 1974). The existence of such floaters suggests that there is considerable space competition in many noncooperatively breeding species. We stress that such reproductive competition is probably typical of territorial species, and that competition for some critical resource other than space is probably typical in those that are not territorial. Thus, breeding competition per se can be only part of the story behind the evolution of cooperative breeding, and a large or small "index of breeding space competition," such as we have calculated here, may not tell us much about the behavior of the population surplus.

Instead, there must be additional features that cause the young of certain species, faced with the inevitable competition for breeding opportunities, to remain at home rather than to become nonterritorial floaters (e.g., Smith, 1978) or to live in "bachelor flocks" (e.g., Jenkins et al., 1963; Holmes, 1970; Atwood, 1980a). One possible critical feature is a high ratio of "optimal" versus "marginal" habitats, as discussed in section 1.3 and in Koenig and Pitelka (1981). Such ecological conditions may very well be peculiar and may thus help explain the relative rarity of cooperative breeding. Other predisposing features may be important as well, however (Brown, 1982, 1987). We will return to some of them when we discuss the evolution of delayed dispersal by nonbreeding helpers in Chapter 12.

8.9. SUMMARY

Of banded nestlings, 65.3% survive to be sexed following their postjuvenal molt in the autumn. Survivorship of fledglings varies greatly among years; overall 56.8% survive to their first spring, at which time dispersal may occur. The sex ratio of first-year birds is slightly, but not significantly, biased toward males; we have no evidence that the sex ratio of offspring varies with group size, composition, territory type, or annual productivity.

Approximately 67% of males and 60% of females surviving to their first spring remain as nonbreeders in their natal group their first year; about 24% of individuals of both sexes are still nonbreeders in their second year, after which time a few males but almost no females remain as nonbreeders. On average, males remain as nonbreeders for 1.08 years and females for 0.91 years. The propensity for birds to remain as nonbreeders varies significantly among years.

There are distinct sexual differences in the propensity for males and females to inherit their natal territory, move within the study area, or disappear. By the end of four years, 17.3% of males and only 7.4% of females have inherited, whereas 67.6% of females compared to only 50.6% of males have disappeared from the study area or died. These differences appear to follow from the general pattern of group composition (which results in more female vacancies than male vacancies) and a tendency for females to disperse farther than males, thereby disappearing from the study population more frequently. The observed root-mean-square dispersal distance is 0.47 km for males and 0.65 km for females; however, after correcting for the finite size of the study area, a reasonable minimum estimate for the true root-mean-square dispersal distance of the entire population is between 1.42 and 4.88 km.

The mean age of first reproduction is estimated to be 2.08 years for males and 1.91 years for females; males are more likely to disperse in sibling units than are females, but the difference appears to be largely attributable to a bias in the average number of other same-sexed siblings available for cooperative dispersal.

Based on the observed frequency of birds immigrating into the study area, it appears as though virtually all nonbreeders that survive to their first spring eventually obtain a breeding position somewhere. This finding contrasts with the expectation that, because of demonstrably high space competition, a large number of nonbreeding helpers should fail to attain breeding positions.

This paradox indicates that space competition per se is not the critical feature of cooperative breeding systems; rather, the important issue is what alternatives exist for offspring faced with what we believe to be the nearly universal situation (at least among territorial species) of space competition. In most species the alternatives chosen are to become a nonterritorial floater or to join a nonbreeding flock. In contrast, we suggest that a low ratio of marginal to optimal habitat makes such options impossible for some species. When this occurs, and when the resources of the natal territory permit, the "victims" of such space competition may delay dispersal and remain in their natal territories as nonbreeders.

CHAPTER NINE

Consequences of Territorial Inheritance

We have already presented data showing that either male or female acorn woodpeckers may inherit their natal territory (Chapter 8). However, such inheritance is male biased: 17.3% of males banded as nestlings and followed for four years inherited their natal territory compared to only 7.4% of females (Table 8.3). What are the consequences of this pattern of inheritance on the continuity of genetic lineages?

Inheritance of the natal territory may be important to the evolution of cooperative breeding in that it could provide significant advantages to both nonbreeders and breeders: for the nonbreeder, inheritance provides a relatively safe way to acquire breeding status in the population—one requiring no dispersal whatsoever. For breeders, this mechanism provides not only a means by which parents might enhance the survival of their offspring, but a way in which they can pass down a good territory, including the storage facilities, to their descendants.

In fact, ever since the early days of the study of cooperative breeding, it has frequently been assumed that the passing down of a territory through a single genetic lineage is one of the significant advantages of group living. Brown (1974), for example, suggested that in Mexican jays "*the most productive territories of an area are handed down from generation to generation within the ownership of the same genetic lineages*" (italics his) and thus "the probability of handing down a good territory to one's own descendents is nearly 100 percent for a flock of Mexican jays." (Brown's subsequent work has shown this to be overstated [Brown and Brown, 1984], but the general issue remains important.) Territorial inheritance by helpers was similarly invoked as a crucial feature of cooperative breeding systems by Gaston (1978a), who dubbed this effect the "patrimony factor" to emphasize the fact that in many such species only male nonbreeders inherit their natal territory (e.g., Florida scrub jays [Woolfenden and Fitzpatrick, 1978, 1986], groove-billed anis [Vehrencamp, 1978; Bowen et al., MS], and several species of babblers [Gaston, 1978a]). However, the long-term demographic consequences of this pattern have not been examined previously.

We have documented elsewhere (Koenig and Pitelka, 1979; Koenig et al., 1984) that inheritance of the natal territory by a nonbreeding helper occurs following the disappearance of all breeders of the opposite sex that were present when the offspring was born (that is, opposite-sexed birds who might be his or her parents). In such circumstances offspring may subsequently breed even though the parent of the same sex is in the group. Thus, two generations may breed simultaneously in a group. Group offspring that have inherited their natal territory may continue to breed if they outlive their parent, thereby completing the transfer of breeding status to the younger generation.

In two cases, we documented birds of three successive generations of the same sex inheriting a territory; both cases involved males. The first group is Lower Haystack: ♂307, ♂309, and ♂390—all birds born at Lower Haystack in 1976 or 1977—inherited and bred together at Lower Haystack in 1978. Two of these males subsequently disappeared, but one of their offspring from 1978, ♂461, survived and bred along with the remaining male (♂307) starting in 1979, after ♂461's mother disappeared and was replaced by an unrelated female.

The second case was at group Lambert. ♂114 was banded as a nestling in this group in 1972, and may have been a breeder when ♂284 fledged here in 1976. By spring 1977, ♂114 was the only adult male in the group and thus had inherited the territory. In winter 1978, the entire group was forced to abandon because of an acorn crop failure. The following spring, ♂284 returned and bred successfully with a new female. The female disappeared in the winter of 1980–81, but two of their offspring, ♂582 and ♂583, survived and were present in 1981 along with ♂284 and another new female. No breeding attempt was made that year, and the birds abandoned the territory that autumn. Nevertheless, three generations of male offspring apparently inherited this territory, demonstrating the possibility of territories being handed down through several generations, at least through the male line.

9.1. SURVIVORSHIP OF LINEAGES

In order to estimate the significance of this pattern, we analyzed the "survivorship" of lineages using data from all focal study groups, beginning when the histories and breeding status of group members were known with sufficient confidence that lineages could be traced. (Because of polygynandry, such information is often easier to infer than actual parentage.) Lineages were followed only through descendants living on the same territory, and we ignored the few cases in which we knew that immigrants to a territory were in fact distant genetic descendants of the current residents. Lineages were followed through (1) either sex in the group (i.e., beginning with either the breeding males or females, the number of years that genetic continuity could be traced through breeders residing in the group), and (2) either the male line alone or the female line alone, in order to compare the consequences of the sex bias in inheritance. We compiled data from 30 groups including 512 lineage-years.

Cumulative survivorship was estimated by multiplying the probabilities of survival for each prior year, as determined by the proportion of lineages entering a year that survived to the next year. The standard error of the cumulative survivorship values was calculated by standard actuarial methods (Hull and Nie, 1981).

Results from these analyses were compared with two additional data sets. First, survivorship of the appropriate sex-specific lineages is compared with the survivorship of individual breeding males and females (from Chapter 7). Second, using the record of continuous occupation by territories, individual territories were considered to "die" when they went unoccupied for one or more breeding seasons, and otherwise were considered to still be "alive," even if the birds residing on it had changed completely since the prior year. These data provide an estimate of territory longevity independent of the individual birds or genetic lineages residing in them.

Results are shown in Table 9.1 and graphed in Figure 9.1. Survivorship of lineages, irrespective of sex, is 86.6% per year, with a mean survival time of 4.37 years. The survivorship of male-only lineages is virtually identical to these values, whereas that for female lineages is 80.5%, with a mean survival time of 3.45 years. Thus, male-specific genealogical lines appear to be primarily responsible for the longevity of lineages. The considerably greater longevity of male lineages parallels the higher survivorship of breeding males, the greater tendency of males to share breeding opportunities compared to females, and the higher probability of males inheriting their natal territory.

Table 9.1 also permits a comparison of sex-specific lineage survivorship to the survivorship of individual breeders. Lineages, as expected, last longer, but the differences are quite modest. Annual survivorship of male lineages is 86.2% compared to average individual male survivorship of 83.4% over the same period; for females, lineage survivorship is 80.5% compared to individual survivorship of 74.4%. Thus, the increase in sex-specific survivorship as a result of considering genetic lineages rather than individuals is modest for females and negligible for males.

In contrast, the "survivorship" of territories, as defined above, is quite high: annual survivorship is 95.0% per year and the mean survival time is 11.87 years. Thus, the survivorship of territories far exceeds that of both the individual breeders and of the lineages that inhabit them.

TABLE 9.1. Survival of Lineages, Territories, and Individuals

	Either Sex	*Males Only*	*Females Only*	*Territories*
LINEAGES				
Percent annual survivorship	86.6	86.2	80.5	95.0
Mean survival time (years)	4.37	4.28	3.45	11.87
Cumulative percent ($\bar{x} \pm$ SE) surviving				
5 years	41.2 ± 5.2	40.3 ± 7.4	27.7 ± 6.5	78.5 ± 7.3
10 years	26.5 ± 5.6	29.1 ± 7.8	7.1 ± 4.9	58.5 ± 9.5
N lineage-years	512	258	246	273
INDIVIDUAL BREEDERS				
Percent annual survivorship[a]	—	83.4	74.4	—
Mean survival time (years)	—	4.01	2.47	—
Cumulative percent ($\bar{x} \pm$ SE) surviving				
5 years	—	41.3 ± 6.8	20.3 ± 5.9	—
10 years	—	16.3 ± 8.0	$0.0 \pm$	—
N bird-years	—	273	170	—

NOTE: Difference between male and female lineages not significant by χ^2 contingency test (annual survivorship) or Lee-Desu statistic (mean survival time).

[a] Figures include data collected before 1973 and thus differ from values reported in Chapter 7.

The mean generation time for the Hastings population is on the order of five years (see Chapter 10). Thus, the cumulative percent of lineages surviving five and ten years provides an estimate of the likelihood that the breeders in a territory will have descended from birds breeding on that same territory one and two generations in the past, respectively. These values, calculated from the actuarial survival analysis, are given in Table 9.1. The probability that a territory will retain the same lin-

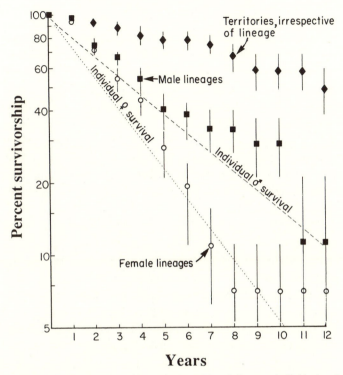

FIGURE 9.1. Survivorship curves for male and female lineages, individual male and female breeders, and for territories irrespective of the relatedness of the inhabitants. For sample sizes, see Table 9.1. Values plotted are the mean percent surviving ± SE.

eage after one generation is 41.2% and after two generations is 26.5%. Once again, the values for the male-specific lineages are virtually identical to those for the sexes combined, whereas the probability of a female breeder successfully bequeathing her territory to descendant females after one generation is 27.7%, and after two generations only 7.1%.

In contrast, the probabilities that a territory will still be continuously occupied five and ten years from any particular date are 78.5% and 58.5%, respectively. As before, a territory has a life expectancy far exceeding that of either the individuals inhabiting it or of their descendants.

Effect of Territory Quality

These analyses suggest that territorial inheritance, although potentially crucial as an alternative to emigration for nonbreeding helpers, nonetheless has only a small effect on the overall demographics of acorn woodpecker populations. There is, however, an important caveat to this conclusion: the probability of inheritance varies significantly with the quality of the territory. As we saw in Chapter 4, a likely candidate for determining whether a territory is good or bad is the size of its storage facilities. Thus, as a test of the importance of territory quality, we analyzed the survivorship of lineages depending on the available storage facilities.

The results are presented in Table 9.2 and graphed in Figure 9.2. Territories were divided into those with less than an estimated 1,000 storage holes, between 1,000 and 2,000 storage holes, and more than 2,000 storage holes. In all comparisons, including sex- and non-sex-specific survivorship as well as the "survivorship" of territories, values increase with increasing number of storage holes, significantly so in all but one case (that of annual survivorship of female lineages).

On territories with over 2,000 holes, mean survival time is over eight years for male lineages as well as non-sex-specific lineages, and over four years for female-specific lineages. Although these values still are considerably lower than territory "survivorship" in comparable areas (where annual survivorship exceeds 98% on territories with over 2,000 storage holes), they suggest that inheritance of territory may be a significant benefit of group living to an important segment of acorn woodpecker populations, specifically, nonbreeding males living in territories with large storage facilities. Lineages on such territories clearly could last fifteen to twenty years; they are unlikely to last longer, at least in areas such as Hastings Reservation, because of the relatively high frequency of crop failures. Females may also benefit, but only through their male descendants; thus, breeding females only stand a reasonable chance of passing their territory on to their sons.

TABLE 9.2. Survival of Lineages and Territories as a Function of Storage Facilities

	< 1,000 Holes	1,000–2,000 Holes	> 2,000 Holes	Comparison
EITHER SEX				
Percent annual survivorship	75.6	82.3	91.4	< 0.001
$\bar{x}$ survival time (years)	2.98	4.07	8.31	< 0.001
Cumulative percent surviving 5 years	0.0±	39.5 ± 10.4	59.7 ± 7.1	—
N lineage-years	84	102	305	—
MALE LINEAGES				
Percent annual survivorship	74.1	82.5	90.8	< 0.05
$\bar{x}$ survival time (years)	3.20	2.87	8.41	< 0.05
Cumulative percent surviving 5 years	0.0±	36.5 ± 14.1	59.3 ± 10.7	—
N lineage-years	42	51	141	—
FEMALE LINEAGES				
Percent survivorship	68.8	78.6	83.8	ns
$\bar{x}$ survival time (years)	2.20	2.70	4.23	< 0.05
Cumulative percent surviving 5 years	0.0±	45.4 ± 13.2	30.2 ± 9.2	—
N lineage-years	38	51	135	—
TERRITORIES				
Percent annual survivorship	88.2	91.1	98.7	< 0.01
$\bar{x}$ survival time	6.60	8.10	> 12.00	< 0.01
Cumulative percent surviving 5 years	62.5 ± 21.3	69.4 ± 14.9	100.0±	—
N territory-years	34	56	154	—

NOTE: Comparison lists the p-value of a χ^2 contingency test (annual survivorship) or Lee-Desu test (mean survival time). Cumulative percent lists $\bar{x} \pm$ SE.

Thus, the potential benefits derived through territorial inheritance may be important to the evolution of sociality in the acorn woodpecker. We will thus consider this feature carefully when we discuss the evolution of delayed dispersal and the retention of offspring in Chapter 12.

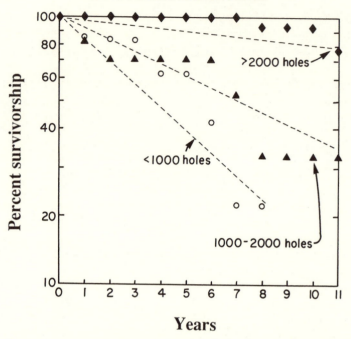

FIGURE 9.2. Survivorship curves for lineages as a function of territory quality as indicated by the number of storage holes. Solid diamonds are territories with >2,000 holes, solid triangles territories with 1,000–2,000 holes, and open circles territories with <1,000 holes. For sample sizes, see Table 9.2. Values plotted are the mean percent surviving; lines were drawn by eye. Differences among groups are significant (Lee-Desu statistic = 10.4, df = 2, $p < 0.01$).

9.2. PARENTAL MANIPULATION OR FACILITATION?

Another consequence of territorial inheritance is that some birds eventually breed at home in a familiar territory, often along with a parent of the same sex, whereas the majority spend their reproductive careers elsewhere, either with siblings or with no relatives at all. Brown and Brown (1984), examining a similar situation in the Mexican jay, provided some evidence that offspring inheriting their natal groups achieved higher reproductive success than those moving to other territories. They suggest that this form of parental facilitation may be important in cooperative breeders, particularly joint-nesting species such as the acorn woodpecker, in which groups consist of more than a pair of breeding individuals. Might such reproductive enhancement be occurring in the acorn woodpecker?

Annual Reproduction and Survivorship

Comparisons of annual reproductive success and survivorship of birds known to have inherited their natal groups (heirs) versus those that dispersed and bred elsewhere (nonheirs) is presented in Table 9.3. Values for annual reproductive success

TABLE 9.3. Comparison of Annual Reproductive Success and Survivorship of Birds Inheriting Their Natal Territory and Moving Elsewhere to Breed

	Inherited	Moved	Statistic[a]	p-value
MALES				
Young fledged per group	2.23 ± 2.09 (53)	2.36 ± 2.18 (101)	0.33	ns
Young fledged per breeding male	1.12 ± 1.27 (49)	1.18 ± 1.18 (97)	0.40	ns
Percent annual survivorship	85.1 (60)	83.5 (121)	0.0	ns
$\bar{x}$ survival time (years)	3.93	3.75	0.22	ns
FEMALES				
Young fledged per group	2.27 ± 2.49 (11)	2.44 ± 2.15 (41)	0.33	ns
Young fledged per breeding female	1.36 ± 1.38 (11)	1.71 ± 1.71 (41)	0.39	ns
Percent annual survivorship	57.1 (14)	75.0 (52)	1.0	ns
$\bar{x}$ survival time (years)	2.00	2.63	0.89	ns

[a] Data are $\bar{x} \pm$ SD (N). Comparisons are by Mann-Whitney U-test (young fledged), χ^2 contingency test (annual survivorship), and Lee-Desu statistic (mean survival time).

assume, in the absence of better information, that there is no bias in reproductive success among cobreeders sharing reproductive status within a group (this assumption is discussed in detail in sections 2.2, 10.3, and 13.4). Data are presented using all available groups, but the results are not changed when controlling for group turnover, territory quality, breeder age, or presence of nonbreeders in the group.

We found no significant differences in annual reproductive success between heirs and nonheirs; indeed, in all four comparisons emigrants do slightly better than heirs. This result thus contrasts with that reported by Brown and Brown (1984) for the Mexican jay.

However, facilitation may manifest itself in ways other than annual reproductive success. Table 9.3 also presents data on the survivorship, from the time of attaining reproductive status, of heirs versus emigrants. Similar to the results for annual reproductive success, there are no significant differences in the survivorship of individual males or females. This suggests that territorial inheritance and emigration yield comparable lifetime fitness.

Lifetime Reproductive Success of Heirs versus Nonheirs

This latter supposition can be tested directly with the lifetime reproductive success of heirs versus nonheirs in the population. (We discuss our determination of lifetime reproductive success in section 10.3.) These data, again calculated assuming no bias in parentage of young among cobreeders, are presented in Table 9.4. No significant differences emerge for either sex in either reproductive success or the average number of cobreeders with which heirs and nonheirs share reproduction.

Thus, we have no direct evidence that reproductive success is enhanced by inheriting and breeding on the natal territory. Lifetime fitness might still be increased if dispersal entails significant likelihood of mortality; however, as shown in section

TABLE 9.4. Lifetime Reproductive Success of Heirs versus Nonheirs

	Heirs	Nonheirs	z-value[a]
MALES			
Group total	5.22 ± 6.36 (9)	4.09 ± 6.51 (32)	0.58
Individual	2.30 ± 2.58 (9)	2.36 ± 3.81 (32)	−0.41
Cobreeders	2.29 ± 1.07 (9)	1.98 ± 1.00 (32)	0.77
FEMALES			
Group total	3.83 ± 6.21 (6)	3.30 ± 4.16 (27)	0.07
Individual	2.17 ± 3.13 (6)	2.93 ± 3.73 (27)	−0.31
Cobreeders	1.39 ± 0.70 (6)	1.31 ± 0.68 (27)	0.47

NOTE: Group total is the total number of young fledged by the individual's breeding group; the individual total estimates lifetime reproductive success by dividing the group total for each year by the number of breeders of that sex present in the group; the cobreeders total estimates the incidence of mate-sharing experienced by individuals averaged over each year they bred. Data include cohorts through 1980 only.

[a] Comparisons by Mann-Whitney U tests; all p-values > 0.05.

8.4, this does not appear to be the case, despite considerable competition for breeding vacancies.

Facilitation by Cobreeding with Parents

An additional form of parental facilitation to consider is that resulting from parents of the same sex being present in the group following inheritance by group offspring. If a significant difference exists between the success of heirs with and without parents, it would suggest that facilitation may occur when new heirs have older, experienced breeders from which to gain mutualistic benefits. Because of joint-nesting by females and the high incidence of mate-sharing by breeding males, the potential for such facilitation is considerable.

Data testing this hypothesis, based on yearly reproductive success, are presented in Table 9.5. For males, there are no significant differences between the two groups of heirs. For females, however, groups containing daughters breeding with

TABLE 9.5. Reproductive Success of Birds Inheriting Their Natal Territory as a Function of Whether a Parent of the Same Sex Was Still Present in the Group

	Parent Present	Parent Gone	z-value	p-value[a]
MALES				
Young fledged	2.23 ± 2.22 (39)	2.21 ± 1.76 (14)	0.16	ns
Young fledged per breeding male	0.91 ± 0.97 (35)	1.64 ± 1.75 (14)	1.15	ns
FEMALES				
Young fledged	4.00 ± 2.55 (5)	0.83 ± 1.33 (6)	2.11	<0.05
Young fledged per breeding female	2.00 ± 1.27 (5)	0.83 ± 1.33 (6)	1.35	ns

[a] Comparisons by Mann-Whitney U tests.

their mother fledge significantly more young than daughters breeding in the absence of their mother; a considerable difference remains when considering young fledged per breeding female, although the difference is no longer significant.

There are at least two reasons to be cautious about these results. First, the sample size of female heirs is quite small, and the difference could very well disappear were it possible to control for age and turnover among breeders of the opposite sex. Second, we have few data to affirm the assumption of equal maternity, and none to affirm equal paternity, among cobreeders. However, these data do at least suggest that this form of parental facilitation might be acting for the few females that inherit their natal territory and breed with their mothers.

Facilitation and Territory Quality

Parental facilitation of breeding by offspring could result from the retention of young in high-quality, familiar territories (Brown, 1974; Brown and Brown, 1984). Indeed, as predicted by Brown (1974), territorial inheritance appears to be of con-

TABLE 9.6. Transfer within and between Territory Types by Birds Immigrating within the Study Area and Inheriting Their Natal Territory

	Territory Immigrated To		
	Always occupied	Unoccupied ≥ 1 year	Total
TERRITORY EMIGRATED FROM			
Always occupied	55 (56.7%)	30 (30.9%)	85
Unoccupied ≥ 1 year	7 (7.2)	5 (5.2)	12
Total	62 (63.9)	35 (36.1)	97
HEIRS	23 (85.2)	4 (14.8)	27

siderable importance in the best territories (i.e., those with large storage facilities; see section 9.1). However, we have no evidence that territorial inheritance results in an increase in the fitness of inheriting offspring compared to those breeding elsewhere.

One possible explanation for the lack of this form of parental facilitation would be if emigrants tended to move to better territories than where they were born. This does not appear to be true: although the majority of emigrants move to territories of the same type as those in which they were born (56.7% to other territories occupied all years, 5.2% to territories that are occasionally unoccupied; Table 9.6), a higher proportion move from good territories to poor ones (30.9%) than the reverse (7.2%; Table 9.6).

Consequently, there is a slight advantage for heirs, 85.2% of which inherited territories that were always occupied, whereas only 63.9% of emigrants moved to territories of similar high quality (Table 9.6). This difference is not, however, significant ($\chi^2 = 3.5$, ns).

Parental Facilitation and Variance Enhancement

In summary, we have no evidence to support the hypothesis that the lifetime reproductive success of individuals remaining

241

at home exceeds that of those moving elsewhere to breed, and thus this form of parental facilitation does not appear to be significant in our population. The only potential exception is among females inheriting their natal territories and breeding with their mothers; however, even if facilitation is real in this situation, it affects only a small proportion of offspring, as less than 10% of females inherit their natal teritories. Our failure to find a significant effect suggests that parental enhancement of their offspring's reproduction is not a factor promoting philopatry and cooperative breeding in this species.

In the absence of such facilitation, the variance enhancement model of Brown (1978, 1987; Brown and Brown, 1984; Brown and Pimm, 1985) appears inapplicable to acorn woodpeckers. This model proposes that parents increase the variance in fitness among their offspring by enhancing the reproductive success of some while suppressing that of others (who then presumably remain as nonbreeding helpers).

In its most general formulation, the variance enhancement model of Brown and Pimm (1985) requires that the effectiveness of individuals in rearing young as nonbreeding helpers must exceed their effectiveness in rearing young as independent breeders. Considerable empirical evidence indicates that, at least on an annual basis, this condition does not hold for the majority of cooperative breeders (Koenig, 1981b, and references therein). Per capita reproductive success almost invariably decreases with increasing number of helpers. Thus, other things being equal, fitness of parents and offspring is likely to be maximized when all offspring breed independently (that is, when variance in offspring reproductive potential is absent). For acorn woodpeckers, at least, this conclusion generally holds even when enhancement of breeder survivorship by nonbreeders (Chapter 7) is taken into account (see Chapter 12).

Alternatively, the widely accepted model of habitat saturation and ecological constraints for the evolution of the retention of young (see Chapters 1 and 12) is consistent with data indicating that both parents and young benefit by offspring breeding on their own whenever possible. Delayed reproduction and helping by offspring should occur only so long as some offspring are unable to obtain breeding positions elsewhere. Thus, variance in offspring reproductive potential is imposed by ecological factors rather than parental behavior (the "variance utilization" model of Brown and Brown [1984] and Brown and Pimm [1985]).

Under conditions of ecological constraints, parents and offspring are not usually in conflict: both benefit from retention when these offspring are unable to acquire breeding positions on their own. Nonetheless, once some young are retained, the interests of parents and offspring frequently conflict over limited food resources, the filling of reproductive vacancies, and parentage of young (Mumme et al., 1983a, 1983b; Hannon et al., 1985, 1987; Koenig et al., in press). Parent-offspring relations in acorn woodpeckers are thus characterized by both conflict (e.g., Emlen, 1982b; Vehrencamp, 1983) and cooperation.

9.3. SUMMARY

Acorn woodpeckers may inherit their natal territory and breed following the demise of the parent(s) of the opposite sex; such heirs may then share mates with their parents and sometimes other offspring of the same sex. Either sex may inherit, but, because of demographic patterns, inheritance by males is much more common than by females.

Here we investigate two potentially important side effects of this phenomenon. First, we calculate the importance of territorial inheritance in extending the tenure of genetic lineages on individual territories (the "patrimony factor" of Gaston [1978a]). Overall, inheritance does not have a major demographic impact on the tenure of lineages: there is little overall difference in the survivorship of lineages compared to individual breeders, and the proportion of territories remaining within the

243

same genetic lineage is small after only a few generations. However, this pattern is quite different in high-quality territories (those with large storage facilities), where the probability of handing down the territory to descendants is high. Thus, as suggested by Brown (1974), territorial inheritance could provide an important advantage to offspring, at least for males living in high-quality territories.

A second possible result of territorial inheritance is parental facilitation whereby parents enhance the reproductive success of some of their offspring, possibly at the expense of others ("variance enhancement" [Brown and Brown, 1984; Brown and Pimm, 1985]). We review data on annual reproductive success, survivorship, and on lifetime reproductive success of acorn woodpeckers inheriting their natal territories compared to those dispersing and breeding elsewhere. We find no support for this form of parental facilitation, with the possible exception of the rare situation in which females inherit and breed with their mothers. Retention of young is thus more parsimoniously explained by ecological constraints than by parentally imposed variance in offspring reproductive potential. Although these constraints result in strongly overlapping interests of parents and their young, both conflict and cooperation between parents and offspring are central to cooperatively breeding societies.

CHAPTER TEN

The Life Table and Lifetime Reproductive Success

Thus far we have discussed individual components of fitness, including reproductive success, survivorship, and fate of offspring. We will now consider the age structure of the Hastings population and combine these analyses to produce a synthetic life table.

This exercise has several goals. First, it allows us to check prior estimates to see if they provide a demographically reasonable picture of the population. Second, it allows us to try out alternative values of critical variables. Third, it extends our prior analyses by providing estimates for a variety of additional demographic parameters. Fourth, in conjunction with the dispersal data discussed in section 8.7, it allows us to estimate effective population size. Finally, it allows us to begin consideration of the life-history consequences of remaining in the natal group as a nonbreeder and of mate-sharing, investigations that we continue in Chapters 12 and 13.

Ultimately, natural selection favors individuals who produce the largest number of surviving offspring (Darwin, 1859; Williams, 1966; Grafen, 1982). Hence, following the life-table analysis we will examine lifetime reproductive success of birds pursuing alternative tactics.

Lifetime reproductive success must be considered in order to adequately determine fitness because of the likelihood that reproduction and survivorship are inversely related. For example, reproductive success per breeding male decreases as the number of breeding males in a group increases (Chapter 6), whereas survivorship of individual males increases with number of breeders (Chapter 7). Thus, annual values may yield misleading information concerning the relative success of males that disperse alone and breed compared to those that disperse and share mates.

10.1. AGE STRUCTURE

A stable age distribution is reached by a population maintaining a constant birth and death rate (Lotka, 1922). Such a distribution is an important assumption of a life-table analysis. Because of the differing survivorship and fecundity schedules of breeders and nonbreeders, it is difficult to derive a precise expected age distribution for acorn woodpeckers. However, the overall age distribution of males and females in the population, plotted in Figure 10.1, is consistent with the assumption that the long-term average age structure is relatively stable.

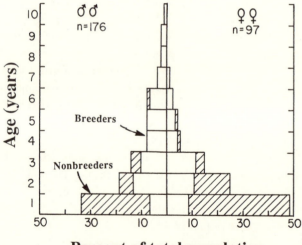

Percent of total population

FIGURE 10.1. Observed age distribution for males and females during the breeding season (excluding young of the year), divided into breeders (open bars) and nonbreeding helpers (hatched bars). Mean age was 3.11 years for males and 2.02 years for females.

Virtually all nonbreeding helpers live in their natal group and are thus of known age. However, because a high proportion of breeder recruitment into the population, especially for females, is by unmarked birds from outside of the study area, only a fraction of the population is of known age. We therefore restricted our analysis to the five breeding seasons of 1978 through 1982, when we had good information concerning ages.

During this period, 47% of 318 breeders in the population were of known age, including 32% of 123 breeding females and 56% of 195 breeding males. Because many birds are of unknown age, it is likely that older individuals are underrepresented in our samples. Thus, our analysis of age structure must be considered only approximate.

The overall age structure is older for males (Figure 10.1), as expected from their lower mortality rate (Chapter 7). Also shown in Figure 10.1 are the proportions of breeders and non-breeders in each age class. As expected, nonbreeders predominate in the younger age classes, particularly in the first year, and drop out later on. The age distribution of breeders alone is shown in Figure 10.2, plotted with the expected age distribution, as determined by age-specific fecundity and mortality schedules (section 10.2) and assuming a steady-state population. As before, the higher mortality rate of breeding females results in a relatively younger distribution for females compared to males.

A comparison of the observed and expected distribution in Figure 10.2 indicates that we are missing many of the older individuals. This is especially true for females, of which a high proportion are of unknown age. Our study has not continued long enough to detect the oldest birds in the population.

The year-to-year range in the age distribution of breeders is indicated by Figure 10.3, which divides the population of known-aged individuals for each year between 1978 and 1982 as well as for all five years combined. Even disregarding the older age classes (>6 yrs), which are biased by the length of

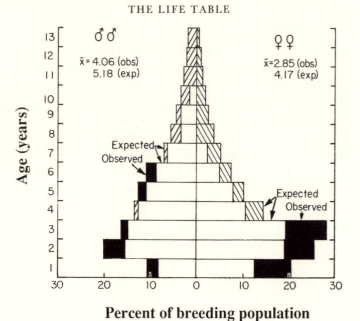

Percent of breeding population

FIGURE 10.2. Observed versus expected age distribution for breeding males and females; expected distributions determined from the life-table analyses shown in Tables 10.4 and 10.5. Excess of expected distribution is indicated by hatched portion of the bars, excess of observed distribution by the solid black portion; white portions are overlapped by the two distributions. Mean ages for the observed and expected distributions are given.

the study, considerable annual variation occurs in the age distribution.

The range in variation shown in this figure probably comes close to representing the extremes, as it includes the 1978 autumn acorn crop failure as well as several years of excellent acorn productivity. This failure, and the subsequent poor over-winter survivorship in 1978–79, resulted in an unusually small cohort of first-year birds entering the breeding population in 1979 as well as a corresponding increase in the proportion of older breeders. One- to two-year-old breeders constituted only 14% of the population in 1979, whereas in 1982, after two good years of reproduction, birds in this category made up 49%

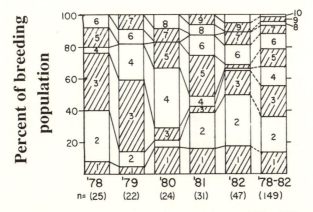

FIGURE 10.3. Yearly observed age distribution of breeders, 1978 to 1982; numbers inside the bars are the ages. The proportions of total breeders whose ages were known were as follows: 1978, 42%; 1979, 40%; 1980, 44%; 1981, 48%; 1982, 57%; all five years combined, 47%.

of the population. In general, the majority of breeders are two or three years old, followed in abundance by first-year birds.

These data suggest that the Hastings population is both in a steady-state and maintaining a stable age distribution, at least on a long-term basis. The major discrepancy between the observed and expected age distribution is in the lack of older age classes, an artifact attributable to the length of our study.

10.2. THE LIFE TABLE

Acorn woodpeckers offer several challenges to a traditional life-table analysis. First, the proportion of individuals breeding cannot be assumed to be a simple step function, as in a species where all surviving individuals reach reproductive maturity and breed at the same age. Second, since more than a single male and female may breed in a group, reproductive success per individual is not congruent with reproductive success per breeding unit. Finally, survivorship of a cohort varies depending on what proportion of individuals are breeders and nonbreeders. However, with appropriate modifications, a synthetic life table can be constructed and the results compared with other species.

Basic Equations

The basic data set from which the life tables are constructed are presented in Table 10.1. As is traditional, we will focus on

TABLE 10.1. Summary of Demographic Parameters Used in Life-Table Analyses

Variable	Value	Source[a]
1. Mean set size	4.82 eggs nest^{-1}	Table 5.3
2. Mean $\mathcal{N}$ breeding males	1.84 group^{-1}	Table 3.6
3. Mean $\mathcal{N}$ breeding females	1.19 group^{-1}	Table 3.6
4. Mean number of males	2.55 group^{-1}	Table 3.6
5. Mean number of females	1.70 group^{-1}	Table 3.6
6a. Sex ratio of offspring	0.531 males	Table 8.1
b.	.469 females	
7. Fledging success	2.12 fledgings group^{-1}	Table 5.2
8. Proportion of groups breeding	0.846	Section 5.4
9. Proportion of groups with 2d nests or renests	0.121	Section 5.1
10. Proportion of groups with autumn nests	0.022	Table 5.2
11. Proportion of eggs hatching	0.680	Table 5.9
12. Proportion of eggs fledging	0.423	Table 5.9
13a. Mean number of ♀ eggs	1.88 (brdg. female)$^{-1}$ year^{-1}	[1 × (8 + 9 + 10) × 6b/3]
b. Mean number of ♂ eggs	1.38 (brdg. male)$^{-1}$ year^{-1}	[1 × (8 + 9 + 10) × 6a/2]
14. Survivorship from fledging to first spring	0.568	Table 8.1
15a. Breeder survivorship	0.801 year^{-1} [males]	Table 7.1[b]
b.	0.691 year^{-1} [females]	
16. Nonbreeder survivorship	0.90 year^{-1}	Section 8.4
17a. Proportion of dispersers obtaining a breeding position	0.680 (males)	Table 8.6
b.	0.774 (females)	

[a] Numbers in brackets refer to values from this table.

[b] Includes birds abandoning their territories; see Section 7.1.

a life table constructed for females; however, we will also construct a life table for males so as to compare the demographic consequences of sex-specific dispersal and survivorship.

The symbols used in these analyses are summarized in Table 10.2. We use f_x as the average fecundity of breeding females of age x and d_x as the fraction of individuals breeding at age x. Hence, $f_x d_x = m_x$, the average fecundity of females of age x, as usual. The parameter x is defined as age class (starting with $x = 1$) rather than age (for a discussion of this procedure, see Murray and Gårding, 1984).

Formulae for the life-table calculations are summarized in Table 10.3. Although we have no evidence for it (see section 7.2), we believe it is realistic to assume that senescence occurs (Botkin and Miller, 1974; Woolfenden and Fitzpatrick, 1984). Consequently, we assume that no acorn woodpecker survives

TABLE 10.2. Symbols Used in Life Table Analyses

Symbol	Meaning
x	Age class (Murray and Gårding, 1984)
ℓ_x	Age-specific annual survivorship
ℓ_{br}	Annual survivorship of breeders
ℓ_h	Annual survivorship of nonbreeding helpers
W	Proportion of dispersing nonbreeders obtaining a breeding position
l_x	Cumulative survivorship to age class x
d_x	Proportion of birds surviving to age class x that are breeders
f_x	Age-specific fecundity of breeders; female eggs per breeding female or male eggs per breeding male of age class x
m_x	Age-specific fecundity; female eggs per female or male eggs per male of age class x
V_x	Reproductive value at age class x
E_x	Expectation of further life at age class x
r	Annual rate of increase
R_o	Net reproductive rate
T	Mean generation time

TABLE 10.3. Formulae for Calculations Contained in Life-Table Analyses

Variable	Formula	Source
1. Proportion of individuals breeding at age x	—	Calculated from Table 8.7
2. Net reproductive rate (R_o)	$\Sigma\,(l_x m_x)$	Ricklefs, 1973
3. Annual rate of increase (r)	Estimated by r such that $\Sigma\,(l_x m_x e^{-rx}) = 1$	Dublin and Lotka, 1925
4. Expectation of further life (E_x)	$(-1/\ln(1-(l_x/\sum l_x)))$	Ricklefs, 1973
5. Mean generation time (T)[a]	$\Sigma\,(x e^{-rx} l_x m_x) - 1$	Caughley 1976
6. Reproductive value (V_x)	$\dfrac{\Sigma\,(e^{-rx} l_x m_x)}{(e^{-rx} l_x)}$	Fisher, 1930

[a] Because x is the age class rather than age, 1 is subtracted from the standard summation for mean generation time (Murray and Gårding, 1984).

beyond age fifteen (see section 7.2). In each case, we present estimated values for net reproductive rate, mean generation time, expectation of further life, and reproductive value.

Because the population sizes at the beginning and end of the study were almost identical (Figure 3.3), the overall net reproductive rate (R_o) was close to 1 and the annual rate of increase (r) was close to 0. However, because the calculations do not employ values that were all derived over the same period of time, we cannot expect them to yield values for these variables exactly equal to those observed during the study. For example, survivorship and reproductive success is estimated from 1972 to 1982, but data for clutch size and other nesting parameters are only from 1976 to 1982.

Results

The life table for females is presented in Table 10.4. Survivorship of nonbreeders is assumed to be 90% year^{-1}, whereas

TABLE 10.4. Synthetic Life Table for Females

Age	x	ℓ_x	l_x	d_x	m_x	$l_x m_x$	$x l_x m_x$	E_x	V_x
Egg	1	0.680	1.000	0.000	0.00	0.000	0.000	2.60	1.06
Nestling	1	0.622	0.680	0.000	0.00	0.000	0.000	2.60	1.57
Fledgling	1	0.568	0.423	0.000	0.00	0.000	0.000	2.90	2.52
1st February	1	0.899	0.240	0.000	0.00	0.000	0.000	3.75	4.43
1 year	2	0.780	0.216	0.330	0.62	0.134	0.268	3.11	4.93
2 years	3	0.726	0.168	0.664	1.25	0.210	0.631	2.85	5.52
3 years	4	0.702	0.122	0.913	1.72	0.210	0.840	2.75	5.89
4 years	5	0.704	0.086	0.959	1.80	0.155	0.773	2.71	5.94
5 years	6	0.694	0.060	0.941	1.77	0.107	0.641	2.66	5.88
6 years	7	0.691	0.042	1.000	1.88	0.079	0.551	2.62	5.93
7 years	8	0.691	0.029	1.000	1.88	0.054	0.435	—	—
8 years	9	0.691	0.020	1.000	1.88	0.038	0.338	—	—
9 years	10	0.691	0.014	1.000	1.88	0.026	0.260	—	—
10 years	11	0.691	0.010	1.000	1.88	0.018	0.197	—	—
11 years	12	0.691	0.007	1.000	1.88	0.012	0.149	—	—
12 years	13	0.691	0.005	1.000	1.88	0.009	0.111	—	—
13 years	14	0.691	0.003	1.000	1.88	0.006	0.083	—	—
14 years	15	0.691	0.002	1.000	1.88	0.004	0.061	—	—
15 years	16	0.691	0.002	1.000	1.88	0.003	0.045	—	—

NOTE: $\ell_{br} = 0.691$, $\ell_h = 0.90$, $W = 0.774$, $f_x = 1.88$. No females are assumed to live beyond age 15. Using these data, $r = 0.012$, $R_o = 1.06$, $T = 3.99$ years.

the proportion of nonbreeders disappearing that succeed in obtaining breeding positions (the probability of successful dispersal) is set at 77.4%.

The overall estimated net reproductive rate given these parameters is 1.064, while the estimated intrinsic rate of change (r) is 0.012. Thus, the life table matches the observed overall stability of the population. Estimated mean generation time for females is 3.99 years.

The expected age distribution of females based on this table is depicted in Figure 10.2. The survivorship curve is given in

Figure 10.4, age-specific reproductive values in Figure 10.5, and the $l_x m_x$ curve, corresponding to the contribution each age class makes to the net reproductive rate, in Figure 10.6.

The analogous life table for males is constructed in Table 10.5. In this case the estimated net reproductive rate is equal to 1.17 and the estimated r is 0.025, suggesting that the table is a reasonably good representation of the demography of males in the population. Because of their higher survivorship as breeders, the estimated mean generation time for males (5.11 years) is considerably longer than that for females. Age-specific survivorship is plotted with that for females in Figure 10.4.

The survivorship curves for both males and females (Figure 10.4) show the "maturational" plus "ecological" survivorship curves typical for birds (Tanner, 1978; this type of curve is characteristic of Deevey's [1947] Type II survivorship). Only

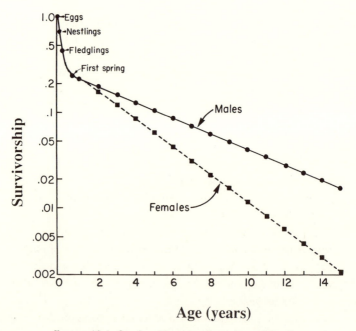

FIGURE 10.4. Survivorship curves for males and females.

254

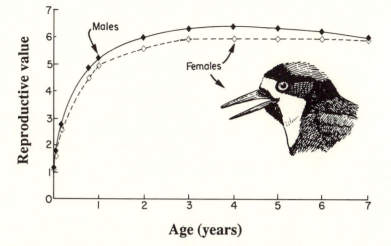

FIGURE 10.5. Age-specific reproductive value (V_x) for males and females. The curves cross at age nine, and the total area under the male line equals that under the female line; however, we have cut the lines off earlier because their shape at older age classes is determined largely by the assumptions of the life table rather than the demography of the sexes. In this case, V_x decreases beyond age four only because we cut the life table off at age fifteen.

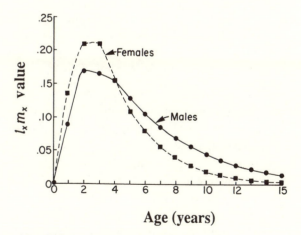

FIGURE 10.6. The observed $l_x m_x$ curves for males and females. The area under the curves indicates the contribution of that age class to the total reproductive output of the population; hence, the total area equals 1.

255

TABLE 10.5. Synthetic Life Table for Males

Age	x	ℓ_x	l_x	d_x	m_x	$l_x m_x$	$x l_x m_x$	E_x	V_x
Egg	1	0.680	1.000	0.000	0.00	0.000	0.000	2.92	1.17
Nestling	1	0.622	0.680	0.000	0.00	0.000	0.000	3.07	1.72
Fledgling	1	0.568	0.423	0.000	0.00	0.000	0.000	3.66	2.76
1st February	1	0.929	0.240	0.000	0.00	0.000	0.000	5.08	4.86
1 year	2	0.811	0.223	0.283	0.39	0.087	0.174	4.43	5.23
2 years	3	0.815	0.181	0.672	0.93	0.168	0.504	4.35	5.97
3 years	4	0.794	0.147	0.799	1.10	0.163	0.650	4.23	6.19
4 years	5	0.808	0.117	0.949	1.31	0.153	0.767	4.20	6.41
5 years	6	0.804	0.095	0.969	1.34	0.127	0.759	4.08	6.31
6 years	7	0.801	0.076	1.000	1.38	0.105	0.735	3.96	6.18
7 years	8	0.801	0.061	1.000	1.38	0.084	0.673	—	—
8 years	9	0.801	0.049	1.000	1.38	0.067	0.607	—	—
9 years	10	0.801	0.039	1.000	1.38	0.054	0.540	—	—
10 years	11	0.801	0.031	1.000	1.38	0.043	0.476	—	—
11 years	12	0.801	0.025	1.000	1.38	0.035	0.416	—	—
12 years	13	0.801	0.020	1.000	1.38	0.028	0.361	—	—
13 years	14	0.801	0.016	1.000	1.38	0.022	0.311	—	—
14 years	15	0.801	0.013	1.000	1.38	0.018	0.267	—	—
15 years	16	0.801	0.010	1.000	1.38	0.014	0.228	—	—

NOTE: $\ell_{br} = 0.801$, $\ell_h = 0.90$, $W = 0.680$, $f_x = 1.38$. No males are assumed to live beyond age 15. Using these data, $r = 0.025$, $R_o = 1.17$, $T = 5.11$ years.

about 20% of eggs live until they are one year old, after which survivorship is relatively high and apparently not strongly age-dependent (see section 7.2).

The plots of age-specific reproductive value (Figure 10.5) indicate the relatively high importance of older offspring. For example, a one-year-old offspring has between four and five times the reproductive value of an egg. Thus, given the choice of investing in either a new reproductive effort or older offspring, we can expect that the older offspring should warrant high priority (see also Woolfenden and Fitzpatrick, 1984;

Fitzpatrick and Woolfenden, 1986), as long as additional investment increases the probability that the offspring will eventually obtain a breeding position. One situation consistent with this prediction occurs when all the breeders of one sex die while several offspring of the same sex are still present in the group. Under these circumstances offspring may keep potential new breeders of the missing sex from joining the group. Remaining breeders, rather than excluding the offspring so that a new mate can join the group, instead allow their offspring to remain, thereby forfeiting a season's reproduction (Hannon et al., 1985).

Finally, the $l_x m_x$ curve shown in Figure 10.6 indicates that the majority of the population's reproductive output is accomplished by two- and three-year-olds, but that first-year birds contribute substantially as well.

These calculations rest on the assumption that the Hastings population is neither an importer nor exporter of young. Nevertheless, it appears plausible that the values we have derived for survivorship and reproduction, as well as for age at first reproduction, are reasonable. In particular, the very high survivorship of nonbreeders (put at 0.90 in the life tables) is at least demographically plausible, as are the relatively high values for the probability of dispersers obtaining successful reproductive positions (Table 10.1).

Comparison with the Florida Scrub Jay

Although other studies of cooperative breeders have yielded important demographic information (e.g., Vehrencamp, 1978; Gaston, 1978a; Ligon, 1981), the only other species for which a comprehensive life-table analysis has been attempted is the Florida scrub jay (Woolfenden and Fitzpatrick, 1984). Survivorship curves derived from their population are similar to ours (cf. Figure 10.4 with Figure 9.14 in Woolfenden and Fitzpatrick, 1984), with the notable exception that adult survivorship in scrub jays is identical for males and females, whereas for acorn woodpeckers it is greater for males (section 7.1). Age-specific reproductive values are also similar for the two species (cf. Figure 10.5 with Figure 9.20 in Woolfenden and Fitzpatrick, 1984). The $l_x m_x$ curves, although similar in shape, are centered differently because of the higher proportion of acorn woodpeckers breeding early in life (cf. Figure 10.6 with Figure 9.19 in Woolfenden and Fitzpatrick, 1984).

Consequences of Delayed Reproduction

What are the demographic consequences of delayed reproduction in this population? This question can be answered by varying the parameters we have used in the above life-table analyses. Specifically, we can vary d_x—the proportion of birds breeding at age x—and observe the results on R_o. This exercise is artificial, as in the real demographic world it is not possible to alter one parameter without influencing the others. However, it serves the heuristic function of allowing us to probe the demographic changes resulting from a hypothetical alteration in the observed pattern of delayed reproduction.

The results, shown in Figure 10.7, indicate that each year reproduction is delayed results in about a 10% decrease in the net reproductive rate. The actual estimated age at first reproduction, calculated from our analysis of the fate of offspring (Table 8.7), is also indicated. Net reproductive rates as estimated from the points indicated by the arrows correspond relatively well to the calculated values, which are ·1.06 for females (Table 10.4) and 1.17 for males (Table 10.5).

The decline in the population growth rate as reproduction is delayed is considerable, but not as great as might be expected from varying such a critical parameter. The reason for this is straightforward: if nonbreeders have survivorship close to one and there is no senescence, the population reproductive rate remains nearly the same regardless of how long first reproduction is delayed (Horn, 1978; Woolfenden and Fitzpatrick, 1984). Indeed, the decline in the net reproductive rates shown in Figure 10.7 is primarily the consequence of the 10% annual mortality

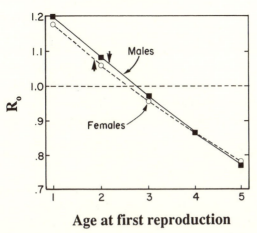

Age at first reproduction

FIGURE 10.7. Net reproductive rate (R_o) as a function of the age at first reproduction. Lines are the trajectories for hypothetical populations of males and females that all delay reproduction until the age given on the x-axis. Arrows represent the estimated average age at first reproduction.

rate suffered by nonbreeders. For additional discussion of the varying effects of delayed reproduction on the rate of population increase, see Wiley (1974).

Effective Population Size and Genetic Variance

Using the above life-table information we now estimate effective population size (N_e) and genetic variance components for the population. These parameters are necessary in order to evaluate virtually any hypothesis of the roles of selection, adaptation, and speciation in evolutionary processes.

We estimated N_e as the neighborhood size (Wright, 1946) devalued by factors correcting for departures from the random transmission of genes caused by the effects of overlapping generations and variance in lifetime reproductive success. We did not correct for the effects of nonnormality of the dispersal distribution (Wright, 1969; Barrowclough and Coats, 1985), because of our uncertainty about the exact shape of this distribution (see section 8.7). (The kurtosis of the observable fraction of the dispersal distribution [see Figure 8.4] is 2.4, a value that results in virtually no change in N_e from that of a population in which dispersal exhibits no kurtosis [Wright, 1969:304].) A summary of our procedures follows. For additional discussion of the estimation of N_e in bird populations, see Barrowclough and Coats (1985).

1. Based on the estimates of mean-square dispersal length derived in section 8.7 and the population-density estimates from Table 3.6 (35 individuals km^{-2}), we calculate neighborhood size (Wright, 1946) as the number of individuals in a circle twice the radius of the root-mean-square (RMS) dispersal length. That is, $N = 4\pi\rho\sigma^2$, where N is neighborhood size, ρ is the population density, and σ^2 is the mean-square dispersal length.

2. Generations of acorn woodpeckers are not discrete, and a substantial proportion of individuals are prereproductive at any one time. N_e can be calculated in such a population using neighborhood size and a life table following the method of Emigh and Pollak (1979). (We will designate this corrected value of effective population size as N_{ep}.)

3. Next, we correct N_{ep} for the unequal progeny production by individuals based on lifetime reproductive success (see section 10.3). This correction is necessary unless the distribution of offspring produced by individuals follows a binomial distribution (Crow and Kimura, 1970); if variance in progeny production is greater than binomial, N_e is correspondingly reduced. We derive this correction separately for the sexes and then recombine them to obtain our ultimate estimate of N_e.

4. We then use our estimate of N_e to derive the among-deme genetic variance (Wright's F_{st}). F_{st} is a measure of inbreeding due to finite population size and an index of the amount of expected differentiation among subpopulations assuming the absence of selection (Wright, 1951; Barrowclough, 1980; for some problems regarding F_{st} estimates, see Felsenstein, 1975).

As discussed in Chapter 8, we are unable to find all dispersers and therefore must estimate the dispersal distances of birds emigrating outside the study area. Using the corrections for the finite study area detailed in section 8.7 and the three estimates for long-distance dispersers discussed in Appendix B, estimates for the true RMS dispersal distance are 1.42 (curve a, Figure B.1), 2.42 (curve b), and 4.88 km (curve c; Table 8.9). These estimates yield values for neighborhood size $(4\pi\rho\sigma^2)$ of 886, 2575, and 10,474 individuals, respectively. Assuming an even sex ratio at birth, Emigh and Pollak's (1979) analysis, based on the life tables presented earlier, yields $\mathcal{N}_{ep} = (0.305)(\mathcal{N})$. That is, because of overlapping generations and other demographic considerations, the effective population size is 30.5% of the neighborhood size. Applying this correction, our estimates for $\mathcal{N}_{ep}$ are 270, 785, and 3,194 individuals.

Our next step is to correct for unequal progeny production with the observed mean and variance of lifetime reproductive success. From Crow and Kimura (1970:362), the general formula for the inbreeding effective population size in a species with separate sexes is:

$$\frac{\mathcal{N}_{t-2}\bar{k} - 2}{\bar{k} - 1 + (V_k/\bar{k})}, \tag{1}$$

where $\mathcal{N}_{t-2}$ is the population size in generation $t - 2$ and $\bar{k}$ and V_k equal the mean and variance in number of progeny per parent, respectively. If we assume that population size is stable over the two-generational time span used by this formula, we can divide by $\mathcal{N}_{t-2}$ to derive the proportional reduction in $\mathcal{N}_e$ due to the nonbinomial distribution of progeny numbers:

$$\frac{\bar{k} - (2/\mathcal{N}_{t-2})}{\bar{k} - 1 + (V_k/\bar{k})}. \tag{2}$$

Assuming $(2/\mathcal{N}_{t-2})$ to be small, we are left with

$$\frac{\bar{k}}{\bar{k} - 1 + (V_k/\bar{k})}. \tag{3}$$

From Table 10.6, the mean and variance of life-time number of young fledged is 2.81 ± 18.32 for males and 3.71 ± 41.22 for females. Substituting these values into equation (3) yields a proportional reduction in $\mathcal{N}_e$ of 0.34 for males and 0.27 for females.

We apply these values as follows. First, we evenly divide the "ideal" population (that is, the one assuming a binomial distribution of progeny numbers) into males and females. We then correct for the nonbinomial distribution of progeny numbers (as well as the uneven sex ratio among adults) by multiplying these values by the ratios derived above (0.34 and 0.27 for males and females, respectively). We then recombine the estimates for the two sexes separately by Wright's (1931) formula:

$$\mathcal{N}_e = \frac{4\mathcal{N}_m\mathcal{N}_f}{\mathcal{N}_m + \mathcal{N}_f}, \tag{4}$$

where $\mathcal{N}_m$ and $\mathcal{N}_f$ are the numbers of males and females, respectively. Using our measures for $\mathcal{N}_{ep}$ corrected by the Emigh and Pollak (1979) analysis, this procedure results in estimated $\mathcal{N}_e$ values of 81, 236, and 961, again depending on whether long-distance dispersers are assumed to follow curves a, b, or c from Figure B.1. These estimates yield F_{st} values of 0.089, 0.031, and 0.008, respectively, relative to a universe of 10^6 demes.

Using values uncorrected for unequal progeny production, our estimates of $\mathcal{N}_{ep}$ (270 to 3,194) are comparable to those derived for a wide variety of other avian species. Estimates of $\mathcal{N}_{ep}$ for eight noncolonial species using the isolation by distance (Wright, 1943) model of gene flow employed here ranged from 176 to 7,679, and $\mathcal{N}_{ep}$ values for eight colonial species, derived by a stepping-stone model (Kimura and Weiss, 1964), ranged between 17 and 64,580 (Barrowclough, 1980). The only other cooperative breeder for which data are available is the Florida scrub jay, for which Woolfenden and Fitzpatrick (1984) derive an estimate of 298 individuals for $\mathcal{N}_{ep}$ almost the same as our estimate of 270 employing the most conservative guess for the dispersal pattern of long-distance emigrants. The lower values we obtain here are also of the same order of magnitude as N_e for avian species derived from electrophoretic surveys (N_e's on the order of 300; Barrowclough, 1983).

Our estimates of N_e (correcting for unequal progeny production) are sufficiently large to indicate that genetic differentiation among subpopulations of acorn woodpeckers is unlikely to occur rapidly. Even our most conservative estimate ($N_e = 81$) is comparable to N_e estimates derived from avian studies of long-term karyotypic change (N_e's ~ 100; Barrowclough and Shields, 1984), reflecting not only unequal progeny production but long-term population fluctuations (i.e., bottlenecks) as well.

Thus, the few available data do not suggest that effective population size for cooperative breeders is particularly small. For both the Florida scrub jay and the acorn woodpecker, N_e values are of the same magnitude as values derived in comparable ways from other birds. Hence, we find no support for the hypothesis that cooperative breeders are prone to particularly intense inbreeding resulting from reduced dispersal and small effective population size.

10.3. LIFETIME REPRODUCTIVE SUCCESS

Thus far our analyses have been on a population level. However, acorn woodpeckers have a complex array of alternative individual life histories. Our goal here is to compare these strategies and to assess their relative impact on individual fitness.

To accomplish this, we focus again on lifetime reproductive success. As discussed in section 9.2, there are theoretical reasons for doing so; in contrast with seasonal or other short-term measures, only lifetime measures have the potential for estimating the fitness of individuals (Cavalli-Sforza and Bodmer, 1961; Grafen, 1982; Clutton-Brock, 1983).

Lifetime reproductive success has rarely been measured in natural populations, especially vertebrates (but see Clutton-Brock, in press). Our ability to determine this important measure is limited not only by the relatively long life span of acorn

263

woodpeckers but also their polygynandrous mating system. This latter feature means that the parentage of offspring cannot be unambiguously determined whenever more than one male or female breed within a group. As a result, some of the more interesting questions we would like to answer using lifetime data must, at least for now, be addressed with demographic models and simulations rather than hard data.

By assuming that parentage of offspring is shared equally among breeders within groups, however, we are able to determine lifetime reproductive success for a reasonable sample of individuals. The assumption of equal parentage is supported for females by detailed data on joint-nesting females (Mumme et al., 1983b; Mumme, 1984). In particular, between-year reciprocal eggs destruction suggests that there is not a significant bias in the number of young fledged by joint-nesting females, at least on a lifetime basis, despite egg destruction and a significant bias in egg ownership in any one year.

For males, however, we do not have comparable information. Joste et al. (1985; see also Mumme et al., 1985) have recently demonstrated multiple paternity; thus, there is no doubt that more than a single male can successfully parent offspring within a brood, even if only a single female lays eggs. Unfortunately, we have no data with which to estimate the bias in paternity within groups. Thus, an assumption of equal paternity is essentially a "null" hypothesis and leads to a minimal estimate of variance in lifetime success of breeding males. Later we will relax this assumption in order to estimate the consequences of increasing bias on variance in yearly and lifetime reproductive success. For a recent discussion of the problem of estimating gamete contribution by males in cooperative breeders, see Craig and Jamieson (1985).

We were able to document the complete reproductive histories of 45 males and 34 females that lived long enough to enter the breeding population. Individuals alive when the study was begun and those still alive in the winter of 1985–86, when the analyses were performed, were excluded. In addition, we excluded individuals entering the breeding population after 1980 in order to minimize the bias resulting from the inclusion of only those birds from the more recent cohorts that died young. (Since at least one individual that became a breeder in 1975 is still alive as of spring 1986, there is still some bias, but by limiting our sample to these older cohorts we minimize it.) Only individuals banded as nestlings or that we are confident were banded prior to successful reproduction were included.

The results of these analyses are given in Table 10.6. The distributions of estimated lifetime number of young fledged are graphed in Figure 10.8, and the distributions of the number of years males and females spent as breeders are shown in Figure 10.9. Also presented in Table 10.6 are data comparing the relative variance ($I =$ variance divided by the mean squared) in lifetime reproductive success and breeding parameters between males and females. Relative variance represents the

TABLE 10.6. Lifetime Reproductive Success and Breeding Data

	$\bar{x} \pm SD$			I		
Variable	Males	Females	♂♂/♀♀	Males	Females	♂♂/♀♀
Breeding life span	2.58 ± 1.95	2.09 ± 1.44	1.23	0.57	0.48	1.19
Total cobreeders	5.44 ± 5.39	2.71 ± 2.01	2.01	0.98	0.55	1.78
Total breeders of opposite sex	2.87 ± 2.45	3.24 ± 3.13	0.89	0.73	0.94	0.78
Estimated young fledged	2.81 ± 4.28	3.71 ± 6.42	0.76	2.23	3.00	0.74
Total young fledged by group	5.02 ± 7.50	4.29 ± 6.86	1.17	2.30	2.55	0.90
Indirect offspring fledged	2.21 ± 3.68	0.59 ± 1.79	3.75	2.77	9.31	0.30
Percent indirect offspring	44.0	14.7	—	—	—	—

NOTE: Based on 45 males and 34 females fledged prior to 1981. I (the opportunity for selection) $= \sigma^2/\bar{x}^2$. For explanation of variables, see section 10.3.

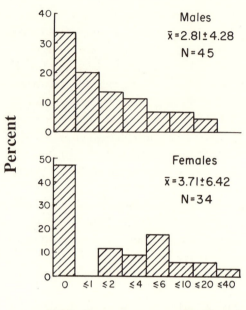

Lifetime young fledged

FIGURE 10.8. Lifetime reproductive success for males and females assuming no bias among cobreeders.

"opportunity for selection" within particular selective events or episodes (Arnold and Wade, 1984a, 1984b).

As expected from the overall higher survivorship of breeding males, the average breeding life span of males is greater than females; furthermore, relative variance in breeding life span is greater among males than females by a similar ratio of about 1.2 to 1. The next measure, total cobreeders, was determined by summing the number of breeders of the same sex, including the individual concerned, for each of the years the bird bred. Similarly, the total number of breeders of the opposite sex was determined by summing the number of opposite-sexed breeders for each year that the individual bred. These measures are thus indices of the degree to which mate-sharing is experienced over the lifetimes of males and females.

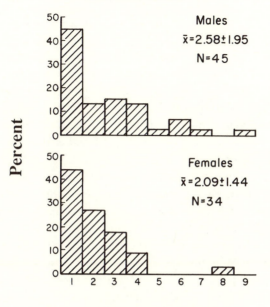

Years as a breeder

FIGURE 10.9. Distribution of survivorship (years as a breeder) among the males and females for which we have lifetime reproductive success.

As expected from the analyses of group composition in Chapter 3, the number of total cobreeders is considerably greater for males than females; both the mean values and I values are nearly twice as great for males. Conversely, the number of females with which a male breeds is only 89% of the number of males with which a female breeds, with a concurrent reduction in relative variance.

Estimates of lifetime number of young fledged are also greater for females than males, reflecting the breeding sex-ratio bias toward males. (The *total* number of young fathered by all males is by necessity exactly equal to the total number mothered by all females; thus, $\bar{x}_m p_m \approx \bar{x}_f p_f$, where $\bar{x}$ equals the mean progeny per bird and p the proportion of adult males (m) and females (f). Using the estimated number of young fledged from Table

267

10.6 and the adult sex ratio from section 3.4, this formula yields $(2.81)(.596) \approx (3.71)(.404)$ or $1.67 \approx 1.50$ and is thus in reasonable agreement to expectation.) Of greater interest are the higher I values for females (3.00) compared to males (2.23), a pattern typical of populations in which males are the limiting sex. This conclusion is also supported by the higher proportion of females entirely failing to reproduce (Figure 10.8) and the greater maximum fecundity of females: the largest number of offspring estimated to have been fathered by a male was 19, while one female, who bred successfully at group Road 3 between 1975 and 1983, mothered a total of 34 young. These results thus contrast with those from polygynous species (Clutton-Brock, 1983; Koenig and Albano, 1987), in which male variance exceeds female variance, as well as monogamous species (such as the Florida scrub jay; Fitzpatrick and Woolfenden, in press) in which relative variance in lifetime reproductive success is equal for the sexes.

However, the sexual difference in the opportunity for selection in number of young fledged found here is relatively minor compared to that observed in the polygynous red deer (*Cervus elaphus*), where I values are 3.4 times greater in males than females (Clutton-Brock, 1983). Furthermore, given our small sample size, the dependence of these figures on a few highly successful individuals, and the unknown degree of bias among reproductive success of male cobreeders, the difference between males and females we observed is probably not significant.

Besides the number of young parented by individuals we can also calculate the total number of young fledged by the group during the individual's lifetime by summing the reproductive success of the group each year it bred. These values are slightly higher for males than females (Table 10.6), probably as a result of their longer average life span. Again, however, relative variance is slightly greater for females.

By subtracting the number of young estimated to have been parented by an individual from the total number of young fledged by groups in which they bred, we are left with the total number of "indirect" offspring produced by males and females; that is, offspring that were parented by cobreeders rather than themselves. This value, given in Table 10.6, is nearly four times greater for males than females, although again the relative variance is greater for females than males. Dividing the average number of such "indirect" offspring by the total number of offspring fledged by the group indicate that, during the lifetime of males, 56.0% the offspring fledged by a breeding male's group is fathered by that male, whereas 85.3% of the offspring are mothered by a particular female. Despite the uncertainty in these values as a consequence of our ignorance concerning the degree of reproductive bias among cobreeders, the differences are great enough to suggest that the evolutionary consequences of mate-sharing and polygynandry (examined in detail in Chapter 13) are greater for males than females.

In order to focus more directly on differences not attributable to differential survivorship between males and females, Table 10.7 presents means and relative variance for reproductive success calculated on a yearly basis; this sample includes individuals for which complete lifetime data are not available. These results show similar trends to those found in Table 10.6. On average, 2.44 males shared breeding status within their groups each year, whereas only 1.42 females did so. Conversely, the mean number of mates per year (obtained by dividing the number of breeders of the opposite sex in a group by the number of breeders of the same sex) is twice as great for females as males.

The estimated yearly reproductive success of breeding females is considerably greater than that of males, due entirely to the greater degree of mate-sharing among males; the total yearly number of young fledged by groups are similar, as expected. Interestingly, the relative variances in young fledged on an annual basis are virtually identical for the two sexes and considerably less than relative variance on a lifetime basis (Table 10.6). This indicates that variance in survivorship (or other factors changing from year to year) reinforces rather than counteracts variance in annual fecundity.

TABLE 10.7. Yearly Reproductive Success and Breeding Data

Variable	$\bar{x} \pm SD$			I		
	Males	Females	♂♂/♀♀	Males	Females	♂♂/♀♀
Mean N cobreeders	2.44 ± 1.55	1.42 ± 0.59	1.71	0.40	0.17	2.35
Mean N breeders of opposite sex	1.38 ± 1.08	2.03 ± 1.44	0.68	0.61	0.50	1.22
Mean mates per year	0.78 ± 1.02	1.58 ± 1.23	0.49	1.72	0.61	2.82
Young fledged	1.11 ± 1.16	1.63 ± 1.66	0.68	1.09	1.04	1.05
Total young fledged by group	2.17 ± 1.99	2.04 ± 1.93	1.06	0.84	0.89	0.94
Indirect offspring fledged	1.05 ± 1.26	0.41 ± 0.84	2.57	1.44	4.16	0.35
Percent indirect offspring	48.4	20.1	—	—	—	—

NOTE: Based on 398 male breeder-years and 263 female breeder-years. I (the opportunity for selection) $= \sigma^2/\bar{x}^2$.

As before, estimates of the number of "indirect" offspring suggest that the importance of indirect fitness is likely to be considerably greater for males (indirect offspring contributing 48.4% to annual fitness) than for females (indirect offspring contributing 20.1% to annual fitness).

10.4. INFLUENCE OF REPRODUCTIVE BIAS AMONG COBREEDERS

It is difficult to determine accurately the influence of reproductive bias on lifetime reproductive success, primarily because of the possibility of tradeoffs or reinforcement between survivorship and reproductive dominance. We can, however, estimate the possible effects of reproductive bias on a yearly basis by imposing an arbitrary average amount of bias on cobreeders. To do this, we compared the "null" situation of no bias with two alternatives: in the first (bias I), each successive cobreeder was given half the success of prior birds; in the second (bias II), each successive cobreeder was given one-fourth the success of prior birds. Yearly reproductive success and relative variance were then calculated as before, using the raw data from Table 10.7.

The results, shown in Table 10.8, indicate that variance in yearly male reproductive success increases considerably as reproductive bias is imposed, whereas variance in female success

TABLE 10.8. Yearly Relative Variance in Young Fledged

Variable	*I*		
	Males	*Females*	♂♂/♀♀
BREEDERS ONLY			
No reproductive bias	1.09	1.04	1.05
Bias I	1.29	1.09	1.23
Bias II	1.66	1.17	1.42
ENTIRE POPULATION			
No reproductive bias	2.06	2.13	0.97
Bias I	2.36	2.21	1.07
Bias II	2.89	2.33	1.24

NOTE: Three different hypothetical degrees of reproductive bias are used: none, each successive breeder one-half as successful as the prior breeder (bias I), and each successive breeder one-quarter as successful as the prior breeder (bias II). Total $N = 398$ male breeder-years (584 total male years) and 263 female breeder-years (404 total female years). Means are 1.11 for males and 1.63 for females (breeders only) and 0.76 for males and 1.06 for females (entire population).

271

changes little, primarily because few females share reproductive status within groups. Thus, if there is no bias among males, relative variance in yearly reproductive success is essentially the same for the two sexes. If reproductive success of cobreeding males is highly biased (bias II), whereas that of cobreeding females is not, males would then experience $1.66/1.04 = 1.60$ times the relative variance as females.

Also shown in Table 10.8 are data on yearly reproductive success including nonbreeders. If there is no bias among cobreeders, the relative variance in yearly success increases to 2.06 for males and 2.13 for females. However, if a modest degree of bias among males occurs (bias I), relative variance among males jumps to 2.36, well above that for females under all but the high-bias regime. Thus, the variance in yearly reproductive success for the total population of males is probably quite similar to that for females.

10.5. PARTITIONING OF THE OPPORTUNITY
FOR SELECTION

Here we partition the opportunity for selection using the lifetime reproductive success data presented above. The methodology for performing these analyses are detailed by Arnold and Wade (1984a). Their analyses provide a valuable means of partitioning and measuring fitness, thereby pinpointing stages in life-history where selection may be greatest.

Following Arnold and Wade (1984a, 1984b), we use relative variance (I values) to measure the opportunity for selection. Two samples were considered. First, we included not only those individuals whose breeding careers we observed, but also the estimated number of individuals dying *prior* to attaining reproductive status. Thus, for males, we assumed 56.8% survivorship from fledging to February (Table 10.5), and that 90% of the survivors obtained some sort of breeding position (see Chapter 8). Hence, we start with 88 males, of which 45 comprise our sample with lifetime data. For females, we assumed 56.8% survivorship from fledging to February, and that all survivors obtain a breeding position (Chapter 8); hence, we start with 60 females, 34 of which comprise our sample with lifetime data. Four episodes of selection are considered: juvenile survivorship, years as a breeder, mates per year, and young fledged per mate. The number of mates was determined by dividing the total number of breeders of the opposite sex by the number of the same sex, as discussed above.

The results are shown in Table 10.9. The total opportunity for selection (W) is 5.45 for males and 6.00 for females, reflecting the slightly greater variance in lifetime reproductive success for females than males. In terms of the proportion of the total opportunity for selection contributed by individual selective episodes, no individual event dominates the others. However, variance in mates per year contributes the least to the total variance in lifetime fitness for both sexes, and juvenile survivorship contributes the most, followed by young fledged per mate.

Dominating the contribution to total I for both sexes, however, is the "cointensity (1234)" term. This composite term represents a sum of covariances between fitness components (see Table 10.9 and Arnold and Wade, 1984a). Thus, we cannot be certain in exactly which episode the greatest variance occurs, as the cause and effect of such covariance effects cannot be determined from this analysis. However, comparing the large I values obtained here with those obtained using only those individuals attaining reproductive status (Tables 10.6 and 10.10) indicates that the primary contributor to this high covariance is high juvenile mortality. A similar conclusion is reached by Fitzpatrick and Woolfenden (in press) for the Florida scrub jay.

In order to partially circumvent this problem and to provide data comparable to studies in which no estimates of juvenile mortality can be made, we have calculated the opportunity for selection using only the 45 males and 34 females that attained reproductive status (Table 10.10). This analysis indicates that

TABLE 10.9. Partitioning of the Opportunity for Selection Using Lifetime Reproductive Success Starting with Fledglings

Source of Variation in Relative Fitness	Contribution to Total Opportunity for Selection		
	Symbol	Value	Percentage
MALES			
Juvenile survival (w_1)	I_1	0.97	18%
Years as a breeder (w_2)	I_2	0.56	10
Cointensities (12)	*	0.54	10
Survival ($w_1 w_2$)	I_{12}	2.07	38
Mates per year (w_3)	I_3	0.10	2
Cointensities (123)	*	0.21	4
Total mates ($w_1 w_2 w_3$)	I_{123}	2.38	44
Young fledged per mate (w_4)	I_4	0.71	13
Cointensities (1234)	*	2.36	43
Young fledged (total selection, W)	I	5.45	100
FEMALES			
Juvenile survival (w_1)	I_1	0.78	13
Years as a breeder (w_2)	I_2	0.47	8
Cointensities (12)	*	0.36	6
Survival ($w_1 w_2$)	I_{12}	1.61	27
Mates per year (w_3)	I_3	0.27	5
Cointensities (123)	*	0.53	9
Total mates ($w_1 w_2 w_3$)	I_{123}	2.41	40
Young fledged per mate (w_4)	I_4	0.66	11
Cointensities (1234)	*	2.93	49
Young fledged (total selection, W)	I	6.00	100

NOTE: Assumes that 56.8% of fledglings of both sexes survive to their first spring and that 90% of surviving males and 100% of surviving females acquire a breeding position (see section 8.4).

* Cointensities include covariance terms between the current episode and all prior episodes; e.g., Cointensities $(12) = COI(1, 2) + COI(12|2) + COI(12, 2|1) - COI(12, 2)$. For formulae, see Arnold and Wade (1984a).

TABLE 10.10. Partitioning of the Opportunity for Selection Using Lifetime Reproductive Success Starting at Reproduction

Source of Variation in Relative Fitness	Contribution to Total Opportunity for Selection		
	Symbol	Value	Percentage
MALES			
Years as a breeder (w_1)	I_1	0.57	25%
Mates per year (w_2)	I_2	0.10	4
Cointensities (12)	*	0.06	3
Total mates $(w_1 w_2)$	I_{12}	0.73	32
Young fledged per mate (w_3)	I_3	0.71	31
Cointensities (123)	*	0.87	38
Young fledged (total selection, W)	I_{123}	2.32	100
FEMALES			
Years as a breeder (w_1)	I_1	0.48	16
Mates per year (w_2)	I_2	0.28	9
Cointensities (12)	*	0.18	6
Total mates $(w_1 w_2)$	I_{12}	0.94	31
Young fledged per mate (w_3)	I_3	0.67	22
Cointensities (123)	*	1.39	46
Young fledged (total selection, W)	I_{123}	3.00	100

NOTE: $N = 45$ males and 34 females fledged prior to 1981.
* See Table 10.9.

variance in young fledged per mate contributes the most to variance in total I for individuals having attained breeding status for both sexes. Second in importance for both sexes is years as a breeder. Variance in the number of mates per year is small for both sexes, but relatively larger for females, as expected given the greater variance in number of male co-breeders and the relatively large influence of additional males on group reproductive success (section 6.2). However, the cointensity terms are still large, again complicating any precise interpretation.

These results indicate that there is considerable opportunity for selection in acorn woodpeckers: values for the total opportunity for selection W from Table 10.10, which includes only individuals attaining reproductive status, are at least as high as values for any of five other species, including three polygynous species (the red deer and two odonates) and three monogamous species (the kittiwake, European sparrowhawk, and Florida scrub jay) for which comparable data exist (Koenig and Albano, 1987; Fitzpatrick and Woolfenden, in press). Unfortunately, considerable additional data from both cooperative and noncooperatively breeding species will be necessary before it will be possible to determine the importance of many of the values calculated here.

10.6. SUMMARY

We present a synthetic life table for acorn woodpeckers, based on the demographic parameters calculated in earlier chapters. We conclude that: (1) values derived for reproduction and survivorship are plausible and yield life tables that are consistent with the observed population stability; (2) survivorship curves for acorn woodpeckers are similar to those for other species of birds, indicating high mortality early in life, and later survivorship apparently age independent, (3) life-table parameters of acorn woodpeckers and Florida scrub jays differ primarily in intersexual patterns of adult mortality (no sex differences in jays, but higher female mortality in woodpeckers) and shape of the $l_x m_x$ curve (comparatively fewer jays breed in their first year); (4) delayed reproduction in acorn woodpeckers probably has relatively little demographic impact due to the high survivorship of nonbreeders.

Based on the life table combined with our estimate of dispersal derived in Chapter 8, we calculate effective population size and among-deme genetic variance. We then correct effective population size for the unequal distribution of lifetime progeny production. Using a conservative estimate for the pattern of dispersal for individuals leaving the study area, $N_e = 81$ individuals; a more liberal estimate of the dispersal pattern yields an $N_e = 961$ individuals. These values are of a similar magnitude as estimates of N_e from other wild populations of birds, but are not comparable inasmuch as prior estimates have assumed a binomial distribution of progeny production. Our estimates of effective population size not incorporating this correction range from 270 to 3,194 individuals and suggest that acorn woodpeckers are not prone to an unusually high degree of inbreeding, despite their social organization.

We next calculate lifetime reproductive success, focusing on sexual differences and on a partitioning of the opportunity for selection into multiplicative episodes. In general, the opportunity for selection is high compared to other species for which data are available. Opportunity for selection is either comparable for males and females or slightly higher in females. However, reproductive bias among male cobreeders could reverse this trend.

Variance in juvenile survivorship appears to contribute substantially to overall opportunity for selection. Considering only birds surviving and obtaining breeding positions, the number of young fledged per mate contributes the largest amount to opportunity for selection in both sexes, followed by survivorship; variance in the number of mates per year is relatively unimportant, particularly for males.

Lifetime reproductive success is difficult to measure, especially in long-lived species such as the acorn woodpecker. Nonetheless, such data are imperative for understanding the evolution of cooperative breeding. Our results, as well as those of Fitzpatrick and Woolfenden (in press), represent a start.

CHAPTER ELEVEN

Population Regulation

We have now discussed the basic demographic processes of acorn woodpecker populations. How do they interact to produce the population characteristics—including size and composition—described in Chapter 3?

As usual, the complex social structure of acorn woodpeckers demands that this question be answered on several levels. First, what limits group size? As a subset of this question, we will consider the problem of what limits the number of breeders within groups. And second, what limits the number of groups? By discussing these two questions separately we can determine the factors that regulate the population as a whole. Several recent papers address population regulation in cooperative breeders (Fitzpatrick and Woolfenden, 1986; Brown, 1985b; Hannon et al., 1987; von Schantz, MS), but in general this phenomenon has been largely ignored since Wynne-Edwards (1962) proposed that cooperative breeders, along with a wide variety of other species, are socially regulated.

11.1. LIMITS TO THE NUMBER OF BREEDERS

Groups rarely contain more than three breeding males and two breeding females, even though they often contain additional nonbreeders (Figure 3.10). We will first consider the role of social behavior in limiting the number of breeders within groups and then discuss the more ultimate ecological factors that are important.

Behavioral Limitation of Breeding

Watson and Moss (1970) discuss the conditions that are necessary and sufficient to show that social behavior, via socially induced mortality or depression of recruitment, is important to limiting breeding populations. We will consider each of their four conditions in turn.

1. A substantial part of the population does not breed. This condition is satisfied by acorn woodpecker populations at Hastings in all years except those following acorn crop failures. Overall, 25% of adults and 39% of the population including first-year birds are nonbreeders (section 3.3).

2. Nonbreeders are physiologically capable of breeding when the more dominant animals are removed. A substantial fraction of nonbreeders attain breeding status in their first year (section 8.3), and virtually all nonbreeders eventually attain breeding positions somewhere (section 8.4). Thus, such individuals are physiologically capable of reproduction. It is furthermore critical that nonbreeders are ecologically constrained from breeding rather than that they choose such a strategy because it directly enhances their fitness. If the latter, then the observed pattern of nonbreeding is not socially induced but is the consequence of ecological factors.

We will discuss the evidence for the importance of ecological constraints in greater detail in Chapter 12; here, however, we note that there is considerable evidence indicating that nonbreeders *are* constrained from breeding (Koenig, 1981a; Hannon et al., 1985; see Chapter 12). In particular, the removal experiments of Hannon et al. (1985) demonstrate that nonbreeders, when given the opportunity, will either fill reproductive vacancies and subsequently breed or inherit and breed in their natal territory.

3. Breeders are not exhausting available food, space, nest sites, or other resources. The existence and high survivorship of nonbreeders within groups sharing territories and storage facilities with breeders suggest, but do not prove, that additional food

279

and other resources are available for breeding but are not being utilized. Furthermore, each acorn woodpecker group usually has several potential nest holes. Unlike the cooperatively breeding red-cockaded woodpecker, which only nests and roosts in holes surrounded by sap in pines infested with heart-rot fungus (Ligon, 1970; Thompson, 1971; Jackson, 1977), acorn woodpecker nest holes have no obvious special features and are not limiting under most circumstances. (Exceptions to this generalization may occur, such as in areas where competition with the introduced European starling is intense [Troetschler, 1976].)

4. *Socially induced mortality or depressed recruitment compensates for mortality from other limiting factors.* We are able to test this condition in part by comparing the proportion of the population that breeds to total population size. The inverse correlation between these two variables, shown in Figure 11.1, indicates a compensatory effect, as specified by this condition. Thus, as the total population declines, a higher proportion of the pop-

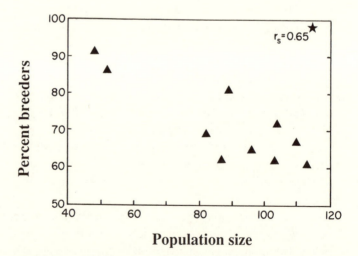

FIGURE 11.1. Relationship between breeding population size and the percent of the population consisting of breeders. $N = 10$ years.

ulation consists of breeders, presumably because socially induced reproductive suppression is no longer so intense.

In summary, social behavior appears to be an important proximate factor limiting the number of breeders within groups of acorn woodpeckers. Incest avoidance (Koenig and Pitelka, 1979; Koenig et al., 1984) and reproductive competition (Hannon et al., 1985) are the proximate mechanisms by which this limitation occurs.

Food and Limitation of Breeding

What ultimate ecological factors limit the number of breeders within groups? At least three analyses suggest that food, specifically stored acorns, are paramount.

First, as shown in Table 4.10, there is a significant positive correlation between the maximum number of acorns stored by the population in the winter and several population characteristics, including the number of breeders per group and the proportion of groups containing joint-nesting females. Thus, groups contain more breeders in years when many acorns are stored.

Second, when analyzed on a territory-by-territory basis, territories with larger storage facilities, as well as those with more species of stored acorns, average more breeders than those with smaller granaries (Table 4.11).

Third, the size of sibling units breeding within groups is indirectly influenced by the acorn crop. As we discuss in detail elsewhere (Koenig et al., 1984; Hannon et al., 1987), reproductive vacancies are usually filled by the largest available unisexual sibling unit; these units usually, but not always (see section 13.1), form the breeding core of the group. Thus, the number of breeders within groups is partly determined by the size of nonbreeding sibling units in the population, which is in turn influenced by the reproductive success of groups. This latter parameter is, as we saw in Chapter 4, determined largely by acorns and acorn storage.

11.2. LIMITS TO TOTAL GROUP SIZE

Fitzpatrick and Woolfenden (1986) show that group size in cooperative breeders can be considered a direct demographic consequence of reproductive success and survivorship. Their conclusion, that the maximum possible average group size in a monogamous population is five to eight individuals, is in accord with the maximum mean group size of 5.24 recorded in any one year from our population (Table 3.6; see also section 8.4). Although acorn woodpeckers are not monogamous, and thus are not necessarily subject to the same constraints modeled by Fitzpatrick and Woolfenden (1986), we agree that group size is primarily a demographic outcome, rather than a strategic optimum. Next, we consider what demographic processes limit total group size.

Survivorship, Reproduction, or Emigration?

Regulation of group size may come about through survivorship, reproductive success, and/or emigration. (In order to distinguish density-dependent processes within groups from those in the population at large we refer to the former as "group-size-dependent.") The first of these, survivorship of breeders, does not decrease with group size (Figure 11.2). Indeed, as discussed in Chapter 7, survivorship of breeding females is independent of group size, whereas that of breeding males increases with group size. However, as discussed below, survivorship may be subject to group-size-dependent effects during poor acorn crop years.

Reproductive success per capita, however, shows strong group-size dependence: larger groups produce relatively fewer offspring (Figure 11.3; see also Koenig, 1981b). Since fledglings nearly always remain in their natal group through their first spring, this group-size dependence of reproductive success is likely to be important in limiting group size. The factors promoting this pattern are discussed in the next section.

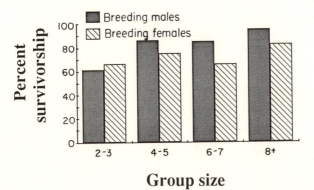

FIGURE 11.2. Survivorship of breeders by total group size. Based on data in Table 7.5.

The third demographic process that may act to regulate group size is emigration by nonbreeders. We examined this possibility by determining the probability that nonbreeders either leave their natal group or disappear as a function of group size in July following the main breeding season. The results, dividing groups arbitrarily into "small" and "large" units, are

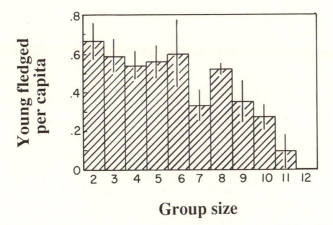

FIGURE 11.3. Young fledged per bird as a function of group size, based on data in Table 6.1. The linear regression (young fledged per capita = −0.049 [group size] + 0.758) is highly significant ($F_{1,217} = 10.7$, $p < 0.01$).

TABLE 11.1. Emigration of Nonbreeders According to Group Size

	Small Groups	Large Groups	Total
TOTAL GROUP SIZE[a]			
Left or disappeared	79	88	167
Stayed	119	140	259
Total	198	228	426
Percent left or disappeared	39.9	38.6	39.2
$\chi^2 = 0.03$, ns			
ADULT GROUP SIZE[b]			
Left or disappeared	67	100	167
Stayed	131	128	259
Total	198	228	426
Percent left or disappeared	33.8	43.9	39.2
$\chi^2 = 4.05$, $p < 0.05$			

NOTE: Adult group size excludes first-year birds.
[a]Small = 3–8, large = 9+
[b]Small = 2–5, large = 6+

shown in Table 11.1. Total group size makes no difference to the probability that nonbreeders leave or disappear, but nonbreeders are significantly more likely to disperse from groups with more adults (Table 11.1). This provides some evidence that emigration may depend on group size.

Several factors might cause this group-size dependence of emigration. We will consider three.

1. *Inheritance of natal territory.* Possibly, when there is only a single breeder of one sex in a group, it would be to the advantage of nonbreeders of the opposite sex to refrain from dispersing, as there would be a relatively high probability that the breeder would die and the offspring would be able to inherit the territory. Conversely, if there are several breeders of one sex in the group, nonbreeders of the *same* sex should disperse, as intra-

284

sexual competition would be intense even if they were able to inherit and breed.

Table 11.2 examines these possibilities by compiling the fates of nonbreeders by sex depending on the number of breeders of the same and of the opposite sex. Fates are determined for each year beginning in July. For nonbreeding females there is no significant difference regardless of the number of breeders of either sex; for nonbreeding males the probability of dispersal tends to *decrease* with increasing number of breeders, significantly so for two compared to one breeding female (the decrease is also significant for groups with one compared to two and three breeding males combined [$\chi^2 = 4.0, p < 0.05$]).

Thus, we find no support for the prediction that dispersal of nonbreeders increases with the number of breeders. This is, however, what we would expect if territorial inheritance does

TABLE 11.2. Emigration of Nonbreeders According to the Number of Breeders in the Group

	Breeding ♂♂			Breeding ♀♀	
	1	2	3	1	2
NONBREEDING MALES					
Left or disappeared	23	28	12	46	16
Stayed	10	30	15	19	32
Total	33	58	27	65	48
Percent left or disappeared	69.7	48.3	44.4	70.8	33.3
	$\chi^2 = 5.0$, df = 2, ns			$\chi^2 = 14.2, p < 0.05$	
NONBREEDING FEMALES					
Left or disappeared	17	22	11	38	12
Stayed	11	16	9	20	11
Total	28	38	20	58	23
Percent left or disappeared	60.7	57.9	55.0	65.5	52.2
	$\chi^2 = 0.2$, df = 2, ns			$\chi^2 = 0.7$, ns	

285

not result in greater lifetime fitness than does dispersal (see Chapter 9).

2. Dispersal and sibling unit size. Hannon et al. (1985) demonstrated that larger sibling units are more likely to win power struggles. Thus, nonbreeders are more likely to disperse from larger groups because they are more likely to have a larger sibling support group with which they cooperate to fill reproductive vacancies.

3. Food limitation. If acorn storage is limited by granary facilities (see Chapter 4), the advantages of remaining at home for a nonbreeder decrease as a function of group size. Thus, it is likely that emigration is influenced directly by food. This is certainly true in years when the crop is poor. In such years, nonbreeders may be forced out of groups by more dominant breeders (Hannon et al., 1987; see below).

In summary, both reproductive success and emigration are probably important in limiting group size in the Hastings population. Survivorship of breeders, however, is either independent of or positively correlated with group size. Comparable analyses have not been conducted on other cooperative breeders, but the tendency for per capita reproduction to decline with increasing group size is widespread (Koenig, 1981b), suggesting that this factor may also limit group size in other cooperatively breeding species.

Factors Promoting Group-Size-Dependent Reproduction

Why does per capita reproductive success decline with group size? As demonstrated in Figure 6.1, larger groups are not more likely to fail completely at nesting than smaller groups; in fact, larger groups tend to fledge more total young than do smaller groups, at least to size eight. Larger groups, although more successful overall, simply do not fledge proportionately more young than do smaller groups.

The ecological factor most likely to produce the relatively poorer success of larger groups is, as usual, stored acorns. As we saw in Chapter 4, the annual success of the whole population correlates strongly with the mean number of acorns stored per bird (Table 4.8). Furthermore, the success of individual territories correlates with the number of storage holes available (Table 4.11), which is relatively fixed. Thus, the number of storage holes per bird, and hence stored acorns per bird, necessarily declines as group size increases.

The regulating effect of acorns on group size is particularly evident during years when the acorn crop is poor. One such acorn crunch occurred in autumn 1983 (Hannon et al., 1987). During that winter, many territories were simply deserted. More interesting were the fates of birds living in territories that were not completely abandoned. Of 60 birds originally present on ten such territories, 31 (51.6%) remained through February. Disappearance was related to dominance: 56.7% of 24 non-breeders disappeared compared to only 42.1% of 19 breeders; overall, the ranking of the percent disappearance by age and status in the population was: breeding males < breeding females < nonbreeding males < nonbreeding females < first-year birds. Observations of dominance interactions suggested that dominant birds were excluding subordinates from the limited acorn supplies, thereby forcing their emigration. Many of these birds excluded from their home territories undoubtedly died.

There is a negative correlation between the percent disappearance of birds in nonabandoned groups in years of poor acorn crops and the number of acorns per bird present in their granaries in December, a pattern shown in both 1978–79 and 1983–84 (Figure 11.4). Thus, the group-size-dependent effect of stored acorns is most evident under conditions of a poor acorn crop. The effect is through a combination of survivorship and emigration; hence, at least under some conditions, survivorship, reproduction, and emigration may all have important regulatory influences.

It is under these same conditions of a poor acorn crop that the density dependence of reproductive success in the

287

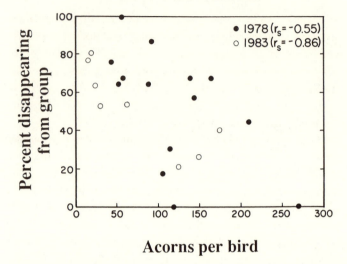

Acorns per bird

FIGURE 11.4. Percent of birds present in August disappearing from their territories by the end of February versus the number of acorns stored per bird in November for groups not abandoned during the poor crop years of 1978–79 and 1983–84. Both correlations are significant ($p < 0.05$). Data from Hannon et al. (1987).

population (Figure 5.4) breaks down. Poor acorn years lead to low population sizes the following spring coupled with poor reproductive success. Otherwise, reproduction is significantly greater in years when the population size is small and the number of acorns stored per bird is high (see section 4.4). Thus, it is possible that the density- and group-size-dependent effects of reproduction regulate the population in most years, but that group-size-dependent survivorship and emigration are important during years when the acorn crashes.

Note, however, that the effects of acorn crop crashes are sometimes catastrophic and at least locally independent of both density and group size. During the 1978–79 crop failure, 7 of 22 territories (32%) occupied during spring 1978 were abandoned. The winter of 1983–84 was even worse: 24 of 34 territories (71%) were abandoned. Many of these abandonments were due to a complete lack of acorns in or adjacent to the territory and thus would have occurred regardless of the number of birds present in the groups involved (see, for example, the discussion of the 1972 acorn crop failure on woodpecker groups in MacRoberts and MacRoberts, 1976:64).

Why the Observed Limit to Group Size?

Although reproductive success and emigration appear to limit group size (see above and Fitzpatrick and Woolfenden, 1986), their effects do not explain why groups of acorn woodpeckers, both at Hastings and elsewhere (e.g., Troetschler, 1976; Swearingen, 1977; Stacey, 1979a; Trail, 1980), are rarely larger than twelve individuals. One plausible hypothesis for this limit is that resources within the home range of groups are never sufficient for the long-term support of more than a dozen or so individuals.

A resource that might provide such a limit is the size of granaries. As we saw in section 4.4, acorn storage is generally limited by the storage facilities available to groups. It has previously been shown by MacRoberts and MacRoberts (1976:72) that maximum group size correlates with size of storage facilities. Thus, as suggested by these authors, the limit to group size may be set by size of granaries.

This hypothesis is testable with data from areas in which granaries contain many more holes than at Hastings, where the average territory contains about 2,516 (SD = 1,683; $N = 24$) holes and the maximum territory contains about 6,200. Localities with granaries much larger than this—on the order of at least 30,000 or more holes—are not unusual (e.g., Dawson, 1923; section 4.1). Unfortunately, no such area has ever been studied in sufficient detail to provide basic group-size information. Presumably this considerable geographic variation in maximum granary size is in turn a consequence of differences in acorn productivity, characteristics of granary substrates, or both (see section 11.3).

Even in areas with considerably larger storage facilities, it is nonetheless likely that there is an upper limit to group size, probably on the order of the maximum size of fifteen observed at Hastings. Such a limit may be set by reproductive competition within groups (e.g., Hannon et al., 1985). In particular, joint-nesting places all potential breeders in direct, intense, reproductive competition (e.g., Mumme et al., 1983a, 1983b). This competition places severe restrictions on breeding opportunities within groups, and thus may lead to group-size-dependent reproduction, emigration, and survivorship.

11.3. LIMITS TO THE NUMBER OF GROUPS

The second component of total size of acorn woodpecker populations is the number of groups present in an area. As discussed in section 3.1 and Chapter 4, groups are closely tied to the granary facilities of a particular territory. Since the number of storage holes is fixed at any one time, it is likely that granaries limit the number of groups in an area, and that limitation is accomplished through density-dependent emigration (see below, "Tests of the Granary Limitation Hypothesis").

Before tackling the ultimate factors determining the number of granaries in an area, we will discuss two tests of the hypothesis that the number of groups is limited by granaries. First, what happens when granaries fall? And second, can new territories be created by the addition of a granary?

Tests of the Granary Limitation Hypothesis: Granary Loss

We did not experimentally remove granaries. However, the primary or exclusive granaries of four territories were lost catastrophically during the study. These losses provided natural experiments testing the consequences of granary removal from previously occupied sites. The histories and results of these four events are summarized as follows.

1. Upper Haystack. Upper Haystack's granary was unusually close to that of the adjacent territory Lower Haystack (see Figure 3.1). Both territories were occupied continuously between spring 1972 and summer 1974. Upper Haystack's granary fell in late summer 1974, following which the residents abandoned; none was ever seen again. The area previously used by Upper Haystack was subsequently annexed by Lower Haystack, and only a single group has used the area since.

2. MacRoberts. This territory was occupied intermittently from the start of the study through winter 1977, when the granary fell. A few limbs with about 300 storage holes remained off the ground, and birds persisted there until summer 1978. The birds then abandoned, and the territory has since remained unoccupied.

3. Big Tree. This territory was occupied continuously between summer 1972 and the poor acorn year of autumn 1978, when it was abandoned. The territory remained unoccupied through summer 1979, when fire destroyed the granary. No group has subsequently settled this area, which has been largely annexed by the adjoining group Arnold 1.

4. Arnold 3. This territory contained a large granary and supported a large, successful group. Two birds were resident during summer 1979 when a fire led to the loss of the main granary; neither resident was seen again. However, a new group (group Fanny Arnold; see Figure 3.1) colonized part of this group's territory in autumn 1980 using some minor storage facilities still present nearby. This group has subsequently been successful in this area.

In summary, the four major granaries lost catastrophically during the study period led to not only immediate abandonment but permanent disappearance of two of the groups dependent on them (cases 1 and 3, above). In case 2, granary loss led to abandonment within six months. Thus, this evidence supports the hypothesis that granaries are an important factor limiting the existence of groups. However, case 4 indicates

291

that other factors, either social or food-related, may limit groups as well (see below).

Tests of the Granary Limitation Hypothesis:
Granary Additions

We created new granary facilities in three areas that had not previously been occupied by acorn woodpeckers. These areas were chosen either because birds had occasionally lived there during the winter (the Pump territory) or because the area appeared to contain suitable habitat yet had never been occupied (the Horsetail and Pipeline territories). Storage facilities added were generally storage limbs that had fallen from other granaries supplemented by holes we drilled. The locations of these territories are shown in Figure 3.1.

1. Pump. This area had been occupied between September 1973 and February 1974 by six birds that stored under loose bark in cracks in a dead willow (MacRoberts and MacRoberts, 1976:104, group 5). No group had ever lived here during the spring. We drilled approximately 1,000 holes in February 1976 and added another 1,250 holes in late March 1977. A group of three birds of unknown origin (one male and two females) occupied the area in October 1977. These birds had 368 acorns stored on 4 December. Unfortunately, the granary fell in a storm on 19 December. We then moved some of the fallen granary facilities to an adjacent valley oak. However, stores were exhausted by 3 February 1978. Two of the birds were seen on 14 March, but the territory was abandoned by 25 March. The granary, still containing about 1,000 holes, has since become overgrown by foliage, and the territory has not been used subsequently.

2. Horsetail. This area was one in which we had often seen birds, but had no permanent group despite being located relatively far away from adjoining territories. We added approximately 1,300 holes in two old storage limbs to a large valley oak on 2 September 1978. (Some additional holes in styrofoam blocks were added to the area in November 1982.) The area remained unoccupied until spring 1985, when two un-banded birds colonized and attempted one nest, which failed in the egg stage. These birds did not store many acorns and subsequently abandoned.

Spring 1985 was exceptional in that coast live oak acorns were present on many trees through the entire breeding season. This group was one of three forming that spring in areas never previously occupied. These three groups either had no access to storage facilities or, as in this case, did not use the storage facilities available to them. Additional details concerning these groups are given below.

3. *Pipeline.* This area had never been occupied despite consisting of excellent riparian habitat. We added 1,900 holes to a sycamore on 2 September 1978. Birds were occasionally seen in the area, but no group occupied the area until September 1982, when it was colonized by a male and two females. They stored acorns in the artificial granary, filling it by March 1983. In spring 1983 they bred and fledged three young. The group abandoned the area in September 1983 during the acorn crop failure. The territory was again reoccupied in the autumn of 1985, and four breeders fledged three young there in spring 1986.

In summary, two of the three artificial granaries we created were eventually used by groups; one of these groups bred successfully. The third territory was eventually inhabited one season, but the storage holes we added were not used for acorn storage. The delay between the artificial granary additions and the colonizations was twenty months for Pump and four years for Pipeline.

Granary Manipulations: Discussion

The above experiments generally support the importance of granaries in determining the number of groups. However, they also suggest that other features besides granaries are important in determining group residency. For example, loss of Arnold 3's primary granary led to the abandonment of the resident group, but the same general area was subsequently recolonized, the new birds focusing on a formerly peripheral section of the Arnold 3 territory. A similar event occurred at territory Bianca, where the main granary largely disintegrated between 1976 and 1978. In this case, a nearby tree, previously a secondary storage facility, became the primary focus for acorn storage and was supporting a group on the identical site by autumn 1981. Thus, in these two examples the loss of a major granary did not ultimately lead to any change in the number of groups.

Similarly, the addition of new granary facilities did not always lead to colonization by a new group, and even when it did there was a considerable time delay between creation of the granary and occupation. Thus, the relationship between granaries and acorn woodpecker groups is not necessarily simple or direct.

Part of the reason for this complex relationship is that granaries are the product of other, more ultimate, factors. Chief among these is certainly food: granaries cannot be made in areas devoid of acorns or other nuts that can be stored in them, whereas highly productive areas will attract and support larger numbers of birds who can, over a period of many years, build storage facilities. We therefore expect that the distribution and size of granaries will reflect the productivity and reliability of the acorn crop.

Other habitat characteristics may also be important in determining granary distribution. Acorn woodpeckers prefer to live in relatively open habitats (either open woodlands or in emergent tall trees above a more continuous forest canopy). It seems reasonable that this preference is partly because it is easier to detect predators in open habitats. In any case, overgrown areas with no emergent trees are likely to have few granaries relative to their acorn productivity. Similarly, existing granaries may be abandoned when they become overgrown.

Characteristics of the storage trees themselves may also influence granary size and distribution. We discussed in Chapter

4 the hypothesis that granary size results from an equilibrium between the drilling of new holes by residents and the loss of old holes through disintegration and storm damage. The rate of both these factors is influenced by characteristics of the trees: for example, the resistance of limbs to rotting and the propensity of the tree to sustain wind damage influence the rate of hole loss, and the ability of birds to drill new holes is affected by the thickness of the bark and the hardness of the wood. Thus, the equilibrium size of granaries is influenced by the trees themselves as well as the productivity of the resources being stored.

Still unexplained is the long delay between the construction of artificial granaries and their eventual use by acorn woodpeckers. Possibly there is "social inertia" that must be overcome in order for a new territory to form. At the minimum, this inertia implies that, given the choice between two equivalent territories, one of which is occupied and one of which is not, birds are more likely to fill a vacancy in the extant group than to colonize the unoccupied territory. At the maximum, this effect suggests that potential colonists are dissuaded from remaining in suitable areas as a consequence of the lack of residents substantiating its viability as a territory.

Direct Effects of Food on Number of Groups

Besides having an indirect effect on the number of territories via granaries, acorns also directly influence the number of groups. In poor years this is obvious: following acorn crop failures, groups may be forced to abandon their territories, regardless of their storage facilities (MacRoberts and MacRoberts, 1976; Hannon et al., 1987; see section 11.2).

Conversely, we have recently found that new groups may form in areas containing no storage facilities whatsoever in years when the acorn crop is unusually good. For example, in 1985, many coast live oak acorns remained on the trees well into June, near the end of the spring breeding season. Two new groups formed, one adjacent to Plaque ("Plaque Annex") and the other on part of Fanny Arnold's territory (for territory locations see Figure 3.1).

The new Plaque Annex group was formed by two former nonbreeding males from Plaque and a female from group 1500. This case constitutes the only clear case of territorial "budding," such as occurs frequently in the Florida scrub jay (Woolfenden and Fitzpatrick, 1978, 1984), that we have observed. Both the new groups attempted nests and produced young of bandable age. (The lone nestling in one of these did not fledge.) Also, both groups abandoned their territories soon after breeding. Interestingly, the Plaque Annex birds, with the exception of the breeding female, merged with group Plaque, forming a large, composite group.

Thus, new groups can form even in the complete absence of storage facilities if acorns are still available on the trees late into the spring. (Three additional new territories not containing granaries occupied in winter 1985–86 and on which successful breeding occurred in spring 1986 indicate that groups can sometimes form without storage facilities even when acorns are not available for a prolonged period.) However, such groups are generally ephemeral and do not persist much beyond the breeding season. They thus mimic the behavior of acorn woodpeckers in some populations in southeastern Arizona, where groups are mostly pairs, storage facilities are small or nonexistent, and the population is migratory (Stacey and Bock, 1978).

How Are Granaries Created?

We conclude that the number of groups is limited usually by the availability of granaries. This poses a problem: given that areas without granaries are uninhabitable by acorn woodpeckers on a long-term basis, how are granaries made in areas where they previously did not exist or had been previously destroyed?

One way is by temporary occupancy of areas by birds forced to abandon their territories in years of crop failures. MacRoberts and MacRoberts (1976) reported that several groups at

Hastings moved to areas where no granary facilities existed during a winter when it was known or suspected that the acorn crop had failed on their original territory. Such groups store acorns in crevices or other natural cavities as best they can. Although they may survive as a group only a short time, during that period they drill new holes that may be used in subsequent years by other displaced groups. After enough time, positive feedback between residency—although temporary—and the production of storage holes can produce a territory with enough holes to support a permanent group of acorn woodpeckers.

At the opposite extreme are the two groups that formed in areas devoid of storage facilities in a year when acorns were available on the trees extremely late in the season. Similar to the birds forced to inhabit areas without storage facilities following crop failures, such birds may drill some holes while present and contribute to the formation of a granary suitable for permanent occupancy.

11.4. TOTAL POPULATION SIZE

Since population size is the number of groups multiplied by mean group size, total population size will be affected by all the factors discussed above and listed in Table 11.3. We have previously documented the overall density dependence of reproductive success (section 5.3) and adult survivorship (Figure 7.3). However, emigration of nonbreeders is not density-dependent: the proportion of surviving offspring remaining as nonbreeders is not correlated with the prior year's breeding population size ($r_s = -0.02$, $N = 9$ years, ns).

As pointed out in Table 11.3, the processes of limitation are often quite different within groups compared to the population as a whole. For example, overall adult male survivorship is density-dependent (Figure 7.3) but inversely group-size-dependent (Table 7.5). These results are not necessarily contradictory: winter survivorship is correlated with per capita acorn

297

Table 11.3. Levels of Population Regulation in the Acorn Woodpecker

Level of Analysis	Limiting Resource	Mechanisms by Which Limitation Occurs	Source
Breeders within groups	Reproductive opportunities and food (acorns)	Dominance and socially induced reproductive suppression	Section 11.1
Total group size	Food (acorns)	Reproductive success (good years) and emigration and/or survivorship (bad years); reproductive competition (?)	Section 11.2
Number of groups	Granaries and food (acorns)	Emigration	Section 11.3
Total density	Granaries and food (acorns)	Reproductive success and survivorship	Section 11.4

storage (Table 4.10); hence, during years of large population size, per capita acorn storage is low, breeder mortality is high, and mortality differentially affects males in smaller groups. Such complications indicate some of the problems in determining the factors regulating populations of cooperative breeders.

What is the relative importance of group size versus number of groups in determining variation in population size? From Table 3.6, variation in group size (C.V. = 55%) contributes far more to determining variance in total population size (C.V. = 25%) than does variance in the number of groups (C.V. = 11%). Similarly, variation in group size is primarily the result of variance in the number of nonbreeders per group (C.V. = 180% for males and 176% for females) rather than the number of breeders per group (C.V. = 54% for males and 37% for females). Thus, differences in population size at Hastings are largely attributable to variation in the number of nonbreeders per group.

11.5. CONCLUSION

Acorn woodpeckers at Hastings are subject to density-dependent forces influencing both reproductive success (Figure 5.4) and survivorship (Figure 7.3). These forces limit the population on several levels, including the number and size of groups. The hierarchy of processes involved in regulating the population at these various levels is complex; those we have discussed here are summarized in Table 11.3 (see also Hannon et al., 1987).

Our data support the importance of food, specifically acorns, in limiting the population at all levels. This limitation is both direct, with acorns and acorn storage influencing group size and number of breeders within groups, and indirect, insofar as acorns ultimately determine the size and distribution of granary facilities, which in turn determine the number of groups.

Processes limiting the population are both density- and group-size-dependent. We have thus far stressed food in this respect, but predation may be important as well, particularly in determining the density-dependent pattern of adult survivorship. Unfortunately, we are generally unable to determine causes of mortality, and thus the role of predation is uncertain. Furthermore, all effects of a poor acorn crop need not be density- or group-size-dependent: complete crop failures in an area lead to abandonment by all birds, regardless of group or population size (see section 11.2).

Catastrophic crop failures occur at intervals of about four to five years (e.g., Hannon et al., 1987). The effects of such acorn crashes vary considerably depending on the geographic scale of the crop failure, from very local (such as that affecting several low-elevation groups at Hastings in autumn 1973; MacRoberts and MacRoberts, 1976:67), through the locally devastating crash of 1983 (Hannon et al., 1987; see section 11.2), to the crash of 1978, which apparently affected an area at least 50 km wide (our unpublished data). Thus, although the effects of the acorn crop are generally density-dependent, acorn crashes can result in large-scale density-independent fluctuations in population size. Additional consequences of acorn crashes include inverse density-dependent reproductive success (e.g., 1979 and 1984; Figure 5.4): following poor crops not only does the population crash but reproductive success is also poor.

The possibility that disease or parasitism may be important to cooperative breeders is suggested by the population crashes, either known or suspected to be due to disease, documented in Florida scrub jays (Woolfenden and Fitzpatrick, 1984:351) and white-fronted bee-eaters (S. T. Emlen, personal communication). Our data do not exclude the importance of disease, but suggest that usually food is the critical factor regulating acorn woodpecker populations.

We also have little evidence that weather is important in limiting the population. We have never witnessed mass mortality due to bad weather, nor is reproductive success correlated, at least in any interpretable fashion, to weather conditions (Table 5.5). Whatever effects weather may have on reproduction appear to be primarily indirect, resulting from the possible effect of weather conditions on acorn production.

11.6. SUMMARY

Population regulation occurs at several levels in acorn woodpeckers, including the number of breeders within groups, total group size, and the number of groups in the population. The processes by which limitation occurs are different at each level; in particular, those operating within groups are distinct from those acting at the population level.

Limitation of the number of breeders within groups is accomplished proximately by reproductive competition and incest avoidance and ultimately by limited reproductive opportunities and limited food. Total group size is also limited by food, as indicated by the density-dependent correlations of reproductive success, survivorship, and dispersal with the acorn crop and acorn storage. These mechanisms vary depending on crop size: density- and group-size-dependent reproduction occur in most years except those with very poor crops, when group-size-dependent survivorship and dispersal act as limiting processes. Because granary facilities are relatively fixed, the number of groups is set proximately by granaries and ultimately by the acorn crop itself.

Of the two variables—group size and number of groups—that produce total population size, average group size varies considerably more from year to year than the number of groups, and the number of nonbreeders per group varies more than the number of breeders. Thus, the ecological constraints that produce nonbreeders—proposed here to be storage facilities—have a strong influence on total population size through their effects on the incidence of nonbreeders.

Population regulation in acorn woodpeckers thus appears to be primarily the result of the limitation of two resources: acorns and granaries. The latter is ultimately determined by acorns as well, but is independent insofar as areas of good acorn abundance generally cannot be used for breeding unless adequate storage facilities exist. The limiting effects of predation, and possibly disease, may also be important, but are generally secondary.

Density-independent effects exist insofar as complete crop failures may result in group abandonment. The primary mechanisms of limitation, however, are density- and group-size-dependent. Weather appears to have little or no direct effect on population size, although it may influence population size indirectly through its effect on acorn production.

CHAPTER TWELVE

Living in Groups: Retention of Offspring

A basic question concerning the evolution of cooperative breeding and sociality in general (e.g., Alexander, 1974; Hoogland and Sherman, 1976) is, why live in groups? Acorn woodpeckers exhibit two phenomena leading to group living: nondispersal of young and mate-sharing (Koenig et al., 1984; see section 1.2). In this chapter we discuss the former.

12.1. INITIAL CONSIDERATIONS

In the majority of avian and mammalian species, newly independent young disperse or are expelled from their natal territory prior to subsequent reproductive attempts by their parents. There are only two basic mechanisms by which a change in this general pattern can occur. First, offspring may gain an intrinsic advantage by deferring dispersal and independent breeding, although adequate space exists for breeding elsewhere. Alternatively, offspring may be ecologically constrained from dispersing and breeding on their own due to limitation of some resource. Such limitation may lead to intense competition for reproductive opportunities. According to this hypothesis, there is an advantage to nondispersal only because of the ecological constraints; in their absence, independent reproduction is superior to remaining as a nonbreeder.

The latter of these hypotheses was not only widely proposed as being important to cooperative breeders in numerous early evolutionary discussions of the phenomenon (e.g., Selander, 1964; Brown, 1969; Verbeek, 1973; see Chapter 1), but is now widely accepted as the single most important factor leading to group living in many cooperatively breeding species (e.g., Stacey, 1979a, Koenig and Pitelka, 1981; Emlen, 1982a, 1984; Woolfenden and Fitzpatrick, 1984; Brown, 1987; for possible exceptions [discussed in Chapter 14] see Rabenold [1984, 1985], Austad and Rabenold [1985], and Stacey and Ligon [1987]). As summarized by Emlen (1982a, 1984), there are several potential "routes" by which ecological constraints might lead to nondispersal of young and cooperative breeding, including a high risk of early dispersal, a shortage of territory openings, a shortage of mates, and a prohibitive cost of independent reproduction.

Although the sex ratio among breeding acorn woodpeckers is biased toward males, the bias is slight and apparently a result rather than a cause of their social system (Koenig et al., 1983; see also Emlen et al., 1986). There is also no physiological barrier to independent reproduction in our population; in fact, the most frequent breeding group size is two (Figure 3.4), and other factors being equal, these are as successful as larger groups at fledging young on a per capita basis (section 6.1). The ecological constraints likely to be significant in acorn woodpeckers are thus two related factors: a high risk of early dispersal and a shortage of territories. We focus on the latter.

Summary of Evidence for Ecological Constraints

We can detail at least nine lines of evidence that are consistent with the hypothesis that ecological constraints are critical to the evolution of group living in the acorn woodpecker. This evidence is primarily related to delayed dispersal and delayed reproduction by offspring, but in some cases is also relevant to the evolution of mate-sharing by males and joint-nesting by females (we discuss the latter two phenomena in Chapter 13).

1. Power struggles. These contests among nonbreeding helpers fighting to fill reproductive vacancies indicate intense competition for reproductive opportunities (Koenig, 1981a; Hannon et al., 1985).

2. Territory occupancy rate. Occupancy rate of territories at Hastings is high and a steep habitat gradient slope (Koenig and Pitelka, 1981; Figure 3.11) exists compared to less social populations of acorn woodpeckers in the Southwest (Stacey, 1979a; see section 3.4).

3. A critical resource. Reproductive success, survivorship, and group composition are strongly correlated with stored acorns and granaries, suggesting that these resources are constraining factors (Chapters 4 and 11). In contrast, there are few correlations between these population characters and other food resources or weather (Chapter 5).

4. Reproductive success per capita. The decrease of per capita reproductive success with an increase in total group size implies a lack of direct reproductive benefits resulting from nondispersal (Koenig, 1981b; see Chapter 6).

5. Annual pattern of nondispersal. The proportion of young remaining as nonbreeders correlates inversely with the number attaining breeding status (section 8.3). The propensity to remain and help as nonbreeders also correlates positively with population size (Figure 8.3). This suggests that the tendency to remain as a nonbreeder is correlated with the degree of constraint, as indexed by population size.

6. Shortage of territories. There are fewer reproductive vacancies in the population than there are offspring available to fill them (section 8.8).

7. Sex bias in tendency to remain as nonbreeders. Territory vacancies are rarer for males than females. Correlated with this sex difference, males have a higher probability of being nonbreeders at each age and tend to remain slightly longer as nonbreeders than females (section 8.8).

8. Constraints and independent reproduction. For acorn woodpeckers, there is a strong correlation between the availability of reproductive opportunities and the tendency for offspring to disperse and breed on their own (Emlen and Vehrencamp, 1983; Emlen, 1984; see also Stacey, 1979a, and section 8.8).

9. Lack of parental facilitation. The absence of evidence that parents facilitate the reproduction of their offspring, either through territorial inheritance or other means, is consistent with the hypothesis that offspring accrue no intrinsic advantage by remaining in their natal group compared to breeding independently (chapter 9).

This evidence supports the hypothesis that ecological constraints, specifically dependence on acorns and granaries, is the primary factor leading to delayed dispersal of young in the acorn woodpecker (for an alternative view, see Stacey and Ligon, 1987). We now assess the impact of deferred dispersal and reproduction on the fitness of the individuals involved.

We approach this problem in three ways: (1) a comparison of lifetime reproductive success of birds delaying dispersal with those breeding in their first year; (2) a simulation of lifetime reproductive success of birds that remain as nonbreeders versus those that do not; and (3) a demographic model determining the indirect fitness received by individuals acting as nonbreeding helpers. Thus, our analyses will address the questions "Why do offspring delay reproduction and remain in their natal groups?" and "What fitness benefits, if any, do nonbreeders gain by remaining and helping to raise young not their own?"

12.2. LIFETIME REPRODUCTIVE SUCCESS OF NONBREEDERS

We compiled data on the lifetime reproductive success of 42 birds (27 males and 15 females) banded as nestlings prior to 1981 for which we were confident of having complete reproductive histories. We divided these birds into those that remained as nonbreeders in their natal territory for at least one year (nondispersers) and those that bred in their first year (dispersers).

TABLE 12.1. Lifetime Reproductive Success and Survivorship of Birds Remaining as Nonbreeders in their Natal Groups for at Least One Year Compared to Those Dispersing and Breeding in Their First Year

	Lifetime Young Fledged	Total Young Fledged by Group	Years as a Breeder	N Birds
ALL MALES				
Bred first year	2.04 ± 2.20	3.75 ± 3.88	2.50 ± 1.93	8
Stayed first year	3.26 ± 4.40	6.26 ± 8.15	3.05 ± 1.93	19
z-value	0.40	0.51	0.71	
ALL FEMALES				
Bred first year	3.33 ± 2.89	6.67 ± 5.77	3.67 ± 0.58	3
Stayed first year	2.92 ± 3.94	3.75 ± 5.29	1.92 ± 1.00	12
z-value	0.46	0.76	2.24*	
MALE NONHEIRS				
Bred first year	3.06 ± 2.19	5.40 ± 3.98	3.40 ± 1.95	5
Stayed first year	3.24 ± 5.11	5.77 ± 8.89	2.85 ± 1.99	13
z-value	0.85	0.70	0.71	
FEMALE NONHEIRS				
Bred first year	5.00 ± 0.00	10.00 ± 0.00	4.00 ± 0.00	2
Stayed first year	3.14 ± 4.60	3.14 ± 4.60	1.71 ± 0.76	7
z-value	0.92	1.54	2.13*	

NOTE: Estimates assume shared paternity and maternity among cobreeders. Only individuals fledged prior to 1981 are used. Mean ± SD shown; tests are by Mann-Whitney U-test.
* $p < 0.05$.

The results are presented in Table 12.1. Presented are data on the estimated number of lifetime young fledged (assuming equal parentage among cobreeders), the total number of young fledged by groups in which the birds bred while they were breeders (this includes both offspring parented by cobreeders as well as their own offspring) and the number of years birds survived as breeders. For males, nondispersers are slightly more successful than dispersers, but the differences are not significant. For females, dispersers are slightly more successful, significantly so for the number of years as a breeder. However, sample sizes are small, and the results should be viewed cautiously.

In order to eliminate complications arising because of territorial inheritance, the bottom half of Table 12.1 presents lifetime success excluding individuals inheriting their natal territory. Results of statistical tests are identical to the prior analyses. Thus, our data suggest a possible survivorship advantage to females that breed in their first year compared to those that remain as nonbreeding helpers. However, we have no evidence to support the hypothesis that individuals remaining as nonbreeding helpers for one or more years experience greater lifetime reproductive success than those dispersing and breeding in their first year.

12.3. A SIMULATION MODEL

A major shortcoming of the above analyses are the small sample sizes, which limit not only the confidence that can be placed in the results but also the extent of the comparisons we are able to perform. Here we use a simulation model to transcend these difficulties. For example, besides their effects on reproduction (Chapter 6) and on breeder survivorship (Chapter 7), nonbreeding helpers have a higher probability of inheriting their natal territory (Chapter 9) and of producing additional siblings with which to disperse and share breeding status (see Chapter 13). Although we have discussed some of these separately, here we model their combined effects so as to determine the overall fitness consequences of nondispersal compared to independent reproduction.

The Model

We use a simulation (Monte-Carlo) model to estimate the components of lifetime fitness for birds that either disperse or remain as nonbreeders for one or two years. A partial flow chart of the model is presented in Figure 12.1.

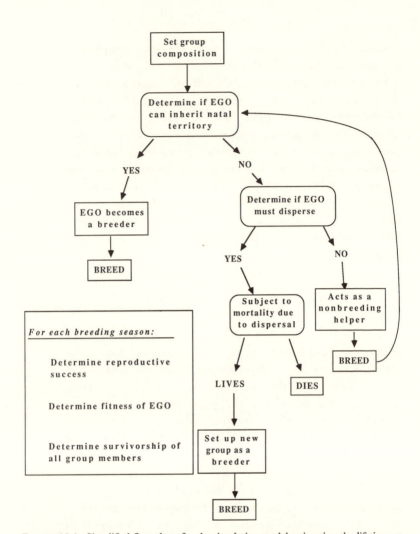

FIGURE 12.1. Simplified flow chart for the simulation model estimating the lifetime reproductive success of birds dispersing compared to remaining as nonbreeders. The model starts in the spring of EGO's first year. Each year EGO can inherit its natal territory, depending on the group composition, or disperse. If EGO inherits, it becomes a cobreeder with remaining birds of the same sex in the parental group. If EGO must disperse (either because all breeders of the same sex died or because he reached the age of forced dispersal), he is subjected to dispersal mortality (either high or low, depending on the initial conditions), after which a new group is set up with EGO as a breeder. Each breeding season, reproductive success of the group is determined and the fitness of EGO is calculated based on his breeding status, relatedness to breeders, and number of cobreeders in the group. If EGO is still a nonbreeder, his indirect fitness is calculated at that time, taking into account his effects on both the survivorship and the reproductive success of the breeders. The model continues until EGO dies or reaches fifteen years of age.

The model focuses on a single offspring (EGO) and begins in its first spring. Three series of runs are made. In the first series, EGO is forced to disperse right away and thus does not breed or help in its natal group. In the second, EGO helps in its first year and then must disperse prior to its second breeding season. In the third, it may remain until its third year. Meanwhile, the patterns of remaining as a nonbreeding helper may be disrupted due to mortality within the group. Specifically, birds may inherit their natal territory and breed if the opposite-sex breeders die; such birds then cobreed along with their parent of the same sex and possibly with siblings, if any are in the group at the time of inheritance. However, if all breeders of the same sex as EGO die, EGO is forced to disperse right away.

Parameters used in the simulation were chosen to reflect the situations observed in our population as realistically as possible. Initial group composition for the model can range from a single breeding pair to three breeding males, two breeding females, and six nonbreeding helpers in addition to EGO. The probability that EGO either disperses or inherits along with siblings depends on the number of other nonbreeders present in the group when these events occur. Cobreeders usually are assumed to share reproductive success equally (see sections 2.2, 10.3, 13.4, and Mumme et al., 1983b), but a slight bias (60:40) was given in successive years to the parent of EGO when EGO inherited his natal territory until EGO was five years old, after which it is assumed to share parentage equally with cobreeders.

Parameters used in the model were the mean adjusted values obtained from an ANOVA of reproductive success (controlling for number of breeding males, breeding females, non-breeders, and experience) and a logit analysis of survivorship (controlling for breeding males, breeding females, and adult nonbreeders). All groups were given a 40% chance of reproductive failure each year (see Table 5.2). Values used, along with the observed values in our population, are summarized in Table 12.2 for reproductive success and Table 12.3 for survivorship.

TABLE 12.2. Summary of Observed and Estimated Values for Young
Fledged per Group as a Function of Group Composition and Experience

| | Observed Values | | |
| | | | |
Group Composition	Mean (N)	Confidence limits	Estimated from ANOVA
PAIRS			
No NB, I	0.88 (26)	1.36–0.41	0.55
No NB, E	1.81 (21)	2.39–1.23	2.02
One NB, E	1.73 (11)	2.68–0.77	2.37
Two+ NB, E	2.33 (15)	3.28–1.38	2.88
TWO MALES, ONE FEMALE			
No NB, I	0.93 (14)	1.62–0.23	1.02
No NB, E	2.67 (12)	4.00–1.34	2.49
One NB, E	3.11 (9)	3.94–2.28	2.84
Two+ NB, E	3.64 (11)	4.70–2.57	3.11
ONE MALE, TWO FEMALES			
No NB, I	1.40 (5)	3.11–0.00	1.11
No NB, E	3.50 (2)	6.44–0.56	2.58
One NB, E	5.00 (1)	— —	2.93
Two+ NB, E	4.00 (1)	— —	3.20
TWO MALES, TWO FEMALES			
No NB, I	0.17 (6)	0.49–0.00	1.58
No NB, E	3.67 (3)	4.32–3.01	3.05
One NB, E	6.00 (1)	— —	3.40
Two+ NB, E	3.57 (7)	4.84–2.30	3.67
THREE MALES, ONE FEMALE			
No NB, I	1.20 (5)	2.77–0.00	1.63
No NB, E	3.50 (2)	4.48–2.52	3.10
One NB, E	2.50 (4)	3.77–1.23	3.45
Two+ NB, E	4.33 (9)	5.87–2.80	3.72
THREE MALES, TWO FEMALES			
No NB, I	3.75 (4)	6.68–0.82	2.19
No NB, E	2.25 (4)	6.03–0.00	3.66
One NB, E	7.00 (1)	— —	4.01
Two+ NB, E	3.83 (6)	5.54–2.12	4.28

NOTE: For observed values, data are the mean and the confidence limits
(1.96 * SE); for the ANOVA, data are mean values adjusted for group composition and turnovers. NB = nonbreeders; I = turnover in group composition (inexperienced); E = no turnover (experienced).

TABLE 12.3. Summary of Survivorship of Breeders as a Function of
Group Composition and Experience

Group Composition	Males		Females	
	% survival (*N*)	Estimated from logit regression	% survival (*N*)	Estimated from logit regression
PAIRS				
No NB	0.63 (40)	0.64	0.71 (34)	0.66
One NB	0.82 (11)	0.76	0.63 (8)	0.57
Two+ NB	0.75 (8)	0.81	0.60 (20)	0.61
TWO MALES, ONE FEMALE				
No NB	0.83 (46)	0.81	0.91 (21)	0.83
One NB	0.87 (23)	0.89	0.69 (16)	0.77
Two+ NB	0.91 (22)	0.91	0.69 (13)	0.80
ONE MALE, TWO FEMALES				
No NB	1.00 (2)	0.77	0.33 (6)	0.72
One NB	1.00 (3)	0.86	0.75 (4)	0.65
Two+ NB	—	0.89	0.67 (3)	0.68
TWO MALES, TWO FEMALES				
No NB	0.92 (12)	0.89	0.93 (15)	0.87
One NB	—	0.94	1.00 (1)	0.82
Two+ NB	0.89 (9)	0.95	0.86 (7)	0.84
THREE MALES, ONE FEMALE				
No NB	0.75 (32)	0.77	0.56 (9)	0.68
One NB	0.83 (18)	0.86	0.67 (3)	0.60
Two+ NB	1.00 (24)	0.89	0.67 (6)	0.64
THREE MALES, TWO FEMALES				
No NB	0.86 (14)	0.86	0.60 (5)	0.75
One NB	1.00 (1)	0.92	1.00 (1)	0.67
Two+ NB	0.90 (10)	0.94	1.00 (6)	0.70

NOTE: For the logit regression, data are mean values adjusted for group composition.

We made two types of runs. In the first set ("low" probability of dispersal), the probability of successful dispersal was set at 0.19 for single males and 0.34 for single females (see Table 8.10; values were arbitrarily doubled for sets of two or more males or females because of the higher success of sibling units in power struggles [Hannon et al., 1985]). In the second set ("high" probability of dispersal), the probability of successful dispersal was set at 1.00. These runs thus compare the consequences of remaining as a nonbreeder when strong ecological constraints against dispersal and independent reproduction are present and absent.

The simulation was run either until EGO died or for fifteen years. Each year, two components of the fitness of EGO were determined. First was EGO's direct fitness (Brown and Brown, 1981b); that is, how many offspring EGO parented, taking into account the sharing of parentage necessitated by mate-sharing. Second was the amount of indirect fitness (Brown and Brown, 1981b) in offspring equivalents due to remaining as a nonbreeder in EGO's natal group. In order to avoid double accounting (Hamilton, 1964), helping fitness was determined as follows: if EGO was a nonbreeding helper its first year, the number of young its natal group fledged was determined. Then the number of offspring fledged for the same natal group *without the presence* of EGO was determined. The difference between these two values was multiplied by a coefficient representing EGO's relationship to siblings (either 0.5 if there was only a breeding pair in the group or 0.45 if there was more than a pair; see Appendix A). Values were then multiplied by 2.0 so as to represent offspring equivalents (West-Eberhard, 1975).

A similar procedure was followed if EGO remained in its second year, with the addition that we also considered its effect on the survivorship of its parents. Following winter survivorship and spring reproduction, the model returned to the group composition prior to the winter and conducted an identical round of hypothetical winter mortality in the absence of EGO. If at

312

least one breeder of each sex remained, reproduction was then simulated so as to yield the number of offspring that "would" have been produced by EGO's group had EGO not remained as a nonbreeder that year. (If at least one breeder of both sexes did not survive, this value was set to zero.) Again, the difference between the two values was multiplied by the coefficient of relationship and then doubled to derive the estimated number of offspring equivalents produced as a result of EGO's presence as a nonbreeder during that year. This procedure thus estimates the indirect fitness effects of birds remaining as nonbreeders as mediated through reproduction and survivorship.

Simulations using a specific set of parameters were run 500 times for each of the three dispersal options (1,500 runs in all). Fitness components were averaged for each year and for all runs combined; indirect and direct fitness values were summed to yield the "total" lifetime fitness for each strategy. We performed runs for EGO as a male and as a female, for a low and a high probability of dispersal, and for zero and two additional nonbreeders in the group prior to the first year (when the simulation begins). This latter alternative was chosen in order to investigate the effects of additional nonbreeding helpers on the relative advantages of staying as a nonbreeder.

Results

Results of simulation runs using groups initially composed of one breeding male and one breeding female are summarized in Table 12.4. Runs employing different initial group compositions yielded similar results. Because the model begins immediately prior to EGO's first spring, birds having to disperse in their first year never remain as nonbreeding helpers (mean years helping = 0); those forced to disperse in their second year always remain one year, whereas those possibly staying until their third year remained an average of 1.38 to 1.42 years, depending on the initial conditions. (This value is less than two because of territorial inheritance.) Longevity of

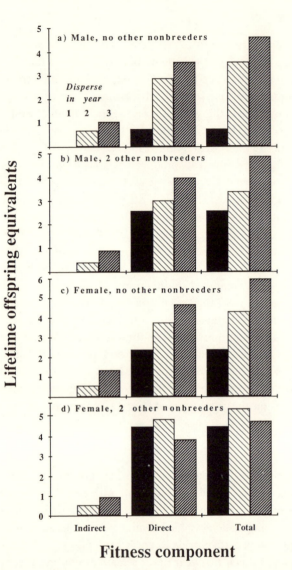

FIGURE 12.2. Estimated lifetime fitness (in offspring equivalents), divided into direct and indirect components, of individuals that disperse in their first year, by their second year, or by their third year. Shown are average values for 500 runs of the simulation model for each condition. (Standard deviations are in general large and all overlap extensively.) In all runs, the probability of successful dispersal was low: 0.19 for single males, 0.34 for single females, and twice these values for coalitions of two or more siblings. (a) EGO is a male and there are initially no additional nonbreeders in the parental group; (b) EGO is a male and there are initially two additional nonbreeders; (c) EGO is a female and there are no additional nonbreeders; (d) EGO is a female and there are two additional nonbreeders.

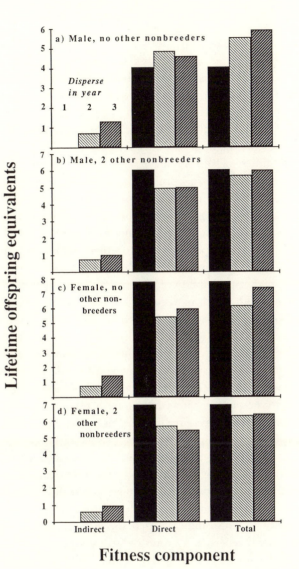

FIGURE 12.3. Estimated lifetime fitness, divided into direct and indirect components, of individuals that disperse in their first year, by their second year, or by their third year. Shown are average values for 500 runs of the simulation model for each condition. In all runs, successful dispersal was certain; that is, there were no constraints and all dispersers obtained territories. Panels as in Figure 12.2.

315

EGO varies considerably depending on the probability of dispersal. As expected, average survivorship is low if the probability of successful dispersal is low and birds are forced to disperse in their first or second years; otherwise, differences in survivorship between runs are not striking. The results also suggest that, under the conditions of the model, inheritance of natal territories is quite high, particularly for males.

The fitness of EGO for these same eight runs are presented in Figures 12.2 (for low probability of dispersal) and 12.3 (for high probability of dispersal). Not surprisingly, when there is a high cost to dispersal, EGO does much better in virtually all respects by remaining at home as a nonbreeder (Figure 12.2). The initial presence of two other nonbreeding helpers in the group moderates this enhancement, primarily because of the higher probability of dispersal given to birds that have a sibling with which to disperse. For females this is enough to equalize the lifetime success of birds dispersing right away compared to those remaining.

In general, however, birds in these simulations gain in both indirect and direct fitness by delaying dispersal and remaining as nonbreeders when the probability of dispersal is low. The majority of gains in direct fitness are a result of the fact that the longer EGO delays breeding the higher the probability that he will inherit his natal territory (Table 12.4). As the only means of acquiring breeding status in acorn woodpeckers that does not entail a significant cost when the probability of successful dispersal is low, the advantages of territorial inheritance under these conditions are considerable.

Figure 12.2 also indicates that the initial presence of additional nonbreeders in a group reduces the potential gains associated with nondispersal. For males, there are still modest advantages to nondispersal in both fitness components when two other nonbreeders are present, but for females no net advantage remains despite the relatively low probability of successful dispersal (note, however, that this probability is higher than for males).

TABLE 12.4. Results of the Simulation Model Estimating the Fitness Effects of Nonbreeders

Initial Group Composition	Year of Dispersal	Probability of Dispersal	$\bar{x} + SD$		Percent Inheritance
			Years helping	Years survived	
EGO = MALE					
1♂, 1♀, 0NB	1	Low	0.00 ± 0.00	1.55 ± 1.71	0.0
	2	Low	1.00 ± 0.00	4.26 ± 3.14	37.6
	3	Low	1.38 ± 0.49	5.14 ± 3.42	52.2
1♂, 1♀, 2NB	1	Low	0.00 ± 0.00	2.69 ± 3.12	0.0
	2	Low	1.00 ± 0.00	4.59 ± 3.50	30.8
	3	Low	1.42 ± 0.49	5.48 ± 3.61	53.2
1♂, 1♀, 0NB	1	High	0.00 ± 0.00	3.96 ± 2.67	0.0
	2	High	1.00 ± 0.00	5.75 ± 3.48	39.0
	3	High	1.41 ± 0.49	5.78 ± 3.36	55.6
1♂, 1♀, 2NB	1	High	0.00 ± 0.00	5.52 ± 3.50	0.0
	2	High	1.00 ± 0.00	6.08 ± 3.45	38.2
	3	High	1.42 ± 0.49	6.23 ± 3.57	50.8
EGO = FEMALE					
1♂, 1♀, 0NB	1	Low	0.00 ± 0.00	2.31 ± 2.47	0.0
	2	Low	1.00 ± 0.00	4.13 ± 3.05	24.2
	3	Low	1.43 ± 0.50	5.16 ± 3.13	31.8
1♂, 1♀, 2NB	1	Low	0.00 ± 0.00	3.73 ± 3.00	0.0
	2	Low	1.00 ± 0.00	4.89 ± 3.03	16.6
	3	Low	1.42 ± 0.49	4.82 ± 2.88	23.8
1♂, 1♀, 0NB	1	High	0.00 ± 0.00	4.77 ± 3.00	0.0
	2	High	1.00 ± 0.00	5.36 ± 2.95	19.4
	3	High	1.41 ± 0.49	5.73 ± 3.12	28.8
1♂, 1♀, 2NB	1	High	0.00 ± 0.00	4.97 ± 3.18	0.0
	2	High	1.00 ± 0.00	5.51 ± 3.08	16.0
	3	High	1.42 ± 0.49	5.76 ± 3.11	25.4

NOTE: NB = nonbreeders. Five hundred runs were made for each set.

By setting the probability of successful dispersal to one, we remove any direct cost to dispersal resulting from ecological constraints. This is done in Figure 12.3, again for groups initially containing either zero or two additional nonbreeders besides EGO. For male nonbreeders in groups initially composed of one breeding male, one breeding female, and no additional nonbreeders, dispersers still fare worse than do nondispersers. This is due to both increased lifetime reproductive success (direct fitness), and the increased indirect fitness accruing to males that remain as nonbreeding helpers in such groups (Figure 12.3).

When two additional nonbreeders are present initially and, for females, even when they are not, there is no advantage to remaining if there is no cost to dispersal. In general, the indirect fitness lost because of immediate dispersal is recouped through increased production of young. Consequently, the total fitness of dispersers is as high or higher than nondispersers under these circumstances.

Implications for the Evolution of Nondispersal

These results support the hypothesis that ecological constraints to dispersal are important to the evolution of nondispersal in this population: in the absence of such constraints, both direct and total fitness for birds that disperse in their first year are usually as great or greater than for those remaining as nonbreeders. The only exception is male offspring in groups containing no other nonbreeders (Figure 12.3a).

Thus, these analyses indicate that there is generally no lifetime benefit to remaining as a nonbreeder as long as dispersal is unconstrained, even considering indirect fitness accruing through increased survivorship and reproduction. It follows that ecological constraints are an important feature selecting for nondispersal of young, at least under the conditions simulated by the model.

318

Index of Kin Selection

Here we use the values derived from the prior simulation model to calculate Vehrencamp's (1979) "index of kin selection" (I_k), a measure designed to quantify the relative importance of individual and kin selection to the maintenance of a particular behavioral trait. This index is the proportion of the total gain in inclusive fitness in the cooperative situation compared to the noncooperative situation due to the kin component (indirect fitness). (Since the simulation includes future indirect fitness effects to the nonbreeders [the i_H of Brown, 1985], our derivation also corresponds to the "index of indirect selection" proposed by Brown [1987].) From Vehrencamp (1979),

$$I_k = \frac{(W_{RA} - W_R)r_{ARy}}{(W_{AR} - W_A)r_{Ay} + (W_{RA} - W_R)r_{ARy}},$$

where

$(W_{RA} - W_R)$ = the change in lifetime reproductive success of the recipient (R) when aided by donor (A);

$(W_{AR} - W_A)$ = the change in lifetime reproductive success of A due to providing aid to R;

r_{ARy} = relatedness of A to R's young;

r_{Ay} = relatedness of A to its own young.

These values can be derived directly from the simulation model (Figures 12.2 and 12.3). Since all fitness values are calculated in lifetime offspring equivalents, the index of kin selection for delayed dispersal can be calculated as the proportion of change in total fitness of nondispersers attributable to change in indirect fitness. Calculation of I_k for nonbreeders in a representative sample of possible initial group compositions is shown in Table 12.5.

When dispersal is constrained, estimates for I_k vary between 0.24 and 0.57. Estimates for the sexes are quite similar;

TABLE 12.5. Calculation of an Index of Kin Selection (I_k) Given One Year as a Nonbreeding Helper

	Fitness of Birds that Disperse in Year 1		Fitness of Birds that Disperse in Year 2		
	Direct	Indirect	Direct	Indirect	I_k
CONSTRAINT TO DISPERSAL PRESENT					
♂♂, no other NBs	0.72	0.00	2.88	0.67	0.24
♂♂, 2 other NBs	2.55	0.00	2.99	0.38	0.46
♀♀, no other NBs	2.36	0.00	3.75	0.54	0.28
♀♀, 2 other NBs	4.50	0.00	4.88	0.50	0.57
NO CONSTRAINTS TO DISPERSAL					
♂♂, no other NBs	4.02	0.00	4.85	0.68	0.45
♂♂, 2 other NBs	6.06	0.00	4.99	0.73	Undefined
♀♀, no other NBs	7.75	0.00	5.40	0.72	Undefined
♀♀, 2 other NBs	6.91	0.00	5.68	0.60	Undefined

NOTE: Calculated from the simulation model (Figures 12.2 and 12.3) for groups with one breeding male and one breeding female. For formulae and additional details, see section 12.3. NBs = nonbreeders.

however, I_k values are slightly over twice as great when two other nonbreeders are initially present in the group as when there are none. These estimates suggest that the indirect benefits to remaining as a nonbreeder are considerable to acorn woodpeckers under these ecological conditions.

Also shown in Table 12.5 are similar calculations assuming there are no constraints to dispersal. Only for lone male nonbreeders is I_k defined; in this case 45% of the benefits to staying accrue through indirect fitness. In the other three cases, the change in lifetime fitness resulting from remaining for one year (ΔW) is negative, and consequently I_k values are undefined. Vehrencamp (1979) suggests that under such circumstances either the actor (the nonbreeder) is being manipulated or group selection explains the observed behavior. In this instance, however, we suggest that the negative values for I_k result from invalid assumptions. The most likely candidate for a faulty assumption is that no constraints to dispersal exist, as has already been discussed in depth in section 12.1.

Values for I_k calculated from Florida scrub jay data by Vehrencamp (1979) yield an estimate of 0.55, whereas I_k values for the splendid wren (Rowley, 1981) range from 0.35 to 0.51 depending on the specific demographic assumptions. Reyer (1984) calculated I_k for the pied kingfisher and derived values of 0.13 for "secondary" helpers (usually unrelated to the birds they are helping) and 1.13 for "primary" helpers (usually offspring). (Values of I_k greater than one indicate that helpers are behaving altruistically in the strict sense of Hamilton [1964] and sacrificing their own personal reproduction for the benefit of relatives.) However, in none of these cases was lifetime reproductive success used; nor did these authors incorporate the future component of indirect fitness (Mumme and Koenig, MS).

12.4. NONBREEDING HELPERS AND INDIRECT FITNESS

Here we use a demographic model to explore the question "What are the indirect benefits of being a nonbreeding helper, and how do they compare to breeding independently?" We begin with the premise that offspring are forced to remain in their natal group and focus on the specific direct benefits accruing to the breeders as a result. We then compare this benefit to that expected if the offspring had not remained as a nonbreeder. This analysis thus links the questions "Why is there delayed dispersal?" and "Why do helpers help?" (see Chapter 1).

A Demographic Model

Consider first a single adult male or female acorn woodpecker that breeds solitarily (i.e., one that does not share a mate or a

nest with other individuals of the same sex). The lifetime reproductive success (W_s) of this individual can be approximated
as follows:

$$W_s = \sum_{i=1}^{n} m_s \ell_s^{(i-1)},$$

where m_s is the annual reproductive output (young fledged per
year) of the solitary breeder, ℓ_s is its annual survivorship, and
n is the number of years of reproductive life. (Hereafter, we
assume that $n = 15$ for acorn woodpeckers.) Note that this expression assumes that the individual always survives to breed
in year one.

Now consider a second solitary breeder that is aided by a nonbreeder only in year one. The nonbreeder disperses immediately prior to the second breeding season in year two. The
lifetime reproductive output of the breeder aided by the nonbreeder in year one can be expressed as

$$W_{sh} = m_{sh} + \sum_{i=1}^{n-1} m_s \ell_{sh} \ell_s^{(i-1)},$$

where m_{sh} is the annual reproductive output of the solitary
breeder assisted by a nonbreeder in year one, and ℓ_{sh} is the survivorship of the breeder (in year one) in the presence of the
nonbreeder. Thus, the overall effect of a nonbreeder, assisting
for one year, on the lifetime reproductive success of the breeder
is $\Delta W_s = W_{sh} - W_s$. Note that ΔW_s is influenced by the effect
of the nonbreeder on both breeder reproductive success (m_{sh})
and breeder survivorship (ℓ_{sh}).

Estimating the effects of nonbreeders on the lifetime reproductive success of two cobreeders sharing mates (males) or
nesting jointly (females) is more complicated. When not aided
by a nonbreeder, the lifetime reproductive success of each
breeder (assuming parentage is shared equally) is

$$W_d = \sum_{i=1}^{n} (1/2) m_d \ell_d^{(2i-2)} + \sum_{i=1}^{n-1} \sum_{j=1}^{i} m_s (1 - \ell_d) \ell_d^{(2j-1)} \ell_s^{(i-j)},$$

where m_d is the reproductive output (per nest) of the group when two breeders are present, and ℓ_d is the corresponding value of breeder survivorship. The first term of the expression represents the lifetime output of the two breeders while both are living. The second term takes into account the probability that one of the breeders will die after one or more years, thereby making the remaining individual a solitary breeder.

If assisted by a nonbreeding helper in year one (but not in subsequent years), the lifetime reproductive output of two individuals that initially share breeding becomes the sum of four terms,

$$W_{dh} = (1/2)m_{dh} + \sum_{i=1}^{n-1} m_s \ell_{dh}(1 - \ell_{dh})\ell_s^{(i-1)}$$
$$+ \sum_{i=1}^{n-1} (1/2)m_d \ell_{dh}{}^2 \ell_d^{(2i-2)}$$
$$+ \sum_{i=1}^{n-2} \sum_{j=1}^{i} m_s \ell_{dh}{}^2(1 - \ell_d)\ell_d^{(2j-1)}\ell_s^{(i-j)}$$

where m_{dh} and ℓ_{dh}, respectively, are the annual reproductive output (per group) and annual survivorship of individuals sharing breeding status. The first term is the reproductive output of the helper-assisted group in year one. The second term represents the probability that one of the breeders dies before the second breeding season (while the nonbreeder is still present), leaving the remaining breeder by himself. The third and fourth terms are similar to those comprising the expression for W_d. They differ only in that they include the probability that both breeders survive (in the presence of the nonbreeder) to breed (unaided) in year two.

Calculation of the lifetime reproductive success of birds that initially breed as trios, either aided or unaided by nonbreeders, is complex. If unaided by a nonbreeder and assuming that parentage is shared equally, W_t (the lifetime reproductive success of birds breeding initially as part of a nest- or mate-sharing trio) is the sum of four separate terms:

$$W_t = T_1 + T_2 + T_3 + T_4,$$

where

$$T_1 = \sum_{i=1}^{n} (1/3) m_t \ell_t^{(3i-3)},$$

$$T_2 = \sum_{i=1}^{n-1} \sum_{j=1}^{i} m_d \ell_t^{(3j-1)} (1 - \ell_t) \ell_d^{(2i-2j)},$$

$$T_3 = \sum_{i=1}^{n-1} \sum_{j=1}^{i} m_s \ell_t^{(3j-2)} (1 - \ell_t)^2 \ell_s^{(i-j)},$$

$$T_4 = \sum_{i=1}^{n-2} \sum_{j=1}^{i} \sum_{k=1}^{j} 2 m_s \ell_t^{(3k-1)} (1 - \ell_t) \ell_d^{(2j-2k+1)}$$
$$(1 - \ell_d) \ell_s^{(i-j)}.$$

In the above expressions, m_t and ℓ_t are the annual values of per group reproductive output and individual survivorship of birds breeding in trios. T_1 represents the lifetime reproductive output of each individual in the nest- or mate-sharing trio while all three individuals are alive. T_2 incorporates the probability that one of the breeders dies after one or more years and calculates the lifetime reproductive output of the resulting duo. If two breeders die in the same year (after one or more years as a trio), the lifetime reproductive success of the remaining breeder is calculated by the expression for T_3. The final equation calculates T_4, the lifetime reproductive success of individuals who begin breeding for one or more years as a trio, breed for one or more years as part of a duo (following the death of one of the breeders), and finally breed as a solitary individual (following the death of the second breeder). Although algebraically complex, these equations are biologically straightforward and capture the basic properties of acorn woodpecker group dynamics (Koenig and Pitelka, 1979; Koenig, 1981a; Koenig et al., 1984; Hannon et al., 1985).

If an initial breeding trio is aided by a nonbreeder in year one (but not subsequently), calculation of the lifetime reproductive success of each individual in the trio (W_{th}) is the sum of eight distinct terms:

$$W_{th} = T_{h1} + T_{h2} + \cdots + T_{h8},$$

where

$$T_{h1} = (1/3)m_{th},$$

$$T_{h2} = \sum_{i=1}^{n-1} m_d \ell_{th}^2 (1 - \ell_{th}) \ell_d^{(2i-2)},$$

$$T_{h3} = \sum_{i=1}^{n-1} m_s \ell_{th} (1 - \ell_{th})^2 \ell_s^{(i-1)},$$

$$T_{h4} = \sum_{i=1}^{n-2} \sum_{j=1}^{i} 2m_s \ell_{th}^2 (1 - \ell_{th})(1 - \ell_d) \ell_d^{(2j-1)} \ell_s^{(i-1)},$$

$$T_{h5} = \sum_{i=1}^{n-1} (1/3) m_t \ell_{th}^3 \ell_t^{(3i-3)},$$

$$T_{h6} = \sum_{i=1}^{n-2} \sum_{j=1}^{i} m_d \ell_{th}^3 \ell_t^{(3j-1)} (1 - \ell_t) \ell_d^{(2i-2j)},$$

$$T_{h7} = \sum_{i=1}^{n-2} \sum_{j=1}^{i} m_s \ell_{th}^3 \ell_t^{(3j-2)} (1 - \ell_t)^2 \ell_s^{(i-j)},$$

$$T_{h8} = \sum_{i=1}^{n-3} \sum_{j=1}^{i} \sum_{k=1}^{j} 2m_s \ell_{th}^3 \ell_t^{(3k-1)} (1 - \ell_t) \ell_d^{(2j-2k+1)}$$
$$(1 - \ell_d) \ell_s^{(i-j)}.$$

The terms m_{th} and ℓ_{th} in the above expressions represent annual measures of per group reproductive success and individual survivorship for birds breeding in trios aided in year one by a nonbreeder.

Fortunately, groups containing more than three breeding birds of the same sex are rare in acorn woodpeckers (Chapter 3; see also Koenig et al., 1984). Thus, we can defer treatment of groups with larger numbers of breeders.

As before, the reproductive success variables necessary to perform the above calculations were estimated from the multiple classification analysis (Nie et al., 1975) resulting from a multiway ANOVA in which we determined the young fledged for groups with one to three breeding males; one or two breeding females; and zero, one, or two or more adult nonbreeders (see Chapter 6; prior group history was not included in this analysis). Similarly, breeder survivorship in groups with these compositions was estimated from the log-linear parameters of

TABLE 12.6. Values of Young Fledged per Group Used in Calculating Lifetime Reproductive Success of Breeders Aided and Unaided by Nonbreeders in the Demographic Model

Number of Breeders		Number of Nonbreeders		
Males	Females	0	1	2+
1	1	1.48	1.83	2.10
	2	2.04	2.39	2.66
2	1	1.95	2.30	2.57
	2	2.51	2.86	3.13
3	1	2.56	2.91	3.18
	2	3.12	3.47	3.74

NOTE: Values of reproductive success estimated from the multiple classification analysis of an ANOVA.

a logit model that evaluated the effect of group composition on survivorship (see Chapter 7). The survivorship and reproductive success values used in this analysis are shown in Tables 12.3 and 12.6.

Results

Using the above model, we calculated lifetime reproductive success of breeders in groups with zero, one, and two non-breeders. These results are presented in Table 12.7. For each type of group, we compared the lifetime reproductive success of breeders that were assisted by an additional nonbreeder for one year (W_h) with that of breeders unassisted by additional nonbreeders (W_{h-1}). For simplicity, we assume that in groups where two adult nonbreeders are already present, additional nonbreeders have no effect on the lifetime reproductive success of the breeders (see Chapters 6 and 7).

From Table 12.7 we can calculate values of $\Delta W\ (=W_h - W_{h-1})$ for groups of various composition. These values con-

TABLE 12.7. Lifetime Reproductive Success of Male and Female Breeders When Aided by a Nonbreeder for One Year (W_h) and When Unaided by a Nonbreeder (W_{h-1})

Number of Breeders		Number of Nonbreeders	Lifetime Reproductive Success						Proportion of Groups
			Males			Females			
Males	Females		W_{h-1}	W_h	ΔW	W_{h-1}	W_h	ΔW	
1	1	1	4.12	4.98	0.86	4.30	4.29	−0.01	0.17
		2	7.66	8.28	0.62	4.28	4.70	0.42	0.10
	2	1	8.55	9.67	1.12	3.93	3.81	−0.12	0.02
		2	15.02	15.75	0.73	3.73	3.98	0.25	0.01
2	1	1	4.67	5.23	0.56	10.76	10.51	−0.25	0.11
		2	8.48	8.83	0.35	9.95	10.45	0.50	0.06
	2	1	9.34	9.98	0.64	9.49	9.28	−0.21	0.02
		2	14.82	15.16	0.34	8.71	9.00	0.29	0.01
3	1	1	4.20	4.66	0.46	8.06	7.76	−0.30	0.05
		2	7.50	7.78	0.28	7.31	7.83	0.52	0.04
	2	1	8.19	8.69	0.50	6.94	6.64	−0.30	0.01
		2	12.49	12.73	0.24	6.11	6.42	0.31	0.01
Groups with 3 or more nonbreeders			12.67	12.67	0.00	8.68	8.68	0.00	0.38
Weighted mean			8.82	9.21	0.39	7.59	7.64	0.05	1.00

stitute the incremental change in lifetime reproductive success realized by each breeder that is directly attributable to the presence of a nonbreeder during one year. Thus, for groups with the compositions shown in Table 12.7, a nonbreeder alters the lifetime reproductive success of each breeder by an amount corresponding to ΔW for each year that it remains as a nonbreeding helper.

The presence of nonbreeders substantially increases the estimated lifetime reproductive success of males breeding in groups with many different compositions. By weighting the values of lifetime reproductive success (W_h and W_{h-1}) by the frequency of group composition observed in the Hastings population (shown in the last column in Table 12.7), we can estimate the average annual effect that nonbreeders have on the lifetime reproductive success of breeders.

On average, each breeding male assisted by a nonbreeder for one year produces 9.21 offspring in his lifetime, compared to 8.82 offspring produced by unassisted males in groups of comparable composition. Thus, by acting as a nonbreeding helper for one year, the average nonbreeder increases the lifetime reproductive success of *each* breeding male in his or her group by 0.39 offspring, an increase of 4.4%. The increase in lifetime reproductive success is much larger in many types of groups. For example, in groups consisting of a lone pair, the lifetime reproductive success of the breeding male increases from 4.12 offspring to 4.98 offspring (an increase of 20.9%) when assisted by a nonbreeder for one year. This parallels the relatively large increase in fitness derived by single nonbreeder males found in the simulation model discussed in section 12.3 (see Figures 12.2a and 12.3a).

The effect that the average nonbreeding helper has on the reproductive success of breeders (ΔW) can be separated into two distinct components. The first (ΔW_M) is the short-term reproductive enhancement attributable to the presence of the nonbreeder in year one (Table 12.6; see also Chapter 6). By subtracting this component from the overall effect, we can calculate the long-term effect on reproductive success of breeders attributable to the nonbreeder's effect on breeder survivorship (ΔW_L). These two components are analogous to the distinction drawn by Brown (1985a, 1987) between the present and future components of indirect fitness.

We estimated the contributions of an average nonbreeder to ΔW_M and ΔW_L by determining the present and future components of ΔW for groups of each composition in Table 12.7 and weighting by the overall frequency of each group. Of the 0.39 offspring fathered by each breeding male attributable to the presence of a nonbreeder for one year, 0.14 (36.8%) represent ΔW_M and 0.25 (63.2%) represent ΔW_L. Thus, the increase in the lifetime reproductive success of male breeders aided for one year by a nonbreeding helper is attributable primarily to improved annual survivorship.

For breeding females, the aid provided by nonbreeding helpers has almost no overall effect on their lifetime reproductive success (Table 12.7). The increased annual reproductive output of females aided by nonbreeders is largely negated by their slightly lower annual survivorship (Tables 12.3 and 7.7). For each year, the average nonbreeding helper increases the lifetime reproductive success of each female in its group by only 0.7% (from 7.59 offspring to 7.64 offspring; Table 12.7). The short-term reproductive component (ΔW_M) of this modest increase in lifetime reproductive success is 0.19, and the contribution of the survivorship component (ΔW_L) is -0.14. Thus, the model suggests that the presence of nonbreeding helpers is of little consequence to the lifetime reproductive success of breeding female acorn woodpeckers.

Calculation of Indirect Fitness

What is the average indirect fitness received by a nonbreeder for each year of aid it provides? To calculate this value, we must first estimate the coefficient of relatedness between nonbreeding helpers and the adults they assist. Because this coefficient of relatedness differs for male and female breeders, we treat the sexes separately.

In 249 nonbreeder-seasons, nonbreeding helpers always assisted either their father or a closely related cobreeder. We have no evidence that breeding males were ever assisted by nonrelatives. Because nonbreeding helpers are always group offspring of prior years, the average coefficient of relatedness between nonbreeders and breeding males is identical to the average coefficient of relatedness between breeding males and offspring, calculated as 0.416 ($= r_{mh}$) in Appendix A.

Breeding females, on the other hand, have been assisted by unrelated birds in 4 of 249 (1.6%) nonbreeding-helper seasons. In all other cases, nonbreeding helpers always assisted their mother or a closely related cobreeder. From Appendix A, the average coefficient of relatedness between breeding females and offspring is 0.469. Thus, the average coefficient of relatedness between female breeders and their nonbreeding helpers (r_{fh}) can be estimated as

$$r_{fh} = (0.469)(245)/249 = 0.461.$$

Using these coefficients of relatedness and the calculations summarized in Table 12.7, we can estimate the average amount of indirect fitness received by a nonbreeding helper through the lifetime reproductive success of the breeders that it assists. Because these calculations are made on a per breeder basis, we weight indirect fitness by the number of breeders each nonbreeder assists.

From this procedure, we estimate that the average nonbreeder increases the lifetime reproductive success of each male breeder in the group by 0.39 offspring for each year of assistance provided. When we repeat these calculations, recognizing that nonbreeders frequently assist more than one breeding male simultaneously, the average nonbreeder increases the lifetime reproductive success of all male breeders in its group by a total of 0.58 offspring. Because the nonbreeder is, on average, related to the male parents of these offspring by 0.416, the total indirect fitness for the nonbreeder is $(0.58)(0.416) = 0.240$ offspring eqivalents realized through the male breeders alone.

The lifetime reproductive success of individual breeding females is, on average, increased by 0.051 offspring in the presence of a nonbreeding helper for one year (Table 12.7). Taking into account the frequency of cobreeding females, the average nonbreeder increases the lifetime reproductive success of all breeding females in its group by a total of 0.048 offspring. Adjusting for the coefficient of relatedness between nonbreeding helpers and the females they assist (0.461), we estimate that the

TABLE 12.8. Average Annual Values of Present and Future Components of Indirect Fitness Received by Nonbreeders in Offspring Equivalents

	Via Male Breeders	Via Female Breeders	Total	Percent of Total
Present	0.083	0.092	0.176	66.9%
Future	0.157	−0.070	0.087	33.1
Total indirect fitness	0.240	0.022	0.263	100.0

total indirect fitness received by nonbreeders through breeding females is (0.048)(0.461), or 0.022 offspring equivalents realized through the female breeders alone.

These results are summarized in Table 12.8. Overall, 0.176 (66.9%) of the 0.263 offspring equivalents making up total indirect fitness is attributable to present reproductive effects of nonbreeders. The remainder (33.1%) is the result of future reproduction attributable to the effect of nonbreeders on breeder survivorship. The future component of total indirect fitness would be considerably greater but for the slightly reduced survivorship, and therefore negative future indirect fitness, of females breeding in the presence of nonbreeders (Table 12.8).

12.5. DISCUSSION

Why Nondispersal?

In this chapter we presented three approaches to determining the demographic and evolutionary effects of nondispersal by offspring and subsequent helping as nonbreeders in the acorn woodpecker. Two of these approaches included estimation of the indirect fitness gained by nonbreeders both through immediate reproductive success of breeders (the present component) and on survivorship of breeders (the future component). All three involved lifetime success determined either directly or indirectly.

331

These analyses and other lines of evidence summarized at the beginning of the chapter, lead to the following conclusions. First, lifetime reproductive success of birds breeding in their first year is not significantly different from that of birds remaining as nonbreeders for at least a year before breeding (section 12.2). Second, nondispersal is generally a strategy superior to independent breeding only if there are ecological constraints to successful dispersal; in the absence of such constraints the fitness of dispersers is usually greater than that of nondispersers (section 12.3). Finally, gains in indirect fitness accruing to nonbreeders are generally small, averaging only 0.240 offspring equivalents via male breeders and 0.022 via female breeders for each year that dispersal is delayed (section 12.4).

These results consistently indicate that remaining as a non-breeder is the result of ecological constraints to dispersal and independent reproduction rather than any intrinsic advantage to nondispersal. In terms of the postdoc analogy (see Chapter 1), young acorn woodpeckers do a postdoc because jobs are scarce, not because postdoctoral positions contribute more to their lifetime success than would a tenure-track job. Similar conclusions have been reached for many other cooperative breeders, including the exhaustively studied Florida scrub jay (Woolfenden and Fitzpatrick, 1984), and pied kingfisher (Reyer, 1984).

Why Do Helpers Help?

Comprehensive examination of the proposed evolutionary reasons that nonbreeders help to raise young not their own requires detailed information on patterns of nest care, dispersal patterns, dominance relationships, and the demographic data we report here. A full discussion of this important issue is beyond the scope of this monograph (see Chapter 1; for a recent comprehensive review see Brown, 1987). However, virtually all nonbreeders take part in cooperative group activities, either through helping to raise young (e.g., Koenig et al., 1983; Mumme, 1984), storing and maintaining acorns and defending the granary (Mumme and de Queiroz, 1985), or both. Thus, the fitness benefits of remaining as a nonbreeder are not independent of those derived from participating in cooperative activities. This dependence allows us to discuss the issue of helping behavior, in contrast with nondispersal, using our demographic results.

First, do nonbreeders enhance their direct fitness through helping? Chief among the proposed mechanisms by which nonbreeders might increase their personal fitness is by increasing their probability of obtaining a territory. Woolfenden and Fitzpatrick (1978, 1984) have argued persuasively that this phenomenon is of significant benefit to male Florida scrub jay helpers. Territory size in this species is correlated with group size; furthermore, helpers in large groups are more likely to later acquire part of their natal territory through budding. Thus, by raising siblings, helpers contribute to their own direct fitness.

The phenomenon of territorial budding per se is not an important benefit of remaining as a nonbreeder in the acorn woodpecker: we observed only a single instance of territorial budding between 1972 and 1985 (see section 11.3). Although analogous benefits could be derived through territorial inheritance, our data show that individuals inheriting their natal territories experience neither increased lifetime reproductive success (section 9.2) nor an increased probability of obtaining a reproductive position (section 8.4). Thus, we have no evidence to suggest that these processes increase the direct fitness of nonbreeding helpers.

A second potentially important route for nonbreeding acorn woodpeckers to increase their probability of acquiring breeding space is by increasing the size of their sibling support unit. As shown elsewhere (Hannon et al., 1985), coalitions of same-sexed siblings significantly increase the probability of successful dispersal (a similar effect has been demonstrated in lions, *Panthera leo* [Bygott et al., 1979; Koenig, 1981c]). Thus, to the extent that nonbreeding helpers help raise younger siblings with which they may later form coalitions, such birds may increase their own likelihood of successfully dispersing. Such a direct benefit is similar to the reciprocity suggested as important to green woodhoopoes by Ligon and Ligon (1978a).

Quantifying this effect requires estimates of the probability of successful dispersal as a function of coalition size and the frequency that siblings from different broods disperse together, as well as conclusive data on the effect of nonbreeding helpers on group reproduction. We defer such an analysis. However, 7 of 13 dispersing male units (54%) and 2 of 9 dispersing female units (22%) consisted of siblings from different broods. Thus, this effect may be an important direct benefit of helping by nonbreeders.

A third way nonbreeders might increase their own direct fitness by helping is by gaining experience that aids them later in life. Lawton and Guindon (1981), for example, showed that young brown jay helpers frequently make mistakes while feeding young; it is likely that the frequency of such errors is reduced as a result of helping-at-the-nest.

Although similar experience may be gained by acorn wood-pecker nonbreeders, we doubt that this hypothesis can explain helping by nonbreeders. The relatively high proportion of birds breeding in their first season (section 8.3), and the finding that the lifetime reproductive success of birds that do and do not help are comparable (Table 12.1), show that prior helping experience is not critical to successful reproduction. In addition, Mumme (1984) found that first-year birds are equally capable as older birds in incubating eggs, brooding young, and feeding young. Nonetheless, given that nonbreeders are trapped in their natal group by ecological constraints, whatever experience they may gain as a result of helping could still contribute to their success. Thus, the role of "experience-enhancement" (Brown, 1987) cannot be entirely discounted.

Helpers also achieve indirect fitness benefits through helping, assuming that their behavior results in increased lifetime reproductive success of nondescendent kin. As shown in sections 12.3 and 12.4, nonbreeders derive substantial indirect fitness from delayed dispersal and helping: between 24 and 57% of the increase in fitness accruing to individuals that delay dispersal for one year is attributable to an increase in the indirect component of inclusive fitness (Table 12.5). However, a major portion of the indirect fitness benefits accruing to nonbreeders comes not from direct aid to younger siblings but from the increased survivorship of male breeders (Table 12.8). Thus, depending on the mechanism by which nonbreeders enhance male survivorship (see section 7.6), alloparental behavior may not be the most important means by which nonbreeders enhance their own indirect fitness.

Combined with our earlier conclusion that the presence of nonbreeding helpers may result in *no* increase in the number of young fledged by groups (see Chapter 6), it is plausible that there is no selective advantage to helping per se, and that this long-debated phenomenon is selectively neutral (see also Woolfenden and Fitzpatrick, 1984). Nest aid contributed by nonbreeders might then be explained as misdirected parental care (e.g., Price et al., 1983; for a thorough discussion of this hypothesis see Jamieson, 1986, and Jamieson and Craig, in press). It is a sobering thought and a challenge to future workers in this field that this null hypothesis for the evolution of helping-at-the-nest by nonbreeders still cannot be rejected for either of these two extensively studied cooperative breeders.

12.6 SUMMARY

In this chapter we discuss the evolutionary consequences of delayed dispersal and helping by nonbreeders. We use three approaches, including (1) a comparison of the lifetime reproductive success of birds dispersing and breeding in their first year with that of birds remaining as nonbreeders for at least one year, (2) a simulation model of lifetime reproductive success as a function of the length of time birds remain as nonbreeders, and (3) a demographic model of the fitness effects of nonbreeding helpers on their own and their parents' fitness. All three analyses support the conclusion that dispersal is generally a superior alternative to nondispersal and helping, and consequently that ecological constraints have been paramount in the evolution of nondispersal and delayed breeding in the acorn woodpecker. Additional evidence supporting this conclusion is also summarized.

We also speculate on the role of indirect selection in the evolution of helping behavior by nonbreeders. In general, the contribution of indirect fitness resulting from nonbreeding to the lifetime fitness of birds appears to be small; nonetheless, as much as 57% of the gain in fitness received by individuals delaying dispersal comes from the indirect component of inclusive fitness. This indicates that indirect selection is important to the maintenance of this behavioral trait.

The major way nonbreeders enhance their indirect fitness is by increasing the survivorship of male breeders. Thus, it is possible that there is little selective advantage to helping-at-the-nest per se.

CHAPTER THIRTEEN

Sharing Nests and Mates: Costs and Benefits

Acorn woodpeckers live and reproduce in a remarkable variety of social situations. Within the Hastings population, individuals breed monogamously, polyandrously, polygynously, and polygynandrously, sometimes assisted by nonbreeding helpers (Chapter 1). How is this diversity of reproductive tactics maintained?

As we have seen in Chapters 6 and 7, each breeding arrangement and group composition carries with it a suite of effects on the major components of fitness. In this chapter we examine the costs and benefits of shared breeding. Ideally, this problem should be addressed using lifetime reproductive success of individuals pursuing different breeding strategies. However, our data on lifetime reproductive success are presently inadequate to address this issue directly (see Chapters 10 and 12).

Instead, we estimate lifetime reproductive success with a demographic model, similar to that employed for nonbreeders in section 12.4. This approach, which uses the life-table parameters derived in the previous chapters, allows us to estimate the fitness consequences of shared versus solitary breeding for both males and females.

13.1. TERRITORY EVICTION, COOPERATIVE DISPERSAL, AND SHARED BREEDING

We first consider two hypotheses for the occurrence of shared breeding in acorn woodpeckers. Both hypotheses propose that there is no inherent direct advantage accruing to birds that share mates. Rather, they suggest that shared breeding is a social strategy forced on acorn woodpeckers because of limited granaries, lack of marginal habitat, and intense competition for breeding opportunities (see Chapter 12).

Hypothesis 1: Individuals sharing breeding status are less likely to be evicted from their territories by intruders than are individuals breeding solitarily. This hypothesis assumes that, because of the intense competition for territories, two or more breeders of the same sex can defend and maintain a territory against competitors where one cannot. This hypothesis is plausible for several reasons: (1) territorial intrusions occur frequently in acorn woodpeckers, and defensive behavior is strongly sex-specific (Mumme and de Queiroz, 1985); (2) most territory defense is performed by breeders rather than nonbreeding helpers (Mumme and de Queiroz, 1985); (3) solitary breeders are occasionally evicted from their territories by units of cooperating intruders of the same sex (see section 3.2).

However, eviction of established breeders by intruders occurs only rarely. We have documented only two certain cases. Both involved a single breeding female that was evicted by a group of three cooperating intruders. In a third possible case, two immigrant females may have forced a single breeding female to leave her territory. These three cases constitute the only evidence of established solitary breeders being evicted from their territories in 160 one-female group-years and 93 one-male group-years documented through 1985. Thus, the vast majority of solitary breeders (98.1% of females, 100% of males) are able to maintain tenure on their territories. The above hypothesis, therefore, can be no more than a minor explanation for the phenomenon of shared breeding.

Hypothesis 2: Mate-sharing and joint-nesting are not directly adaptive but are indirect consequences of cooperative dispersal by unisexual sibling units. Breeding vacancies in acorn woodpeckers are frequently contested by groups of potential immigrants in power struggles (Koenig 1981a). Hannon et al. (1985) showed that power struggles are more likely to be won by unisexual units of cooperating siblings than by individuals that fight independently. Generally, most or all members of an immigrant sibling unit become reproductively active on their new territory and share breeding status (Koenig and Pitelka, 1979; Koenig et al., 1984; Hannon et al., 1985).

Thus, similar to delayed dispersal by offspring (see Chapter 12), intense competition for breeding space may force individuals to cooperate with siblings in order to acquire a breeding territory. Individuals then share reproductive status with these siblings because it is difficult or impossible to prevent them from breeding.

Although this hypothesis is attractive, we believe that it cannot be the entire explanation for the sharing of nests and mates by acorn woodpeckers. First, in many instances, individuals from winning units frequently return (or are forced to return) to their natal territory following a power struggle to resume their roles as nonbreeding helpers. Second, cooperative dispersal is not the only route by which acorn woodpeckers share nests and mates; established breeders that have already secured a territory frequently allow other individuals (usually offspring or other relatives) to acquire shared breeding status, either through territorial inheritance (Chapter 9) or independent dispersal (Koenig et al., 1984).

The relationship between cooperative dispersal and shared breeding is shown for both males and females in Table 13.1. Between 1979 and 1983, we documented 20 nesting attempts of groups following cooperative dispersal by a group of 2 or more immigrant males. At 19 of these (95%), at least some of the dispersing males shared mates in the subsequent nesting attempt. At only 1 nest (5%) did mate-sharing not subsequently occur. For females, 16 nesting attempts followed joint dispersal by 2 or more females. In 12 of these (75%), at least some of the dispersing females remained and nested jointly. At the remaining 4 nests (25%), only 1 female nested while her siblings returned to their natal territory to resume their roles as nonbreeding helpers (Table 13.1).

TABLE 13.1. The Relationship between Cooperative Dispersal and Shared Breeding in Acorn Woodpeckers

	♂♂	♀♀
PROXIMATE RESULT OF COOPERATIVE DISPERSAL		
Shared breeding	16 (80.0%)	10 (62.5%)
Partial shared breeding[a]	3 (15.0)	2 (12.5)
Solitary breeding	1 (5.0)	4 (25.0)
Total	20 (100)	16 (100)
PROXIMATE CAUSE OF SHARED BREEDING		
Cooperative dispersal	14 (26.4)	10 (52.6)
Partial cooperative dispersal[a]	1 (1.9)	2 (10.5)
Inheritance by group offspring	27 (50.9)	2 (10.5)
Independent dispersal	7 (13.2)	5 (26.3)
Cooperative dispersal and inheritance	1 (1.9)	0 (0)
Cooperative and independent dispersal	3 (5.7)	0 (0)
Total	53 (100)	19 (100)

[a] Some but not all members of the dispersing unit remained and shared breeding.

Thus, cooperative dispersal by both males and females often leads to mate-sharing and joint-nesting. However, Table 13.1 indicates that the converse relationship is much weaker. Of 53 cases of mate-sharing by breeding males, only 19 (35.8%) were attributable, even in part, to joint dispersal. Most instances of mate-sharing by males result from inheritance of the natal territory and shared breeding with their father or other relatives (Table 13.1, see also Chapter 9). For females, the relationship between joint dispersal and joint-nesting is stronger. Of 19 instances of joint-nesting, 12 (63.2%) were partially or entirely attributable to cooperative dispersal. The remaining 7 (36.8%) followed either independent dispersal or shared territorial inheritance.

We will consider the relationship between cooperative dispersal in sibling units and shared breeding further in section 13.4. Nonetheless, the above discussion suggests that although cooperative dispersal certainly contributes to the occurrence of mate-sharing and joint-nesting in acorn woodpeckers, it is not a completely satisfying explanation of these phenomena, particularly for males.

The alternative hypothesis is that there is a direct advantage to males sharing mates and females nesting jointly. In order to investigate this hypothesis, we estimate the lifetime reproductive success and inclusive fitness of individuals pursuing solitary and shared breeding strategies.

13.2. SHARING BREEDING WITH ONE OTHER INDIVIDUAL: DUOS

The Model

We first examine the selective consequences of breeding in duos. Following the conclusion, reached in section 13.1, that cooperative dispersal and shared breeding are partially independent, the model considers the situation in which a single male or female acorn woodpecker (the "primary" or "alpha" individual, α) has already successfully secured a breeding territory. The model addresses two questions: (1) Under what conditions should this established breeder allow a second bird (the "secondary" or "beta" individual, β) to join, or remain in, the group and share breeding status? (2) Would this secondary individual achieve higher fitness by joining the established breeder and sharing breeding status or by remaining on (or returning to) its natal territory as a nonbreeding helper and searching for an independent breeding opportunity?

Before we attempt to formulate answers to these two questions, we first define the following terms:

m_s = The mean annual reproductive success (fledged offspring per group) of groups in which males or females breed solitarily.

m_d = The mean annual reproductive success (fledged offspring per group) of groups containing two breeders of the same sex.

ℓ_s = The annual survivorship of individuals breeding solitarily.

ℓ_d = The annual survivorship of individuals that share breeding with one other bird of the same sex (breeding in a "duo").

ℓ_h = The annual survivorship of a nonbreeding helper living on its natal territory.

p = The probability of a nonbreeder in its natal group successfully acquiring a territory and breeding independently in any given year.

b_d = The bias in reproductive success of birds sharing breeding with one other individual. This bias is expressed as the proportion of the offspring of the communal nest parented by the primary (α) individual (see above). The proportion of the offspring parented by the secondary individual (β) is thus $(1 - b_d)$.

r = The coefficient of relatedness between birds sharing breeding.

We first consider the case of two birds, α and β. Assume that the α individual has secured a breeding territory and is alive at the beginning of the breeding season in year one. The β individual, in contrast, has not acquired breeding status and is living on its natal territory as a nonbreeder. Following our life-table analyses (Chapter 10), annual values of reproductive success and survivorship are not age-dependent. If the α individual breeds solitarily, its lifetime reproductive success ($W_{s\alpha}$) is

$$W_{s\alpha} = \sum_{i=1}^{n} m_s \ell_s^{(i-1)},$$

where n is the number of years of potential reproduction (designated as fifteen for acorn woodpeckers). This expression assumes

342

that the α individual survives through at least year one, so that its reproductive output in year one is m_s.

Now consider a second α individual that has also secured a territory and is about to begin breeding. Instead of breeding solitarily, however, this bird allows (or recruits) a second individual (β) to join the group and share breeding status as a duo. The lifetime reproductive success of the α individual sharing mates with one other bird is

$$W_{d\alpha} = \sum_{i=1}^{n} b_d m_d \ell_d^{(2i-2)} + \sum_{i=1}^{n-1} \sum_{j=1}^{i} m_s (1 - \ell_d) \ell_d^{(2j-1)} \ell_s^{(i-j)}.$$

The first term represents the lifetime production of offspring by α while both α and β are alive and sharing breeding. The second term incorporates the probability that the β individual will die after one or more years, making the remaining α individual a solitary breeder.

Now we consider the lifetime reproductive success of the β individual. We assume that β has not secured a breeding territory and is still living on its natal territory, assisting its parents and other relatives as a nonbreeding helper. If the β individual attempts to find an independent breeding opportunity on its own and breeds solitarily, its lifetime reproductive success can be estimated as

$$W_{s\beta} = \sum_{i=1}^{n} \sum_{j=1}^{i} pm_s (1 - p)^{(j-1)} \ell_h^{(j-1)} \ell_s^{(i-j)}.$$

The double summation is necessary because when independent breeding opportunities are restricted ($p < 1$), the β individual may be forced to delay breeding, spending one or more years as a nonbreeding helper before acquiring breeding space on its own.

Now consider the possibility that the β individual, instead of attempting to disperse and breed independently, is able to join the α individual on the latter's territory. In this situation, the β individual is not forced to delay breeding and there is no cost of dispersal. Thus, the lifetime reproductive success of β, sharing breeding as part of a duo with α, is

$$W_{d\beta} = \sum_{i=1}^{n} (1 - b_d) m_d \ell_d^{(2i-2)} + \sum_{i=1}^{n-1} \sum_{j=1}^{i} m_s (1 - \ell_d) \ell_d^{(2j-1)} \ell_s^{(i-j)}.$$

Like the expression for $W_{d\alpha}$, the first term represents the lifetime reproductive output of β while both α and β are alive and sharing breeding [devalued by $(1 - b_d)$ for β, and b_d for α]. The second term is the lifetime reproductive output of the β individual in the event of α's death, making β a solitary breeder.

These equations estimate the lifetime reproductive success of individuals pursuing solitary and shared breeding strategies. However, because cobreeding acorn woodpeckers are typically closely related (Koenig et al., 1984; see also Appendix A), estimates of lifetime reproductive success by themselves are inadequate to estimate the inclusive fitness of individuals pursuing different reproductive strategies. The next step, then, is to employ a modified version of Hamilton's rule (Hamilton, 1964; Michod, 1982; Grafen, 1984; Brown, 1987) to evaluate the selective costs and benefits of solitary and shared breeding.

First, however, we must perform an additional calculation. Our estimate of the lifetime reproductive success of a β individual that attempts to disperse and breed solitarily includes the probability that, when independent breeding opportunities are limited $(p < 1)$, β may spend one or more years as a nonbreeding helper. Although not contributing to its direct fitness during this time, β is nonetheless supplementing its indirect fitness by increasing the reproductive success and survivorship of the breeders it assists. As shown in Chapter 12, this increase amounts to 0.26 offspring equivalents for each year it delays dispersal. Thus, the overall indirect fitness gained by a β individual waiting for an opportunity to breed independently (I_{he}) is

$$I_{he} = \sum_{i=1}^{n} m_{he} (1 - p)^i \ell_h^{(i-1)},$$

344

where m_{he} is the increment in indirect fitness achieved by the β individual for each year it stays in its natal group. As discussed above (see also Chapter 12), $m_{he} = 0.26$ in acorn woodpeckers.

This increase in indirect fitness realized by the β individual while helping also has implications for the α individual. If, as is nearly always the case in acorn woodpeckers, the α individual is closely related to both β and to the birds β assists, α's indirect fitness may actually be reduced by allowing β to join its group and share breeding. If, instead, β remains in its natal territory as a nonbreeding helper until it finds an independent breeding opportunity, β enhances the lifetime indirect fitness of α by an amount corresponding to I_{hc}, where

$$I_{hc} = \sum_{i=1}^{n} m_{hc}(1-p)^i \ell_h^{(i-1)}.$$

In this expression, m_{hc} is the annual increment in indirect fitness received by potential cobreeders (that is, the α individual) through β's action as a nonbreeding helper.

We can now calculate the overall inclusive fitness of α and β individuals breeding solitarily and sharing breeding. If both breed solitarily, the relative inclusive fitness of the α and β individuals are, respectively,

$$IF_{s\alpha} = W_{s\alpha} + I_{hc} + rW_{s\beta}$$

and

$$IF_{s\beta} = W_{s\beta} + I_{he} + rW_{s\alpha}.$$

If α and β instead share breeding as a duo, their relative inclusive fitnesses are, respectively,

$$IF_{d\alpha} = W_{d\alpha} + rW_{d\beta}$$

and

$$IF_{d\beta} = W_{d\beta} + rW_{d\alpha}.$$

Thus, for the α individual, shared breeding would be selectively favored over solitary breeding if $IF_{d\alpha} > IF_{s\alpha}$, or

$$W_{d\alpha} + rW_{d\beta} > W_{s\alpha} + I_{hc} + rW_{s\beta}.$$

Similarly, for the β individual, shared breeding would be advantageous if $IF_{d\beta} > IF_{s\beta}$, or

$$W_{d\beta} + rW_{d\alpha} > W_{s\beta} + I_{he} + rW_{s\alpha}.$$

These equations specify the conditions under which sharing breeding with a second bird is favored by natural selection. Our next objective is to determine if these conditions are met by acorn woodpeckers at Hastings Reservation.

Application of the Model: Males

The values of the model parameters used to estimate lifetime reproductive success and relative inclusive fitness are summarized in Table 13.2. We assume that m_{hc}, the annual increment in indirect fitness received by a potential cobreeder as a consequence of β's actions as a nonbreeding helper, equals m_{he}, the annual increment in indirect fitness achieved by β while a nonbreeder in its natal group. We do this for the following reason. Consider a β individual that does not form a nest- or mate-sharing duo with a closely related α individual and, instead, spends one or more years assisting its parents as a nonbreeder. Because potential cobreeders (α and β) in acorn woodpeckers are usually close relatives, α is likely to be as closely related to the birds β assists as is β. Thus, α's indirect fitness increases to the same degree as does β's from the aid provided by β to their mutual relatives.

The results of the analysis for male acorn woodpeckers are shown in Table 13.3. Despite inferior annual reproductive success per male for males sharing mates (when $b_d = 0.5$, $m_d = 1.13$ fledged young for males sharing mates, and $m_s = 1.73$ fledged young for males breeding solitarily), the higher survivorship of males breeding in duos compared to males breeding solitarily (0.866 versus 0.703) more than compensates for the loss in annual breeding success. Both males in a mate-sharing duo can expect to produce 6.99 offspring over their lifetime, compared to 5.80 and 3.94 offspring for α and β breeding solitarily. In

346

TABLE 13.2. Values of Key Demographic Variables Used in Analysis of Costs and Benefits of Shared Breeding in Acorn Woodpeckers

Variable Symbol	Description	Values Used Males	Values Used Females	Reference
n	Breeding life span (years)	15	15	Chapter 7
r	Coefficient of relatedness between same-sexed breeders	0.45	0.45	Appendix A
m_s	Annual reproductive success of groups with 1 breeder	1.73 (1.84)	2.08 (2.08)	Tables 6.3, 6.6
m_d	Annual reproductive success of groups with 2 breeders	2.25 (2.31)	2.81 (2.64)	Tables 6.3, 6.6
m_t	Annual reproductive success of groups with 3 breeders	3.31 (2.92)	—	Table 6.3
m_{he}	Annual indirect fitness gain to ego from helping	0.26	0.26	Chapter 12
m_{hc}	Annual indirect fitness gain to cobreeders from helping	0.26	0.26	Chapter 12
ℓ_s	Annual survivorship of birds breeding singly	0.703 (0.731)	0.700 (0.706)	Table 7.6
ℓ_d	Annual survivorship of birds breeding in duos	0.866 (0.865)	0.792 (0.764)	Table 7.6
ℓ_t	Annual survivorship of birds breeding in trios	0.834	—	Table 7.6
ℓ_h	Annual survivorship of nonbreeding helpers	0.90	0.90	Section 8.4
p	Annual probability of nonbreeders acquiring independent breeding space	0.19	0.34	Table 8.10
b_d	Bias in reproductive success of duo nests	0.50	0.50	Section 10.3
b_t	Bias in reproductive success of trio nests	0.33	—	Section 10.3

NOTE: Values of reproductive success and survivorship corrected by ANOVA are shown in parentheses.

TABLE 13.3. Lifetime Reproductive Success and Total Fitness of α and β Male Acorn Woodpeckers Breeding Singly and Breeding as Part of a Mate-Sharing Duo, Using Observed Values of Reproductive Success and Survivorship

Component of Fitness	Breeding Singly		Breeding in Duo	
	α	β	α	β
Lifetime reproductive success	5.80	3.94	6.99	6.99
Offspring equivalents via helping by β male	0.78	0.78	—	—
Offspring equivalents via breeding by cobreeder	1.77	2.60	3.14	3.14
Total fitness	8.35	7.32	10.13	10.13
TOTAL FITNESS IF . . .				
α and β unrelated ($r = 0$, $m_{hc} = 0$)	5.80	4.72	6.99	6.99
Breeding opportunities unrestricted ($p = 1$)	8.40	8.40	10.13	10.13
Strong reproductive bias ($b_d = 0.75$)	8.35	7.32	11.35	8.91

terms of total inclusive fitness, the advantage for mate-sharing males is equally pronounced (Table 13.3).

Thus, male acorn woodpeckers appear to benefit directly from living and breeding in mate-sharing duos. This benefit is measurable in both lifetime reproductive success and inclusive fitness. The primary factor contributing to this advantage is the higher survivorship of males living in groups with two or more breeding males (section 7.4).

In fact, the main problem posed by the results presented in Table 13.3 is why males should ever *not* share mates. Even when the α and β individuals are unrelated ($r = 0$ and $m_{hc} = 0$), when independent breeding opportunities are unrestricted ($p = 1$), and when the reproductive success of shared nests is biased strongly in favor of α ($b_d = 0.75$), shared breeding remains advantageous for both the α and β individuals (Table 13.3). However, 45.6% of groups contain only a single breeding male and 25.8% of all breeding males breed solitarily (Table 3.4). Furthermore, mate-sharing coalitions between unrelated males

348

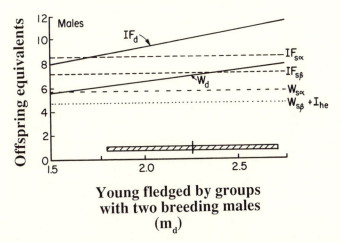

**Young fledged by groups
with two breeding males
(m_d)**

FIGURE 13.1. Estimated lifetime reproductive success (W) and inclusive fitness (IF) of α and β males breeding in a mate-sharing duo (subscript "d") and breeding solitarily (subscript "s") in relation to the annual reproductive success of groups with two breeding males (m_d). W_d and IF_d are identical for the α and β male. The 95% confidence limit around the estimate of m_d (2.25) is shown. For the β male breeding solitarily, we show the sum of his lifetime reproductive success ($W_{s\beta}$) and the amount of indirect fitness he receives as a helper (I_{he}) while waiting for the opportunity to acquire breeding space and breed independently.

are rare in the Hastings population (only one case observed). Why do these solitary-breeding males not join and breed with other males?

Errors in our estimates of key fitness parameters could provide one possible explanation. As shown in Figures 13.1 and 13.2, small changes in the reproductive success and survivorship of duo-breeding males (m_d and ℓ_d) relative to the reproductive success and survivorship of solitary-breeding males (m_s and ℓ_s) result in large changes in lifetime reproductive success and inclusive fitness. However, within the 95% confidence limits of our estimate of m_d, mate-sharing duos consistently outproduce lone males (Figure 13.1). For survivorship, only at the extreme lower limit of the confidence interval around our estimate of ℓ_d are solitary males doing as well as mate-sharing males (Figure 13.2). Thus, our conclusion that males breeding

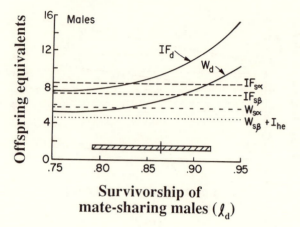

FIGURE 13.2. Estimated lifetime reproductive success (W) and inclusive fitness (IF) of α and β males breeding in a mate-sharing duo (subscript "d") and breeding solitarily (subscript "s") in relation to annual survivorship of males in mate-sharing duos (ℓ_d). W_d and IF_d are identical for the α and β male. The 95% confidence limit around the estimate of ℓ_d (0.866) is shown. $W_{s\beta}$ and I_{he} as in Figure 13.1.

in mate-sharing duos have higher lifetime reproductive success and inclusive fitness than males breeding solitarily is sound.

A second possible resolution of the paradox suggested by Table 13.3 is that our estimates of the annual reproductive success and survivorship of solitary and duo-breeding males are confounded by correlations with other variables. For example, males breeding in mate-sharing duos may have higher survivorship than males breeding solitarily because groups with two breeding males are more likely to contain nonbreeders than groups with one breeding male. Because nonbreeders greatly enhance the survivorship of males (section 7.4), the estimates of male survivorship used in Table 13.2 could be biased. Similar biases may interfere with the reproductive success data (see Chapter 6).

To correct for these potential biases, annual reproductive success of solitary and mate-sharing males was also estimated from the multiple classification analysis of a four-way ANOVA in

350

TABLE 13.4. Lifetime Reproductive Success and Total Fitness of α and β Male Acorn Woodpeckers Breeding Singly and Breeding as Part of a Mate-Sharing Duo, Using Corrected Values of Reproductive Success and Survivorship

	Breeding Singly		Breeding in Duo	
Component of Fitness	α	β	α	β
Lifetime reproductive success	6.78	4.58	7.48	7.48
Offspring equivalents via helping by β male	0.78	0.78	—	—
Offspring equivalents via breeding by cobreeder	2.06	3.05	3.36	3.36
Total fitness	9.62	8.41	10.84	10.84
TOTAL FITNESS IF . . .				
α and β unrelated ($r = 0$, $m_{hc} = 0$)	6.78	5.35	7.48	7.48
Breeding opportunities unrestricted ($p = 1$)	9.83	9.83	10.84	10.84
Strong reproductive bias ($b_d = 0.75$)	9.62	8.41	12.08	9.59

which three group composition variables (the number of breeding males, breeding females, and nonbreeders) and prior group history (breeder turnover) were factors. The resulting corrected estimates of m_s and m_d are the average annual reproductive success of groups with one and two breeding males, respectively, controlling for the effects of the other variables (see Chapter 12). Similarly, annual survivorship of solitary-breeding and duo-breeding males was estimated from the log-linear parameters of a logit model in which the number of breeding males, breeding females, and adult nonbreeding helpers were factors. The corrected values of reproductive success and survivorship are shown in parentheses in Table 13.2.

Using these corrected values, we recalculated lifetime reproductive success and inclusive fitness. Again, however, males in mate-sharing duos have substantially higher lifetime reproductive success and inclusive fitness than males breeding singly (Table 13.4). Again, the advantage of mate-sharing holds even if the males are unrelated ($r = 0$, $m_{hc} = 0$), if independent breeding opportunities are unrestricted ($p = 1$), and if reproductive success of mate-sharing males is strongly biased in favor of the α individual ($b_d = 0.75$; Table 13.4).

Thus, we are faced with a paradox: even though our estimates of lifetime fitness suggest that males sharing mates do better than solitary males under all reasonable circumstances, over a quarter of all males breed solitarily.

Application of the Model: Females

Estimated lifetime fitness of solitary and joint-nesting females is shown in Table 13.5. The results show that females initially breeding in a joint-nesting duo have lifetime reproductive success and inclusive fitness comparable to that of females breeding solitarily. Both the α and β female of a joint-nesting duo can expect to produce 6.77 offspring during their lifetimes compared

TABLE 13.5. Lifetime Reproductive Success and Total Fitness of α and β Female Acorn Woodpeckers Nesting Singly and Nesting as Part of a Joint-Nesting Duo, Using Observed Values of Reproductive Success and Survivorship

	Nesting Singly		Nesting in Duo	
Component of Fitness	α	β	α	β
Lifetime reproductive success	6.90	5.74	6.77	6.77
Offspring equivalents via helping by β female	0.43	0.43	—	—
Offspring equivalents via breeding by cobreeder	2.58	3.10	3.04	3.04
Total fitness	9.91	9.27	9.81	9.81
TOTAL FITNESS IF . . .				
α and β unrelated ($r = 0$, $m_{hc} = 0$)	6.90	6.17	6.77	6.77
Breeding opportunities unrestricted ($p = 1$)	10.01	10.01	9.81	9.81
Strong reproductive bias ($b_d = 0.75$)	9.91	9.27	10.85	8.78

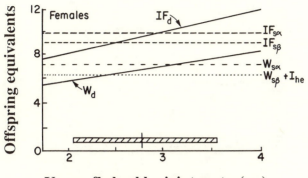

Young fledged by joint nests (m_d)

FIGURE 13.3. Estimated lifetime reproductive success (W) and inclusive fitness (IF) of α and β females breeding in a joint-nesting duo (subscript "d") and breeding solitarily (subscript "s") in relation to the annual reproductive success of groups with two joint-nesting females (m_d). W_d and IF_d are identical for the α and β female. The 95% confidence limit around our estimate of m_d (2.81) is shown. For the β female breeding solitarily, we show the sum of her lifetime reproductive success $(W_{s\beta})$ and the amount of indirect fitness she receives as a helper (I_{he}) while waiting for the opportunity to acquire breeding space and breed independently.

to 6.90 offspring for the α female and 5.74 offspring for the β female if both breed independently. Inclusive fitness values are also similar (Table 13.5).

However, our conclusion that the fitness of solitary and joint-nesting females is comparable must be taken with caution. Because only 21.6% of groups contain two or more breeding females and only 35.5% of breeding females nest jointly (Figure 3.6), our estimates of reproductive success and annual survivorship of joint-nesting females are based on relatively small samples (Tables 6.4 and 7.6). Thus, the 95% confidence limits around our estimates are large (Figures 13.3 and 13.4). Within these confidence limits of m_d and ℓ_d, it is possible that joint-nesting could be disadvantageous for both the α and β females, advantageous for β but disadvantageous for α, or advantageous for both females (Figures 13.3 and 13.4).

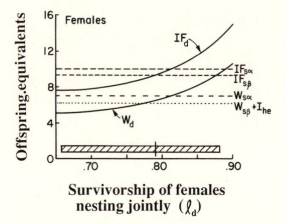

FIGURE 13.4. Estimated lifetime reproductive success (W) and inclusive fitness (IF) of α and β females breeding in a joint-nesting duo (subscript "d") and breeding solitarily (subscript "s") in relation to annual survivorship of joint-nesting females (ℓ_d). W_d and IF_d are identical for the α and β female. The 95% confidence limit around our estimate of ℓ_d (0.792) is also shown. $W_{s\beta}$ and I_{he} as in Figure 13.3.

Table 13.5 also illuminates several other aspects of female reproductive behavior. For example, if independent breeding opportunities are unrestricted ($p = 1$), females nesting solitarily have higher fitness than females nesting jointly. Thus, females should be better off breeding solitarily whenever independent breeding opportunities are available. Second, if the reproductive output of joint nests are strongly biased in favor of the α female ($b_d \gg 0.5$), the β female achieves higher fitness by breeding solitarily. Thus, joint-nesting is likely to occur in acorn woodpeckers only if reproductive bias is weak and independent breeding opportunities limited.

Third, although a β female that has not secured a breeding territory achieves higher fitness if she forms a joint-nesting association with any established female, α females suffer a slight loss in relative fitness by forming coalitions with nonrelatives. Thus, joint-nesting between nonrelatives should be rare. All three of these expectations are met by female acorn woodpeckers

TABLE 13.6. Lifetime Reproductive Success and Total Fitness of α and β Female Acorn Woodpeckers Nesting Singly and Nesting as Part of a Joint-Nesting Duo, Using Corrected Values of Reproductive Success and Survivorship

Component of Fitness	Nesting Singly		Nesting in Duo	
	α	β	α	β
Lifetime reproductive success	7.04	5.85	6.18	6.18
Offspring equivalents via helping by β female	0.43	0.43	—	—
Offspring equivalents via breeding by cobreeder	2.63	3.16	2.78	2.78
Total fitness	10.10	9.44	8.96	8.96
TOTAL FITNESS IF . . .				
α and β unrelated ($r = 0$, $m_{hc} = 0$)	7.04	6.28	6.18	6.18
Breeding opportunities unrestricted ($p = 1$)	10.21	10.21	8.96	8.96
Strong reproductive bias ($b_d = 0.75$)	10.10	9.44	9.83	8.09

at Hastings Reservation (Koenig and Pitelka, 1981; Mumme et al., 1983b; Hannon et al., 1985; see also Chapter 3 and Appendix A).

As in our analyses for mate-sharing males, we adjusted our estimates of reproductive success and survivorship via ANOVA and logit analyses. These corrected values are shown in parentheses in Table 13.2. The recalculated estimates of fitness for solitary and joint-nesting females are shown in Table 13.6.

Using the corrected estimates, females nesting jointly are at a slight disadvantage compared to those nesting solitarily (Table 13.6). Thus, other things being equal, joint-nesting is apparently inferior to solitary nesting in terms of both lifetime reproductive success and inclusive fitness. However, because other things are *not* equal, joint-nesting females actually have roughly the same fitness as solitary females (Table 13.5). We will return to this issue in section 13.4.

355

13.3. SHARING BREEDING WITH TWO
OTHER INDIVIDUALS: TRIOS

The Model

Thus far, we have considered males and females that share their breeding status with only one other individual. However, 35.3% of all breeding males are part of a unisexual trio that shares one or two breeding females (Chapter 3). Here, then, we focus on the costs and benefits of males breeding in mate-sharing trios relative to those of males breeding singly or in duos. Although females may also form nest-sharing coalitions of three or more breeders, such groups are quite rare (Chapter 3). Thus, our discussion will concern only males.

The model we use to estimate lifetime reproductive success and inclusive fitness of males breeding in trios is an extension of the demographic model developed in the previous section for mate-sharing duos. As before, first consider a single acorn woodpecker (the α individual) that has successfully secured a breeding territory, and two other individuals (the β and γ individuals) that are still living as nonbreeding helpers on their natal territory. We address two questions: (1) Under what conditions should the α individual allow a second or third bird to join the group and form a mate-sharing duo or trio? (2) Would the β and γ individuals achieve higher fitness by forming a duo or trio with α, or by remaining in their natal territories as nonbreeding helpers, where they search for independent breeding opportunities on their own? Additional variables needed to address these questions include:

m_t = The mean annual reproductive success (fledged offspring per group) of groups containing three breeding birds of the same sex.

ℓ_t = The annual survivorship of individuals that share breeding with two birds of the same sex, thus forming a mate-sharing trio.

$b_t =$ The bias in reproductive success of mate-sharing trios, expressed as the proportion of the offspring of the trio nest that is produced by the α individual. For simplicity, we assume that the proportion of the offspring produced by the β and γ individuals are the same and therefore equivalent to $(0.5)(1 - b_t)$.

The lifetime reproductive success of the α individual, $W_{t\alpha}$, given that he initially breeds in a trio, is the sum of four separate terms:

$$W_{t\alpha} = T_{t\alpha} + T_{d\alpha} + T_{st} + T_{sd},$$

where

$$T_{t\alpha} = \sum_{i=1}^{n} b_t m_t \ell_t^{(3i-3)},$$

$$T_{d\alpha} = \sum_{i=1}^{n-1} \sum_{j=1}^{i} 2b_d m_d \ell_t^{(3j-1)}(1 - \ell_t)\ell_d^{(2i-2j)},$$

$$T_{st} = \sum_{i=1}^{n-1} \sum_{j=1}^{i} m_s \ell_t^{(3j-2)}(1 - \ell_t)^2 \ell_s^{(i-j)},$$

and

$$T_{sd} = \sum_{i=1}^{n-2} \sum_{j=1}^{i} \sum_{k=1}^{j} 2m_s \ell_t^{(3k-1)}(1 - \ell_t)\ell_d^{(2j-2k+1)}(1 - \ell_d)\ell_s^{(i-j)}.$$

$T_{t\alpha}$ is the lifetime reproductive output of the α individual while all three birds are alive. $T_{d\alpha}$ represents the possibility that the β or γ individual will die after one or more years, leaving a mate-sharing duo. If the β and γ individual die during the same year, α becomes a solitary breeder. T_{st} calculates the reproductive output of α under these conditions. The final term, T_{sd}, incorporates the probability that α spends one or more years breeding as a duo with either β or γ, and then outlives β or γ and breeds solitarily for the remainder of its life.

For the β individual that breeds initially in a trio, its lifetime reproductive success is

$$W_{t\beta} = T_{t\beta} + T_{d\beta} + T_{st} + T_{sd},$$

357

where

$$T_{t\beta} = \sum_{i=1}^{n} 0.5(1 - b_t)m_t\ell_t^{(3i-3)}$$

and

$$T_{d\beta} = \sum_{i=1}^{n-1} \sum_{j=1}^{i} m_d\ell_t^{(3j-1)}(1 - \ell_t)\ell_d^{(2i-2j)}.$$

In the expression for $T_{d\beta}$, we assume that if the α individual dies, the proportion of young of the duo nest parented by β becomes b_d. If, on the other hand, the γ individual dies, β's contribution to the communal nest is $(1 - b_d)$.

The lifetime reproductive success of the γ individual breeding initially in a trio is

$$W_{t\gamma} = T_{t\gamma} + T_{d\gamma} + T_{st} + T_{sd},$$

where

$$T_{t\gamma} = \sum_{i=1}^{n} 0.5(1 - b_t)m_t\ell_t^{(3i-3)}$$

and

$$T_{d\gamma} = \sum_{i=1}^{n-1} \sum_{j=1}^{i} 2(1 - b_d)m_d\ell_t^{(3j-1)}(1 - \ell_t)\ell_d^{(2i-2j)}.$$

We now use these expressions to calculate the relative inclusive fitness of the α, β, and γ individuals pursuing trio-breeding, duo-breeding, and solitary breeding strategies. If all three nest as a trio, the relative inclusive fitness values of the α, β, and γ individuals, respectively, are

$$IF_{t\alpha} = W_{t\alpha} + rW_{t\beta} + rW_{t\gamma},$$
$$IF_{t\beta} = W_{t\beta} + rW_{t\alpha} + rW_{t\gamma},$$

and

$$IF_{t\gamma} = W_{t\gamma} + rW_{t\alpha} + rW_{t\beta}.$$

In contrast, if the α and β females breed as a duo, and the γ breeds solitarily, relative inclusive fitness for the α individual is

$$IF_{d\alpha} = W_{d\alpha} + rW_{d\beta} + rW_{d\gamma} + I_{hc},$$

where $W_{d\gamma} = W_{s\beta}$ (see section 13.2). For the β and γ individuals, relative inclusive fitness is, respectively,

$$IF_{d\beta} = W_{d\beta} + rW_{d\alpha} + rW_{d\gamma} + I_{hc}$$

and

$$IF_{d\gamma} = W_{d\gamma} + rW_{d\alpha} + rW_{d\beta} + I_{he}.$$

If all three birds breed solitarily, values of relative inclusive fitness of the α, β, and γ individuals are, respectively,

$$IF_{s\alpha} = W_{s\alpha} + rW_{s\beta} + rW_{s\gamma} + 2I_{hc}$$

(where $W_{s\gamma} = W_{s\beta} = W_{d\gamma}$),

$$IF_{s\beta} = W_{s\beta} + rW_{s\alpha} + rW_{s\gamma} + I_{he} + I_{hc},$$

and

$$IF_{s\gamma} = W_{s\gamma} + rW_{s\alpha} + rW_{s\beta} + I_{he} + I_{hc}.$$

We can now use the above expressions to estimate the fitness consequences of each breeding strategy.

Application of the Model: Males

The values of parameters used in the analysis of mate-sharing by male trios are shown in Table 13.2. As discussed above, we estimated lifetime fitness using both the observed values of reproductive success and survivorship and values adjusted for the effects of the other variables using ANOVA and logit analyses.

The results of these analyses are summarized in Table 13.7 and 13.8. Based on the observed values of male reproductive output and annual survivorship, males breeding in mate-sharing trios are more successful than either duo-breeding or solitary males (Table 13.7). This analysis also demonstrates that the

TABLE 13.7. Lifetime Reproductive Success and Total Fitness of α, β, and γ Male Acorn Woodpeckers Breeding Singly, Breeding as a Mate-Sharing Duo, and Breeding as a Mate-Sharing Trio, Using Observed Values of Reproductive Success and Survivorship

Component of Fitness	Breeding Singly			Breeding in Duo			Breeding in Trio		
	α	β	γ	α	β	γ	α	β	γ
Lifetime reproductive success	5.80	3.94	3.94	6.99	6.99	3.94	7.06	7.06	7.06
Offspring equivalents via helping by β and γ	1.56	1.56	1.56	0.78	0.78	0.78	—	—	—
Offspring equivalents via cobreeders	3.54	4.38	4.38	4.91	4.91	6.29	6.36	6.36	6.36
Total fitness	10.90	9.88	9.88	12.68	12.68	11.01	13.42	13.42	13.42
TOTAL FITNESS IF . . .									
Males unrelated ($r = 0$, $m_{hc} = 0$)	5.80	4.72	4.72	6.99	6.99	4.72	7.06	7.06	7.06
Breeding opportunities unrestricted ($p = 1$)	11.01	11.01	11.01	12.74	12.74	12.08	13.42	13.42	13.42
Strong reproductive bias ($b_d = 0.75$, $b_t = 0.75$)	10.90	9.88	9.88	13.90	11.46	11.01	16.16	12.38	11.71

TABLE 13.8. Lifetime Reproductive Success and Total Fitness of α, β, and γ Male Acorn Woodpeckers Breeding Singly, Breeding as a Mate-Sharing Duo, and Breeding as a Mate-Sharing Trio, Using Corrected Values of Reproductive Success and Survivorship

Component of Fitness	Breeding Singly			Breeding in Duo			Breeding in Trio		
	α	β	γ	α	β	γ	α	β	γ
Lifetime reproductive success	6.78	4.58	4.58	7.48	7.48	4.58	6.63	6.63	6.63
Offspring equivalents via helping by β and γ	1.56	1.56	1.56	0.78	0.78	0.78	—	—	—
Offspring equivalents via cobreeders	4.12	5.11	5.11	5.42	5.42	6.73	5.96	5.96	5.96
Total fitness	12.46	11.25	11.25	13.68	13.68	12.09	12.59	12.59	12.59
TOTAL FITNESS IF . . . Males unrelated ($r = 0$, $m_{he} = 0$)	6.78	5.36	5.36	7.48	7.48	5.36	6.63	6.63	6.63
Breeding opportunities unrestricted ($p = 1$)	12.88	12.88	12.88	13.89	13.89	13.50	12.59	12.59	12.59
Strong reproductive bias ($b_d = 0.75$, $b_t = 0.75$)	12.46	11.25	11.25	14.92	12.44	12.09	14.85	11.79	11.12

advantage of sharing breeding in trios holds even if mate-sharing males are unrelated ($r = 0$ and $m_{hc} = 0$), if independent breeding opportunities are unrestricted ($p = 1$), and if the reproductive success of males sharing mates is strongly biased in favor of the α male ($b_d = 0.75$ and $b_t = 0.75$; Table 13.7).

The above conclusions are altered if the values of reproductive success and survivorship are statistically controlled for the effects of other variables. As shown in Table 13.8, when these new values are used, α and β males breeding in mate-sharing trios do worse than α and β males breeding in duos. The γ male, however, would achieve higher fitness by forming a trio with α and β than by breeding solitarily while α and β bred as a duo (Table 13.8). Thus, for a coalition of three males, our results suggest a possible conflict of interest (Emlen, 1982b). Other things being equal, we would expect α and β to form a mate-sharing duo and exclude γ's attempts to acquire shared breeding status.

In conclusion, our results suggest that males breeding in mate-sharing trios have higher fitness than solitary-breeding or duo-breeding males. However, when the effects of confounding variables are statistically controlled, males breeding in mate-sharing trios have fitness greater than that of solitary-breeding males, but less than that of duo-breeding males.

13.4. WHY SHARE BREEDING?

Shared versus Solitary Breeding

Why do acorn woodpeckers share breeding opportunities with others? Our analyses suggest that the answer differs for males and females and for groups with different numbers of mate-sharing individuals. Nonetheless, a few general conclusions can be drawn.

First, males that breed in mate-sharing duos achieve higher lifetime fitness than males that breed solitarily. This advantage

is not simply an artifact of other variables correlated with reproductive success and survivorship. When the effects of potentially
confounding variables are statistically controlled, duo-breeding
males still achieve higher fitness than solitary males.

Second, males are not forced to share mates in order to avoid
being evicted from ecologically restricted breeding territories by
coalitions of competitors (e.g., Craig, 1984). In 93 group-years,
we have never observed a single male breeder being evicted by
a coalition of male intruders.

Nor is mate-sharing by males simply an indirect consequence
of selection for cooperative dispersal. Although cooperative dispersal is an important component of mate-sharing, only about
one-third of all instances of mate-sharing by males are a direct
or indirect result of cooperative dispersal (Table 13.1). Most
instances of mate-sharing are attributable to either territorial inheritance or independent dispersal, whereby male acorn woodpeckers join (or remain with) one or more established breeding
males with whom breeding is then subsequently shared.

For females our conclusions are quite different. Female duos
have about the same overall fitness as female singletons (Table
13.5). However, after controlling for confounding variables,
joint-nesting is inferior to solitary nesting (Table 13.6).

Thus, the explanation for joint-nesting by female acorn woodpeckers lies primarily in factors correlated with the presence of
two breeding females rather than joint-nesting per se. The most
important of these related factors is probably the intense competition among females for breeding vacancies and the increased
competitive ability of coalitions of females in acquiring and
defending high-quality breeding territories. We base this conclusion on three types of evidence:

1. Cooperative units of dispersing females are more likely to
win power struggles and successfully acquire breeding territories
than are singletons (Hannon et al., 1985). Power struggles are
particularly intense for female vacancies (Koenig, 1981a), and
the intensity of power struggles is positively correlated with
territory quality (Hannon et al., 1985).

2. Nearly two-thirds (63.2%) of all joint nests are partially or entirely attributable to cooperative dispersal. This is almost twice the proportion of cases of mate-sharing by males attributable to this phenomenon (35.8%; Table 13.1). Although joint-nesting is not an obligate byproduct of cooperative dispersal, it is possible that joint-nesting by females is largely a consequence of cooperative dispersal.

3. Although established breeders are only rarely evicted from their territories, all the observed cases involve single females being evicted by a cooperative unit of two or more females. We have never observed eviction of male breeders (section 13.1).

Based on these data, we propose that females who disperse and breed in sibling units are more likely to acquire and maintain a breeding position in a high-quality territory (Hannon et al., 1985) with a larger number of breeding males (Chapter 3) and where the females will have higher reproductive success (Chapters 5 and 6) and perhaps higher survivorship (Chapter 7). Once such a position has been acquired, however, such females could still achieve higher fitness by excluding their siblings and breeding solitarily (Table 13.6). Presumably this does not occur because the costs of preventing other females from breeding generally exceed the potential benefits.

There is another possible explanation for the occurrence of joint-nesting by females despite the apparently lower fitness such females suffer: sexual conflict. As suggested by Davies (1985; Davies and Lundberg, 1984; Davies and Houston, 1986) for the similarly variable mating system of the dunnock, such variation may result when there are conflicting costs and benefits to the differing strategies for each sex. In the dunnock, for example, individuals do best by nesting singly and by being shared by several individuals of the opposite sex. A similar result holds for female acorn woodpeckers: although females appear to have higher fitness nesting singly, the slightly higher reproductive success of groups containing more than one female (section 6.2), means that it may be to the breeding males advantage to share two joint-nesting females rather than a single female. Such conflicting interests could very well be important in producing the observed diversity in breeding systems. A similar suggestion has been made by Brown (1987).

For males that breed in mate-sharing trios, our conclusions are similar but not identical to those of joint-nesting females. The fitness of trio-breeding males is slightly higher than that of duo-breeding males, and considerably higher than that of solitary-breeding males (Table 13.7). However, when the effects of confounding variables are removed (Table 13.8), the advantage of breeding in trios compared to duos is eliminated, although trios still outperform males breeding solitarily.

Thus, as with joint-nesting females, a major component of fitness accruing to males breeding in mate-sharing trios is related not to trio breeding per se, but to factors correlated with trio breeding (e.g., greater numbers of breeding females and nonbreeders). However, because cooperative dispersal is not the predominant route by which males form mate-sharing coalitions (Table 13.1), the benefits of cooperative dispersal, described above for females, can only play a small role in selecting for mate-sharing trios. However, sexual conflict may still be important: mate-sharing male trios produce more total offspring than duos (Figure 6.3) and thus should be favored by breeding females. Hence, female preference for larger coalitions of male breeders could help to explain why large sets of mate-sharing males are relatively common despite the apparent disadvantage of trios compared to duos.

Our analyses leave us with a paradox: if the fitness of males breeding solitarily is inferior to those breeding in mate-sharing duos or trios, why do a quarter of males breed solitarily? Ironically, this paradox is the opposite of that originally suggested by data on annual reproductive success, in which males breeding solitarily produce more offspring than males in mate-sharing duos (section 6.2).

Unfortunately, the hypothesis of sexual conflict fails to resolve this dilemma. As discussed above, total group reproductive success increases with additional males; consequently, females gain by breeding with two or more mate-sharing males rather than a singleton. Thus, this pattern exacerbates, rather than negates, the problem posed by the existence of a substantial proportion of singleton male breeders.

Our calculations suggest that even unrelated males breeding solitarily would profit by forming mate-sharing units. Coalitions composed of unrelated males are virtually nonexistent at Hastings. Interestingly, however, coalitions of unrelated males may be common in at least some populations of acorn woodpeckers, as suggested by Stacey's (1979b) work in Water Canyon, New Mexico. In contrast, Stacey did not observe females nesting jointly in the New Mexico population. Thus, although the results of our analyses conflict with the observed behavior of acorn woodpeckers at Hastings, it is perhaps significant that the expected results—namely that, other things being equal, females should not share mates, whereas males should do so even if it means cobreeding with an unrelated individual—are observed elsewhere within their range.

Equitable versus Biased Breeding

Here we have tried to answer the question "Why do acorn woodpeckers share breeding opportunities with others of the same sex rather than breed solitarily?" Vehrencamp (1979, 1983) and Stacey (1982) divide this question into two related questions: (1) Why do unisexual coalitions of potential breeders form? (2) Why is breeding shared more or less equitably among breeders rather than biased strongly in favor of a single individual?

The distinction between these two questions can be made clear by contrasting the social organization of the California acorn woodpecker with that of the green woodhoopoe, an African cooperative breeder (Ligon and Ligon 1978a, 1978b, 1982; Ligon, 1981). Like acorn woodpeckers, breeding vacancies in green woodhoopoe groups are often filled by unisexual

366

sibling units. All members of these unisexual units are capable of reproduction. However, unlike acorn woodpeckers, only the dominant individual of each woodhoopoe unit breeds: neither joint-nesting nor mate-sharing has been observed. Thus, why are green woodhoopoes despotic and acorn woodpeckers egalitarian?

Vehrencamp (1979, 1983) and Stacey (1982) propose that the answer may be determined by the costs and benefits of forming unisexual coalitions and by the reproductive options available to subordinate individuals. For example, Vehrencamp's (1983) analysis assumes that the degree of reproductive bias is controlled by a dominant individual, and that this dominant individual will enforce the amount of bias that maximizes its inclusive fitness. However, as she notes, an unknown difficulty in the analysis is the "cost of despotism": how difficult is it for dominant individuals to enforce strong reproductive bias on subordinates? And how frequently do the costs of enforcing bias outweigh the potential benefits?

The difficulty of enforcing reproductive bias may in part explain the relatively nondespotic nature of joint-nesting in the acorn woodpecker. In many species of cooperative breeders in which only one female breeds per group, only that female incubates. Furthermore, incubating females in some species are fed on the nest by breeding males (e.g., the Florida scrub jay: Woolfenden and Fitzpatrick, 1984). Because females in these species are constantly on the nest during egg laying and incubation, it is relatively easy for them to deny other female group members access to the nest and thereby enforce singular nesting.

As in all woodpeckers, however, both sexes of acorn woodpeckers incubate and males perform virtually all nocturnal incubation (MacRoberts and MacRoberts, 1976; Short, 1982; Koenig et al., 1983). Nor are incubating females fed on the nest. Thus, it is probably difficult, if not impossible, for female acorn woodpeckers to completely deny other female group members access to the nest during egg laying and incubation.

The degree of difficulty with which females can maintain exclusive access to the nest may be a factor in explaining the occurrence of joint-nesting in other species as well: for example, both sexes incubate in the groove-billed ani (Vehrencamp, 1977; 1978), pukeko (Craig, 1980), and white-winged chough (Rowley, 1978). At least two exceptions to this hypothesis exist, however: both the brown jay (Lawton and Lawton, 1985) and the grey-crowned babbler (Counsilman, 1977a; Brown and Brown, 1981b) exhibit joint-nesting and predominantly female incubation. Furthermore, this hypothesis provides, at best, only a proximate answer to the problem of joint-nesting; it does not explain why females in joint-nesting species have not been selected to defend their nests more vigorously, thereby potentially thwarting egg laying by additional females.

In joint-nesting acorn woodpeckers, aggression and interference, including destruction of rival eggs, is intense during egg laying. This behavior notwithstanding, behaviorally dominant females usually do not prevent (or are unable to prevent) subordinates from laying eggs, and the observed reproductive bias between joint-nesting females is small (Mumme et al., 1983b).

In contrast with females, estimating the degree of reproductive bias among mate-sharing males is difficult (e.g., Joste et al., 1985; Mumme et al., 1985). Conceivably, however, in groups with just one breeding female, dominant males could enforce despotism through a combination of mate guarding (Mumme et al., 1983a) and intense aggression toward subordinates. Our analyses also indicate that dominant males could impose considerable reproductive bias before it would be to the advantage of subordinates to leave the group.

How much bias dominant males actually impose is unknown. From behavioral data it appears that, with some exceptions (Joste et al., 1982), breeding males may contribute more or less equally to parentage of offspring (Stacey, 1979b; Mumme et al., 1983a; for a discussion of the difficulties in estimating gamete contributions of males in mate-sharing species, see Craig and

Jamieson, 1985). Experimental data also support the hypothesis that reproductive bias between cobreeding males cannot be absolute if nesting is to be successful (Koenig, unpublished data). Unfortunately, attempts at answering this problem using starch-gel electrophoresis have been unsuccessful (Joste et al., 1985; Mumme et al., 1985). As discussed earlier (section 2.2), we consider it unlikely that a satisfying resolution to this problem will be made until better genetic techniques are developed.

The one situation in which we do know the degree of bias is for offspring in their natal group. As described in Chapter 1 and elsewhere (Koenig and Pitelka, 1979; Koenig et al., 1984), breeding status of both males and females within acorn woodpecker groups is generally determined by a fairly simple behavioral rule: "If the breeding individuals of the opposite sex are parents or other close relatives, do not breed. If the opposite-sexed breeders are unrelated, breed." Although we have observed several exceptions to this incest avoidance rule (see Koenig et al., 1984), it nonetheless is a useful generalization (Mumme et al., 1983a; Hannon et al., 1985; for contrasting views, see Shields, 1987, and Craig and Jamieson, MS). Thus, whether a subordinate male or female acorn woodpecker in a group breeds or refrains from breeding (that is, whether bias is complete or not) is determined primarily by the identity of opposite-sexed breeders and only secondarily by the aggressive and manipulative behaviors of dominant individuals.

13.5. SUMMARY

A substantial fraction of acorn woodpeckers share breeding status with one or more close relatives. Breeders are rarely evicted from territories by intruders even when nesting solitarily; thus, territorial defense does not provide a major impetus for mate-sharing by males or joint-nesting by females. Nor is mate-sharing a simple byproduct of cooperative dispersal by sibling units, as a substantial proportion of cobreeding units arise by means other than cooperative dispersal. Consequently, we explore the possibility that there is a direct fitness advantage to birds sharing mates by means of a demographic model estimating the lifetime reproductive success and fitness of birds pursuing differing breeding strategies.

The results of these models suggest that males in mate-sharing duos and trios enjoy enhanced fitness over males nesting singly, even when using reproductive success values corrected for the correlated effects of group composition and history. Mate-sharing trios have slightly higher fitness than those in duos; however, using the corrected data, trios do slightly worse than duos but still considerably better than singletons. In contrast, joint-nesting females appear to have lifetime fitness approximately equal to that of females breeding singly using observed demographic data. Using corrected data the fitness of joint-nesting females is lower than that of females breeding singly.

These results suggest that, relative to males breeding solitarily, male acorn woodpeckers benefit directly from living and breeding in mate-sharing duos and trios. The enhancement is such that even unrelated male cobreeders would experience higher lifetime fitness than singletons. Why a substantial proportion of males still breed solitarily is unresolved by our model.

Joint-nesting females, however, do not appear to reap direct fitness benefits. Thus, joint-nesting can be expected under conditions of intense competition for breeding vacancies. Furthermore, there should be relatively little reproductive bias between joint-nesters, who should generally be related to each other. These predictions are met by females in our population.

The egalitarian relationship between joint-nesting female acorn woodpeckers, with each female contributing approximately equally to the communal nest, is of particular interest. This pattern is in sharp contrast to the strong despotism seen in species such as the green woodhoopoe, and could be a proximate consequence of the pattern of shared incubation found in woodpeckers. Such sharing makes it difficult or impossible for females to maintain exclusive access to the nest during egg laying and incubation.

In contrast, cobreeding males could conceivably monopolize matings through behavioral means, at least when sharing a single female. However, there is no evidence that such monopolization occurs. The available evidence, both behavioral and experimental, suggests that reproductive bias among cobreeding males is not pronounced. However, such evidence is weak, and the issue is not likely to be resolved until new genetic techniques to determine parentage are applied to the problem.

CHAPTER FOURTEEN

Acorn Woodpecker Sociality: Conclusions and Unanswered Questions

Our goal in the preceding two chapters has been to investigate the factors responsible for the evolution of the complex social organization of acorn woodpeckers. We have approached this problem by focusing on delayed dispersal by offspring and on mate-sharing by siblings.

Here we take a broader view of sociality in acorn woodpeckers and in cooperative breeders generally. Drawing upon the analyses presented in prior chapters, we set forth a scenario of the major steps and critical features leading to the evolution of cooperative breeding in this species. In addition, we will identify some of the problems that need to be solved before a satisfactory understanding of this unusual bird is achieved.

14.1. ACORNS AND SOCIALITY

A major conclusion from the prior chapters is that ecological constraints play a premier role in the evolution of group living in acorn woodpeckers. We have also presented evidence that the primary ecological constraint is the availability of storage facilities in which birds cache acorns; evidence for this conclusion includes the major influence that acorn storage has on virtually all aspects of demography (Chapter 4) and on population regulation (Chapter 11).

Unique Qualities of Acorn Storage

Numerous species of birds store food (e.g., Roberts, 1979b; Andersson, 1985). The unique feature of the acorn woodpecker system of food storage is the use of individually drilled holes that the birds make themselves and use over and over again (Figure 4.1). Although the recovery rate of items stored in the ground or natural cavities may be high (Swanberg, 1951) as a result of the spatial memory and visual cues used for cache recovery (e.g., Balda, 1980; Vander Wall, 1982; Balda and Turek, 1984; Kamil and Balda, 1985), it is this granary-based system that allows acorn woodpeckers to recover close to 100% of the acorns they store.

Such efficiency, however, is costly: granary holes are almost entirely bird-made and are the product of generations of woodpeckers each drilling at most a few hundred holes a year in a tree that may serve dozens of generations of birds. Many limbs used for storage are dead and suffer high attrition; hence, a good fraction of the holes drilled merely replace those which fall. (Acorn woodpeckers, like the Red Queen, work as hard as they can just to stay in the same place insofar as their storage facilities are concerned.)

Because acorn woodpeckers do not drill a hole all at once, we do not have a good estimate of the time necessary to create a useable storage hole. MacRoberts and MacRoberts (1976) estimated that each granary hole represents 30 to 60 minutes of work. Using an estimate of 45 minutes for each hole, an average granary at Hastings containing about 2,500 storage holes represents 1,875 hours of drilling, and a large granary, such as the one with 30,000 holes reported by Dawson (1923), represents 22,500 hours of work. Mumme and de Queiroz (1985) report that breeders spend a minimum of 2.7% and nonbreeders 1.4% of their time building storage holes. Given 12-hour days, no attrition of holes, and an average group consisting of 3 breeders, 1 adult nonbreeder, and 1.5 first-year nonbreeders (see Chapter 3), the average Hastings granary represents 1,347 group-days of work (3.7 group-years), and Dawson's large granary 16,164 group-days of effort (44.3 group-years)!

The cumulative investment in time and energy devoted to creating granaries is obviously extraordinary. Considering this effort, it is perhaps not surprising that storage facilities themselves, rather than availability of acorns per se, appear to be the primary factor limiting the number of acorns that can be stored by woodpeckers (see Chapter 4). But why, given the apparent importance of storage facilities, do acorn woodpeckers spend only a small percentage of their time drilling holes?

We can propose two answers to this question. First, there is often only limited substrate available for new storage holes in a granary, and thus in some cases the drilling of new holes may be constrained by the granary itself. Second, although the percentage of time birds spend drilling holes may seem small, it is possible that competing activities, more immediately necessary to the survival of birds than the long-term benefit reaped by additional holes, may render the fraction of time spent on granaries relatively large. (A useful analogy is the time spent by many academics writing research papers. Although critical to long-term academic fitness, other more immediate concerns inevitably consume a large amount of time and energy, leaving only a small proportion of time left for the production of research papers.) Indeed, the majority of the benefits derived from drilling additional holes are likely to be experienced only many years in the future, and so not by the bird actually doing the drilling or, in many cases, even its descendants. This is especially true given the frequency of crop failures (Hannon et al., 1987). Thus, faced with the uncertain future benefits of daily life, acorn woodpeckers may be spending as much time as is evolutionarily feasible building storage holes.

The investment these birds put into acorn storage does not end with drilling holes. Acorns must be found and individually fit into the available holes; this process often takes several minutes of testing several possible sites (see MacRoberts and MacRoberts, 1976, for a description of acorn storage and acorn

374

stores). Later in the season, storage facilities are defended both intra- and interspecifically, and granaries must be tended; on average, breeders spend close to 40% and nonbreeding helpers almost 20% of their time in granaries during the winter (Mumme and de Queiroz, 1985). Much of this time is spent in granary maintenance; birds may move up to 1% of their acorns every week to new, better-fitting holes as the acorns dry and shrink (Koenig, unpublished data).

There is, however, a big payoff for groups whose granaries are sufficiently large to supply stored acorns throughout the subsequent spring: not only do such birds have increased over-winter survivorship, but their reproductive success is greatly enhanced. Granary facilities are, in effect, a kind of real estate for acorn woodpeckers, fashioned largely by the birds themselves and, in California, virtually indispensable for survival and reproduction.

Granaries and Group Living

Although a variety of woodpeckers are known or suspected to be cooperative breeders (e.g., Short, 1970), the family Picidae is not generally known for its social tendencies (Short, 1982). Why, then, have acorn woodpeckers evolved their extraordinary social behavior?

One possible set of relationships, beginning with the oaks themselves, is shown in Figure 14.1. A first step in this relationship is that between oak species diversity and acorn productivity. This relationship is as yet unproven, but is supported by the correlation between acorn woodpecker density and oak species diversity found by Bock and Bock (1974) in California and Roberts (1979a) in the Southwest (see Chapter 4). As suggested by these authors, as well as by Stacey and Bock (1978) and Trail (1980), a relatively high oak species diversity may lead to an acorn supply stable enough to support a population of permanently resident acorn woodpeckers.

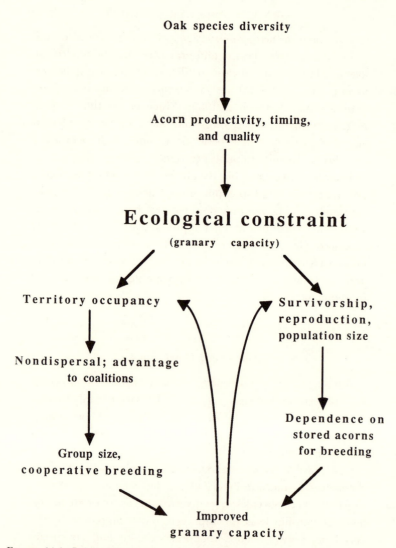

FIGURE 14.1. Relationships between oaks, acorns, granaries, and group living in the acorn woodpecker. Granaries, by limiting acorn storage, are the primary ecological constraint in this system. Constraints lead to high territory occupancy and nondispersal of young (the left feedback loop) and to a population size that exceeds the carrying capacity of the environment in the absence of acorn storage—hence the strong dependence on stored acorns for breeding (the right feedback loop).

As we also saw in Chapter 4, several features besides the absolute size of the acorn crop, including the timing and the quality of the acorns, may be important to determining how good the crop is for acorn woodpeckers. It is the overall availability of acorns, taking these factors into consideration, which ultimately allows acorn woodpeckers to live in one place long enough to create granaries.

But because of the basic problem posed by the temporal occurrence of acorns, granaries per se become a critical feature in this scheme. Except in extraordinary years when acorns remained on some trees into the subsequent summer (see section 11.3), acorn woodpeckers cannot make use of acorns during the spring breeding season without granaries.

Such facilities are not available to young dispersers because of the high occupancy rate of territories with granaries (that is, such territories are saturated with birds). Areas lacking granaries are unavailable for occupancy, regardless of the productivity of the oaks. Marginal territories with inadequate granaries are only intermittently occupied and are uncommon; hence the habitat gradient slope (Koenig and Pitelka, 1981) is steep (see section 3.4). Years of acorn production sufficient to support breeding the following spring without granaries occur, but are rare.

These conditions provide strong incentives for young birds that are unable to secure adequate territories to delay dispersal and remain on their natal territories, if their parents will let them. Furthermore, they also provide a strong advantage in joining sibling coalitions when competing for reproductive vacancies (Hannon et al., 1985). Thus, these conditions could well provide a major evolutionary impetus for both types of sociality seen in acorn woodpecker societies. As group size increases, either through retention of young or the formation of mate-sharing coalitions, there would be an advantage for improving granary capacity. Larger granaries, in turn, would decrease the probability of territory abandonment in poor years, thereby increasing the territory occupancy rate and completing the left feedback loop depicted in Figure 14.1.

The formation of extensive storage facilities results in a relative surplus of food during the winter within groups. Consequently, there is relatively little cost to parents who retain offspring or to individuals who share breeding status. Construction of storage facilities minimizes the problem of resource depletion (Waser, 1981; Brown, 1982), which otherwise would act to select against group formation. However, once groups are formed, dependence on stored acorns is reinforced, completing the second feedback loop drawn in Figure 14.1.

Thus, there are two critical features leading to group living and cooperative breeding: first, there must be some ecological constraint to independent reproduction. Second, the problem of resource depletion, an unavoidable consequence of group formation (Alexander, 1974; Hoogland and Sherman, 1976; Waser, 1981; Brown, 1982) must be reduced. The particularly advanced form of acorn storage practiced by acorn woodpeckers accomplishes both these features, linking their ecology, demography, and social organization. Although it seems likely that some similar set of interdependences may be found for other cooperative breeders, there are few other examples that are likely to be as clear.

14.2. UNANSWERED QUESTIONS

Why Granaries?

A critical detail is why acorn woodpeckers employ the unique habit of storing in individually drilled holes in centralized granaries. Several species of *Melanerpes* woodpeckers besides the acorn woodpecker are known to store food. Red-headed woodpeckers use cracks or cavities (Kilham, 1958), Lewis's woodpeckers use cracks in dead wood (Bock, 1970), and red-bellied woodpeckers store food in natural cracks or crevices (Kilham,

1963). (For a discussion of the evolutionary significance of avian food-storing behavior in general, see Roberts, 1979b.) Here we are concerned with the question of why only the acorn woodpecker drills holes specifically for individual food items in a manner leading to massive, centralized storage facilities used year after year.

This distinctive style of acorn storage results in a system very different from that of most other food-caching species, as exemplified by Clark's nutcrackers storing pine nuts (Vander Wall and Balda, 1977). Because acorn woodpeckers store acorns in a special substrate, they can store only enough food to supply a small fraction of their total energetic needs, compared to the massive scale of storage practiced by the nutcrackers storing in unspecialized and dispersed caches in the ground (see section 4.5). Unfortunately, this observation simply begs the question of why they use granaries at all: acorn woodpeckers could store many more acorns if they were not limited by the supply of holes to put them in.

Granaries have at least three advantages over other types of caches. First, because acorns are usually wedged tightly into storage holes, they provide sites relatively safe from marauders, such as other woodpeckers, jays, and squirrels, that are heavy acorn predators (MacRoberts, 1970; MacRoberts and MacRoberts, 1976; Mumme and de Queiroz, 1985). Second, they provide a substrate in which, unlike the ground, virtually all acorns stored can be retrieved and used. Third, storage holes provide a substrate that allows acorns to dry out and remain mold- and rot-free despite the generally wet winter conditions prevalent at Hastings and throughout much of the range of acorn woodpeckers. That these considerations are important is suggested by the observation that granaries are almost always either dead limbs or the bark of relatively porous wood—substrates that allow acorns to dry without damage. Even acorns stored en masse in cavities, such as described in section 4.4, are generally removed and put into regular storage holes when space is available.

Possibly these advantages make acorn storage in individual holes a necessity if acorns are to be dependably available for later use. Certainly no alternative would provide for as high a recovery rate of intact acorns.

Ecological Constraints and Group Living

Although the acorn woodpecker's granary system is unique, other species have analogous requirements that may turn out to provide equally important insights into the details of the ecological factors leading to cooperative breeding. One such bird is the red-cockaded woodpecker of the southeastern United States, a species in which nonbreeding helpers are relatively common (Ligon, 1970; Gowaty and Lennartz, 1985). Cooperative breeding in this species may be related to its extraordinary dependence on specially made roost holes built in pines infected with heart-rot fungus and surrounded by massive quantities of pine sap (Ligon, 1970; Thompson, 1971; Jackson, 1977). Holes are relatively safe from predators such as snakes (Jackson, 1974) and, like granaries, are used for decades and serve as foci for much of the activity of birds within groups (J. R. Walters, personal communication). Although the chain of causality between safe roosting sites and group living still remains to be elucidated, it seems likely that access to such sites may provide a key to cooperative breeding in this species similar to that provided by granaries for acorn woodpeckers.

It is considerably more challenging to provide similar scenarios for other cooperative breeders. For example, for the Florida scrub jay, Woolfenden and Fitzpatrick (1984; Fitzpatrick and Woolfenden, 1986) conclude, as we do, that cooperative breeding is primarily a result of ecological constraints. The constraints, in their case, are related to their obligatory reliance on the specialized fire-maintained oak-scrub habitat found in central Florida (Fitzpatrick and Woolfenden, 1986). Such habitat specificity indicates a sharp division between the optimal habitat represented by Florida oak scrub and the poor habitat anywhere else.

Studies on noncooperatively breeding populations of scrub jays provide important comparative evidence supporting this hypothesis. On Santa Cruz Island off the coast of California, as in Florida, breeding is delayed for up to several years. However, rather than remain in their natal group, young jays disperse and live in flocks in marginal habitat (Atwood, 1980a, 1980b). At Hastings, young scrub jays may remain in their natal territory for much of the year, but are forced out during the breeding season, when they spend their time either on the periphery of other occupied territories or in unoccupied areas of good food productivity (W. J. Carmen, personal communication). Thus, in both these areas, young nonbreeders have the alternative of leaving their natal group and dispersing to marginal habitats.

Despite the support these studies offer for the hypothesized relationship between the relative abundance of marginal habitat and group living, there remains the important question of why scrub jays in Florida are restricted to fire-maintained oak scrub. Potential answers to this problem include food, competition, and predation.

First, the habitat specificity of scrub jays in Florida may be the result of a correspondingly restricted food supply. Second, some predator may be devastatingly efficient at capturing scrub jays or destroying their nests whenever they stray outside of the oak scrub in Florida, thereby effectively resulting in a sharp habitat gradient. Yet a third possibility is interspecific competition with the blue jay, which replaces the scrub jay in areas where the scrub becomes dominated by pine (Woolfenden and Fitzpatrick, 1984). Even after explaining group living in a species such as the Florida scrub jay proximately with the concepts of habitat saturation and relative lack of marginal territories, there remains the general problem of determining what ultimate ecological factors are responsible for their habitat specificity.

One approach to this problem is examination of habitat characteristics (e.g., Brown et al., 1983; Zack and Ligon, 1985a, 1985b; Koford et al., 1986; see section 4.10). Specific habitat requirements may provide clues as to the exact nature of the ecological constraint.

Social Facilitation and Group Living: The Alternative

We have stressed the role that ecological constraints play in the evolution of group living, not only in acorn woodpeckers but also in other cooperative breeders (see Emlen, 1982a). Once such constraints, facilitated by the other ecological conditions discussed earlier in section 14.2, result in the formation of groups, it is reasonable to expect strong selection to overcome the inherent disadvantages of group living (Alexander, 1974; Hoogland and Sherman, 1976). Such adaptations might result in higher reproductive success per capita and survivorship of birds living in cooperative breeding groups, thereby obscuring the original ecological conditions leading to group living.

However, recent evidence suggests that, in some species, direct benefits accrue to nonbreeding helpers that are so pronounced that group living is unlikely to have evolved in response to ecological constraints. The clearest examples come from the *Campylorhynchus* wrens studied by Rabenold and co-workers (Rabenold, 1984, 1985; Wiley and Rabenold, 1984; Austad and Rabenold, 1985). In the stripe-backed wren, for example, groups with four or more adults produce six times as many young per year than trios or pairs and nearly three times as many offspring per capita (Rabenold, 1984). This reproductive facilitation provides clear indirect fitness advantages to offspring remaining as nonbreeding helpers until their third year, by which time the average relatedness with the breeders has fallen sufficiently to make dispersal equivalent to remaining as a helper (Rabenold, 1985). Similar advantages to helping may exist in the closely related bicolored wren, in which the presence of a single helper triples group reproductive success through better nest defense (Austad and Rabenold, 1985).

The implication is that pairs in these species are less efficient breeding units than are larger groups. Thus, similar to winter foraging flocks (e.g., Caraco, 1979a, 1979b; Sullivan, 1984) and some social carnivores (Caraco and Wolf, 1975; Nudds, 1978), group living in these species, rather than a result of ecological constraints, is the result of social facilitation occurring among individuals within groups (see also Gowaty, 1981; Brown, 1987; Stacey and Ligon, 1987).

For acorn woodpeckers, our data indicate that social facilitation is not an important factor leading to delayed dispersal and reproduction by offspring (Chapter 12). We have, however, shown that direct reproductive advantages accrue to males that share mates, and thus that social facilitation occurs in this context (Chapter 13). The primary proximate advantage to such cobreeders is higher survivorship (Chapter 7). The ultimate ecological factors responsible are unknown, but are presumably the result of either cooperative sharing of tasks, improved predator detection, or both. Thus, both ecological constraints and social facilitation are necessary to explain cooperative breeding in the acorn woodpecker.

Ultimately, however, the hypothesis of social facilitation leads back to a form of ecological constraint. For systems in which social facilitation selects for group living, constraints are working against independent reproduction per se; that is, individuals are unable to breed independently because there is some ecological constraint (e.g., predation) that makes it *directly* impossible for them to do so successfully, even though they may have the opportunity to try. In contrast, the ecological constraints model hypothesizes that individuals are perfectly able to breed independently, but that some constraint restricts their ability to gain the opportunity (e.g., acquire access to the resources necessary for them to be successful). Thus, at least in this sense, these two alternative hypotheses for the evolution of group living in cooperative breeders tie back into each other: in both cases reproduction by pairs is restricted (see Emlen, 1982a).

The documentation of social facilitation in cooperative breeders presents new challenges to the field. No longer can it be said that habitat saturation, stressed here as well as by numerous other workers, is the sole "modus operandi for ecological thinking concerning the evolution of helping behavior" (Emlen, 1982a). A full synthesis of the ecological factors leading .o group living in cooperative breeders is yet to be produced.

Helping Behavior

This treatise has been primarily concerned with the ecological and demographic factors that have contributed to the evolution of the unusual social organization of the acorn woodpecker. We agree with Fitzpatrick and Woolfenden (1981) that such an analysis is a necessary prelude to a thorough analysis of the phenomenon of helping-at-the-nest. (For a recent review of this complex phenomenon, see Brown, 1987.) Here we briefly discuss the analyses that are relevant to the controversy (e.g., Zahavi, 1974; Brown, 1974, 1975, 1978, 1987; Ligon and Ligon, 1978a; Woolfenden and Fitzpatrick, 1978, 1984; Clarke, 1984) surrounding this phenomenon.

From the analyses in Chapter 12 we conclude that the fitness resulting from being a nonbreeder is relatively small. However, much of the benefit received by nonbreeders that delay dispersal and act as helpers is attributable to the indirect component of inclusive fitness (Table 12.5). This suggests that indirect selection has played an important role in the evolution of delayed dispersal and helping within acorn woodpecker groups. However, a considerable fraction of the indirect benefits received by nonbreeding helpers is not from the increased reproductive success of the individuals they assist (the present component), but rather from the increased survivorship of closely related breeding males (the future component; Table 12.8). Depending on the mechanism by which nonbreeders enhance the survivorship of breeding males, much of what indirect benefits exist may not be the result of helping behavior per se.

Reciprocity may also be important to the full spectrum of helping behavior observed in acorn woodpeckers, particularly among cobreeders (Chapter 13). Because of mixed paternity (Joste et al., 1985; Mumme et al., 1985) and maternity (Koenig and Pitelka, 1979; Mumme et al., 1983b) such individuals help to rear each other's young, and are thus engaged in more or less immediate reciprocity. However, such cobreeders—at least at Hastings—are almost always close relatives, and the indirect fitness benefits to sharing breeding status may be substantial (see Chapter 13). It is no simple matter to disentangle the role of reciprocity independent of kinship, although the apparent lack of relatedness between cobreeders in New Mexico (Stacey, 1979b) and recent experimental results (Koenig, unpublished data), suggest that kinship is not a feature universally important to this phenomenon. Furthermore, because both parties benefit directly by feeding young in communal nests, this phenomenon is most likely an example of by-product mutualism rather than the rarer and more theoretically interesting score-keeping mutualism (or reciprocal altruism) involving delayed repayment of aid (Brown, 1987). For a possible example of communal feeding of young in a cooperative breeder involving reciprocal altruism, see Caraco and Brown (1986).

Reciprocity as a consequence of nonbreeders raising offspring who may later aid in their own reproductive efforts (e.g., Ligon and Ligon, 1978a) is not a tenable hypothesis in acorn wood-peckers: younger siblings rarely act as nonbreeding helpers to older siblings. However, siblings from successive broods can and do disperse together (Koenig et al., 1984; see Chapter 13). Thus, reciprocity may be a factor in helping behavior to the extent that nonbreeders help to raise individuals that may later repay them by aiding in their own dispersal attempts. Again, however, this is more likely a by-product mutualism than true reciprocal altruism (Brown, 1987; Koenig, in press).

In addition, are there direct selective advantages to helping behavior? Territorial budding, important in the Florida scrub jay (Woolfenden and Fitzpatrick, 1978, 1984) has been observed only once in acorn woodpeckers at Hastings. Territorial inheritance follows the death of opposite-sexed breeders and is not facilitated by helping behavior. Thus, neither of these phenomena can provide significant direct advantages to helping behavior in our population. Furthermore, our analyses indicate there is little direct lifetime fitness advantage to serving as a nonbreeding helper (section 12.2). Nonbreeding helpers certainly gain direct benefits as a consequence of having a safe place to live and not being forced to disperse, but we consider it unlikely that helping by nonbreeders is payment to breeders for this luxury (e.g., Gaston, 1978b).

Thus, we are unable to point to any unambiguous fitness benefits, direct or indirect, for the phenomenon of helping by nonbreeders in acorn woodpeckers. However, this conclusion is tentative and begs for more detailed analyses of the patterns and ecological consequences of helping behavior. We hope that such analyses will benefit substantially from the demographic and ecological background provided by this book.

14.3. SUMMARY

Group living in the acorn woodpecker is the result of a series of interrelated ecological factors, beginning with the population ecology and evolutionary history of the oaks. Fruiting phenology of oaks over much of the range of acorn woodpeckers dictates that acorns must be cached in order to be available during the winter and the spring breeding season. Acorn woodpeckers have solved this problem by the construction of granaries. Because of their unique, bird-built origin, granaries have a direct limiting effect on the number of social units present in an area. However, granaries are able to support more than two individuals, thereby overcoming the potential constraint of resource depletion experienced by birds living in groups. Thus, granaries provide both a highly permissive ecological situation for groups, and simultaneously restrict the availability of useable territories to young birds.

This scenario may provide a paradigm for the evolution of group living in other cooperative breeders as well. However, recent work on *Campylorhynchus* wrens, as well as our own results concerning mate-sharing by male acorn woodpeckers, indicate that social facilitation, rather than ecological constraints, may sometimes be responsible for group living in cooperative breeders. Thus, ecological constraints, as put forth by Koenig and Pitelka (1981) and Emlen (1982a), appear to be critical to the evolution of cooperative breeding in the acorn woodpecker, but do not provide a complete explanation for the evolution of this phenomenon even within this species.

The primary focus of this book has been the evolution of group living rather than the evolution of helping behavior per se. At this time, we cannot be certain that nonbreeders gain either direct or indirect fitness benefits by helping to raise young not their own. The demographic and ecological analyses presented here nonetheless provide a framework for the understanding of cooperative breeding in this extraordinary species.

Relatedness between Siblings and between Parents and Offspring

Following the general procedure of Bertram (1976), we estimate the mean degree of relatedness between siblings (equal to the mean relatedness between cobreeders) and between parents and offspring in the Hastings acorn woodpecker population. Breeding group composition of the Hastings population is shown in Table 3.4. We initially assume that offspring are always sired by breeders in their own groups, that cobreeders are always siblings that share parentage equally, and that breeding males are unrelated to breeding females.

All these assumptions are violated in some cases. First, the occurrence of kleptogamy in the Hastings population has been documented by Mumme et al. (1985). Second, cobreeders are sometimes parents and offspring and in rare cases completely unrelated (Koenig et al., 1983). Third, inbreeding does occur in our population (Koenig et al., 1983). Finally, although there is little or no bias between female cobreeders (Mumme et al., 1983b; see also section 13.4), the degree of reproductive bias between male cobreeders is unknown. The recent work of both Joste et al. (1985) in Water Canyon, New Mexico, and Mumme et al. (1985) at Hastings, indicates that broods can be multiply sired. However, there are currently no data indicating the average degree of bias, although such bias could be substantial (e.g., Joste et al., 1982). We consider the effects of these factors on our estimates of relatedness in section A.3.

APPENDIX A

A.1. MEAN RELATEDNESS BETWEEN SIBLINGS

Let the relatedness of a typical pair of siblings $r_{sib} = z$. Of all breeding groups, 40% contain only a breeding pair. In these groups, siblings are related by $r_{11} = 0.5$ (the subscripts represent the number of cobreeders present in the group for the two sexes). Thirty percent of breeding groups contain two breeders of one sex and a single breeder of the other (two males and one female or vice versa). Half of the time in such groups, siblings will share both parents, hence $r = 0.5$. The other half, they share one parent and the other parent is related to their own parent by z. Hence, mean relatedness between siblings in such trios is

$$r_{12} = (0.5 \times 0.5) + (0.5 \times (p_1 + p_2)), \tag{1}$$

where $p_1 = $ the contribution to r of the one parent and p_2 the contribution by the second parent. Let p_1 be the contribution by the single parent. Then $p_1 = 0.25$ and $p_2 = 0.25 \times z$. Substituting into (1)

$$r_{12} = (0.25) + (0.5 \times (0.25 + 0.25z)) \tag{2}$$
$$= 0.375 + 0.125z. \tag{3}$$

Following identical reasoning, similar equations can be derived for all observed group compositions. Groups with three breeders of sex A and one of sex B (occurs in 11% of groups; these groups always contained three males and one female):

$$r_{13} = 0.334 + 0.167z. \tag{4}$$

Groups with four breeders of sex A and one of sex B (occurs in 2% of groups; these groups aways contained four males and one female):

$$r_{14} = 0.313 + 0.188z. \tag{5}$$

Groups with two breeders of both sexes (occurs in 9% of groups):

$$r_{22} = 0.25 + 0.25z. \tag{6}$$

The derivation for equation (6) differs from that for (3) only in that siblings may share both, one, or neither parent. If they share neither parent, r is derived based on the assumption that cobreeders of both sexes are related by z.

Groups with three males and two females (occurs in 7% of groups):

$$r_{23} = 0.209 + 0.292z. \tag{7}$$

Finally, 1% of groups contained three breeders of both sexes:

$$r_{33} = 0.167 + 0.333z. \tag{8}$$

Mean relatedness of siblings is calculated by summing the r values of all observed group compositions multiplied by their frequencies (f) and then solving for z.

$$z = \sum_{j=1}^{4} \sum_{i=1}^{3} (r_{ij}f_{ij}) \tag{9}$$

$$z = 0.401 + 0.106z \tag{10}$$

$$= 0.449.$$

Hence, given the initial assumptions, average siblings are related by $r_{\text{sib}} = 0.449$, only 10% less than full siblings ($r = 0.5$).

A.2. MEAN RELATEDNESS BETWEEN PARENTS AND OFFSPRING

Let $r_{m(n)o}$ be the relatedness between n male cobreeders and nonbreeders of either sex, $r_{f(n)o}$ between n female cobreeders and nonbreeders of either sex, and r_{po} the mean relatedness between breeders and their offspring. For males, 46% of the time they are the sole breeder, hence $r_{m(1)o} = 0.5$. Males share breeding with one other male in 34% of groups. As before, each male fathers half the offspring, and the cobreeders are related by z. Hence

$$r_{m(2)o} = (0.5 \times 0.5) + (0.5 \times (0.5 \times z)). \tag{11}$$

Replacing z with 0.449 and solving yields $r_{m(2)o} = 0.362$. Similarly, $r_{m(3)o} = 0.317$ and $r_{m(4)o} = 0.293$. Multiplying each r value by their frequency and solving yields an overall average $r_{mo} = 0.416$. Identical calculations for females, using the observed frequencies of groups with one, two, and three breeding females, yields $r_{fo} = 0.469$. Hence, offspring share slightly fewer genes with the average breeding male in their natal group than breeding female. On average, r_{po} between breeders and offspring in groups is $(0.416 + 0.469)/2 = 0.443$.

A.3. RELAXING THE ASSUMPTIONS

Extra-group Copulations and Parentage

Mumme et al. (1985), based on starch-gel electrophoresis, documented that 4 of 186 nestlings (2.2%) in the Hastings population were fathered by a male other than a breeder in the group; the genotypes of all offspring tested were consistent with one of the breeding females. Thus, we have no evidence for the occurrence of intraspecific nest parasitism (see, for example, Yom-Tov [1980], Gowaty and Karlin [1984]), but kleptogamy apparently occurs.

Because kleptogamy can only be detected when the genotypes of the offspring conflict with those of the breeders, its actual incidence could be considerably greater than these estimates. If 5% of offspring are fathered by males outside of their groups (approximately twice the observed incidence), the mean degree of relatedness of siblings will decrease 2.5% (to $r_{sib} = 0.439$). Mean relatedness of breeding males to offspring in their group decreases 5% to 0.395, whereas the mean relatedness of breeding females to offspring in their group is, of course, unchanged.

Mate-Sharing of Parents with Their Offspring

Of males cobreeders of known relationship, 33% were parent-offspring and another 9% included both sibling and parent-offspring relationships; for females, 14% were parent-offspring (Koenig and Mumme, unpublished data). Despite this relatively high frequency, the effect of cobreeders being parents and offspring rather than siblings is minimal, inasmuch as $r_{sib} \approx r_{po}$ (0.449 versus 0.443).

Unrelated Cobreeders

Unrelated cobreeders reduce the average r between siblings and between breeders and nonbreeders within groups. Of male cobreeders of known relationship between 1976 and 1983, one set of two (1% of groups) were apparently unrelated; similarly two sets of two females (7% of groups) consisted of nonrelatives (see Chapter 12). Using these figures, new r values are readily calculated by modifying the equations used in section A.1. The new r_{sib} equals 0.447, similar to the earlier figure. The estimated value of r_{po} is also virtually unchanged.

Reproductive Bias between Cobreeders

We currently have no data on the degree of reproductive bias among male cobreeders within groups. In the above calculations, we have assumed no bias. If bias is complete, then only a single male effectively reproduces regardless of group composition. This changes r_{sib} to 0.493, considerably greater than the earlier estimate of 0.449 and tantamount to making $r_{sib} = 0.5$, the value for full siblings. Given complete reproductive bias between male cobreeders, r_{mo} is 0.5 and r_{po}, the mean degree of relatedness between parents and offspring in their group, is 0.485. Thus, reproductive bias among cobreeders tends to increase relatedness between siblings and, to a lesser extent, between breeders and nonbreeders within groups, thus countering the effects of kleptogamy and the occasional unrelated cobreeders. We are unable to objectively assess the extent of the effect at this time.

Inbreeding

The final factor potentially influencing the average related-ness among group members is inbreeding, which is not uncommon in our population (Koenig et al., 1984). Using evidence from genealogies, we attempted to derive a value for F, the mean inbreeding coefficient of relatedness occurring in the population (Koenig and Mumme, unpublished data). The estimated mean F is 0.022 ($N = 82$ breeding units), including each "mated" set of birds as a sample, and 0.012 ($N = 183$ years), including each breeding core for each year they bred together. Modifying r_{sib} and r_{po} by multiplying our estimates by $(1 + F)$ yields new values of 0.459 and 0.453 using the first estimate for F and 0.453 and 0.448 using the lower estimate for F. These revised estimates remain only slightly below the values of 0.5 expected in a monogamous society.

Estimation of
Long-Distance Dispersal

We observed 77 male and 52 female dispersers (Figure 8.4). From the estimated percent of nonbreeders obtaining territories within the study area in Table 8.6, these figures represent 59.5% of all male and 31.1% of all female dispersers. Applying Barrowclough's (1978) correction, we estimate that 111.4 males and 93.9 female nonbreeders dispersed a distance less than the diameter of the study area (2 km). Thus, we estimate that 86.1% of all male and 56.2% of all female nonbreeders dispersed less than 2 km; e.g., $(111.4)(59.5/77)$ for males, $(93.9)(31.1/52)$ for females.

Of the 40.5% (e.g., $100\% - 59.5\%$) male and 68.9% ($100\% - 31.1\%$) female nonbreeders that do not obtain breeding positions within the study area, we estimate from Table 8.6 that 90% of the males ($= 36.5\%$) and 100% of the females ($= 68.9\%$) survive and obtain territories elsewhere. Thus, for males, 86.1% disperse less than 2 km and 96.0% successfully obtain territories somewhere (e.g., $59.5\% + 36.5\%$); for females, 56.2% disperse within 2 km and 100% obtain territories. Subtracting, we conclude that 9.9% of nonbreeder males and 43.8% of nonbreeder females disperse a distance greater than 2 km.

Rescaling back to the observed values, an estimated 12.8 male and 73.4 female dispersal events involved movements an unknown distance greater than 2 km. (Note that these values added to the number of dispersers already theoretically present exceeds the known number of nonbreeders with known fates [Table 8.6, line 1], considerably so for females. This is because all dispersal events, even those by birds of unknown origin and "secondary" dispersal events, are included in the former figures, whereas Table 8.6 includes only group offspring banded as nestlings.)

In order to derive reasonable estimates of the actual RMS dispersal distance, we plotted the distance of dispersal (log transformed) versus the cumulative frequency of dispersal (arcsin transformed; Figure B.1). We then chose three plausible tails for the distribution. The first (curve a) assumes an exponential shape to the curve and provides a conservative estimate for the dispersal pattern. The other two curves assume sigmoidal distributions; the farthest dispersers in curve b travel 10 km, whereas those in curve c go 100 km.

Although we cannot objectively choose among these curves, or even know whether any of them are realistic, acorn woodpeckers are relatively strong flyers easily capable of traveling well over 10 km in a short time. We thus suspect the true distribution is likely to be somewhere between curves b and c, probably closer to the latter.

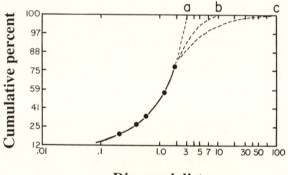

Dispersal distance

FIGURE B.1. Distance of dispersal (log transformed) versus the cumulative frequency of dispersal. Cumulative frequency was arcsin transformed ($y = \arcsin \sqrt{x}$); numbers listed are the back-transformed values. Solid circles and solid line indicate the observed dispersal distribution. Dotted lines represent three plausible dispersal patterns beyond the observable range: (a) assumes an exponential distribution, (b) assumes a logistic curve with a maximum dispersal distance of 10 km, and (c) assumes a logistic curve with a maximum dispersal distance of 100 km.

Scientific Names of Birds

TABLE C.1. Scientific Names of Birds Mentioned in the Text

Family	Common Name	Scientific Name
Ciconiidae	White stork	*Ciconia ciconia*
Accipitridae	Cooper's hawk	*Accipiter cooperi*
	European sparrowhawk	*Accipiter nisus*
	Galápagos hawk	*Buteo galapagoensis*
	American kestrel	*Falco sparverius*
Tetraonidae	Willow ptarmigan	*Lagopus lagopus*
Rallidae	Pukeko	*Porphyrio porphyrio*
	Purple gallinule	*Porphyrula martinica*
Laridae	Kittiwake	*Rissa tridactyla*
Cuculidae	Groove-billed ani	*Crotophaga sulcirostris*
Strigidae	Northern pygmy owl	*Glaucidium gnoma*
Alcedinidae	Pied kingfisher	*Ceryle rudis*
Meropidae	European bee-eater	*Merops apiaster*
	White-fronted bee-eater	*Merops bullockoides*
Phoeniculidae	Green woodhoopoe	*Phoeniculus purpureus*
Picidae	Red-bellied woodpecker	*Melanerpes carolinus*
	Acorn woodpecker	*Melanerpes formicivorus*
	Lewis' woodpecker	*Melanerpes lewis*
	Red-cockaded woodpecker	*Picoides borealis*
Tyrannidae	Ash-throated flycatcher	*Myiarchus cinerascens*
Troglodytidae	Bicolored wren	*Campylorhynchus griseus*
	Stripe-backed wren	*Campylorhynchus nuchalis*
Mimidae	Galápagos mockingbird	*Nesomimus parvulus*
Prunellidae	Dunnock	*Prunella modularis*
Muscicapidae	Superb blue wren	*Malurus cyaneus*
	Splendid wren	*Malurus splendens*
	Grey-crowned babbler	*Pomatostomus temporalis*
	Hall's babbler	*Pomatostomus halli*
	Arabian babbler	*Turdoides squamiceps*

(*continued*)

Tᴀʙʟᴇ C.1. *(continued)*

Family	Common Name	Scientific Name
Laniidae	Common fiscal shrike	*Lanius collaris*
	Grey-backed fiscal shrike	*Lanius excubitorius*
Sturnidae	European starling	*Sturnus vulgaris*
	Chestnut-bellied starling	*Spreo pulcher*
Paridae	Black-capped chickadee	*Parus atricapillus*
	Blue tit	*Parus caeruleus*
Ploceidae	House sparrow	*Passer domesticus*
	White-browed sparrow weaver	*Plocepasser mahali*
Emberizidae	Northern oriole	*Icterus galbula*
Grallinidae	White-winged chough	*Corcorax melanorhamphus*
Corvidae	Scrub jay	*Aphelocoma coerulescens*
	Mexican (gray-breasted) jay	*Aphelocoma ultramarina*
	Blue jay	*Cyanocitta cristata*
	Pinyon jay	*Gymnorhinus cyanocephalus*
	Brown jay	*Psilorhinus morio*
	Common crow	*Corvus brachyrhynchos*
	Clark's nutcracker	*Nucifraga columbianus*
	Yellow-billed magpie	*Pica nuttalli*

Literature Cited

Alexander, R. D. 1974. The evolution of social behavior. *Ann. Rev. Ecol. & Syst.* 5:325–83.

Andersson, M. 1985. Food storing. In *A dictionary of birds,* edited by B. Campbell and E. Lack, pp. 235–37. Vermillion, S. Dak.: Buteo Books.

Arnold, S. J., and M. J. Wade. 1984a. On the measurement of natural and sexual selection: Theory. *Evol.* 38:709–719.

———. 1984b. On the measurement of natural and sexual selection: Applications. *Evol.* 38:720–38.

Ashmole, N. P. 1961. The biology of certain terns. Ph.D. diss., Oxford University, Oxford, England.

———. 1963. The regulation of numbers of tropical oceanic birds. *Ibis* 103:458–73.

Assoc. of Offic. Agric. Chemists. 1965. *Official methods of analysis.* 10th ed., Washington, D.C.

Atwood, J. L. 1980a. Social interactions in the Santa Cruz Island scrub jay. *Condor* 82:440–48.

———. 1980b. Breeding biology of the Santa Cruz Island scrub jay. In *The California islands: Proceedings of a multidisciplinary symposium,* edited by D. M. Power, pp. 675–88. Santa Barbara, Calif. Santa Barbara Museum of Natural History.

Austad, S. N., and K. N. Rabenold. 1985. Reproductive enhancement by helpers and an experimental inquiry into its mechanism in the bicolored wren. *Behav. Ecol. & Sociobiol.* 17:19–27.

Axelrod, R., and W. D. Hamilton. 1981. The evolution of cooperation. *Science* 211:1390–96.

LITERATURE CITED

Balda, R. P. 1980. Recovery of cached seeds by a captive *Nucifraga caryocatactes*. *Z. Tierpsychol.* 52:331–46.

Balda, R. P., and R. J. Turek. 1984. The cache-recovery system as an example of memory capabilities in Clark's nutcracker. In *Animal cognition*, edited by H. L. Roitblat, T. G. Bever, and H. S. Terrace, pp. 513–32. Hillsdale, N.J.: Lawrence Erlbaum Associates.

Barrowclough, G. F. 1978. Sampling bias in dispersal studies based on finite areas. *Bird-Banding* 49:333–41.

———. 1980. Gene flow, effective population sizes, and genetic variance components in birds. *Evol.* 34:789–98.

———. 1983. Biochemical studies of microevolutionary processes. In *Perspectives in ornithology*, edited by A. H. Brush and G. A. Clark, Jr., pp. 223–61. New York: Cambridge University Press.

Barrowclough, G. F., and S. L. Coats. 1985. The demography and population genetics of owls, with special reference to the conservation of the spotted owl (*Strix occidentalis*). In *Ecology and management of the spotted owl in the Pacific Northwest*, edited by R. J. Gutiérrez and A. B. Carey, pp. 74–83. Portland, Oreg.: Pacific Northwest Forest and Range Experiment Station General Technical Report PNW-185.

Barrowclough, G. F., and G. F. Shields. 1984. Karyotypic evolution and long-term effective population sizes of birds. *Auk* 101:99–102.

Bate-Smith, E. C. 1973. Haemanalysis of tannins: The concept of relative astringency. *Phytochemistry* 12:907–912.

Beal, F.E.L. 1911. Food of the woodpeckers of the United States. *U.S. Dept. Agr. Biol. Surv. Bull.* 37.

Bekoff, M., and J. A. Byers. 1985. The development of behavior from evolutionary and ecological perspectives in mammals and birds. *Evol. Biol.* 19:215–86.

Bendire, C. 1895. Life histories of North American birds, II. *U.S. Natl. Mus. Bull.* 3.

Bent, A. C. 1939. Life histories of North American woodpeckers. *U.S. Natl. Mus. Bull.* 174.

Bertram, B.C.R. 1976. Kin selection in lions and in evolution. In *Growing points in ethology,* edited by P.P.G. Bateson and R. A. Hinde, pp. 281–301. Cambridge: Cambridge University Press.

————. 1978. Living in groups: Predators and prey. In *Behavioural ecology: An evolutionary approach,* 1st ed., edited by J. R. Krebs and N. B. Davies, pp. 64–96. Oxford: Blackwell.

Bock, C. E. 1970. The ecology and behavior of the Lewis woodpecker (*Asyndesmus lewis*). *Univ. Calif. Publ. Zool.* 92:1–91.

Bock, C. E., and J. H. Bock. 1974. Geographical ecology of the acorn woodpecker: Diversity versus abundance of resources. *Amer. Nat.* 108:694–98.

Bock, C. E., and R. B. Smith. 1971. An analysis of Colorado Christmas counts. *Amer. Birds* 25:945–47.

Bolton, H. E. 1933. *Font's complete diary.* Berkeley: University of California Press.

Botkin, D. B., and R. S. Miller. 1974. Mortality rates and survival of birds. *Amer. Nat.* 108:181–92.

Bowen, B. S., R. R. Koford, and S. L. Vehrencamp. MS. Dispersal in communally nesting groove-billed anis.

Boyd, R., and R. J. Richerson. 1980. Effect of phenotypic variation on kin selection. *Proc. Natl. Acad. Sci.* 77:7506–7509.

Brown, J. L. 1969. Territorial behavior and population regulation in birds. *Wilson Bull.* 81:293–329.

————. 1970. Cooperative breeding and altruistic behavior in the Mexican jay, *Aphelocoma ultramarina. Anim. Behav.* 18:366–78.

————. 1972. Communal feeding of nestlings in the Mexican jay (*Aphelocoma ultramarina*): Interflock comparisons. *Anim. Behav.* 20:395–402.

———. 1974. Alternate routes to sociality in jays—with a theory for the evolution of altruism and communal breeding. *Amer. Zool.* 14:63–80.

———. 1975. Helpers among Arabian babblers, *Turdoides squamiceps. Ibis* 117:243–44.

———. 1978. Avian communal breeding systems. *Ann. Rev. Ecol. & Syst.* 9:123–56.

———. 1982. Optimal group size in territorial animals. *J. Theor. Biol.* 95:793–810.

———. 1983. Cooperation—a biologist's dilemma. In *Advances in behavior*, vol. 13, edited by J. S. Rosenblatt et al., pp. 1–37. New York: Academic Press.

———. 1985a. The evolution of helping behavior—an ontogenetic and comparative perspective. In *The comparative development of adaptive skills: Evolutionary implications*, edited by E. Gollin, pp. 137–71. Hillsdale, N.J.: Lawrence Erlbaum Associates.

———. 1985b. Cooperative breeding and the regulation of numbers. *Proc. XVIIIth Int. Ornith. Congr.*, pp. 774–82.

———. 1987. *Helping and communal breeding in birds: Ecology and evolution.* Princeton: Princeton University Press.

Brown, J. L., and R. P. Balda. 1977. The relationship of habitat quality to group size in Hall's babbler (*Pomatostomus halli*). *Condor* 79:312–20.

Brown, J. L., and E. R. Brown. 1980. Reciprocal aid-giving in a communal bird. *Z. Tierpsychol.* 53:313–24.

———. 1981a. Extended family system in a communal bird. *Science* 211:959–60.

———. 1981b. Kin selection and individual selection in babblers. In *Natural selection and social behavior: Recent research and new theory*, edited by R. D. Alexander and D. W. Tinkle, pp. 244–56. New York: Chiron Press.

———. 1984. Parental facilitation: Parent-offspring relations in communally breeding birds. *Behav. Ecol. & Sociobiol.* 14:203–209.

Brown, J. L., E. R. Brown, S. D. Brown, and D. D. Dow. 1982. Helpers: Effects of experimental removal on reproductive success. *Science* 215:421–22.

Brown, J. L., D. D. Dow, E. R. Brown, and S. D. Brown. 1978. Effects of helpers on feeding of nestlings in the grey-crowned babbler (*Pomatostomus temporalis*). *Behav. Ecol. & Sociobiol.* 4:43–59.

———. 1983. Socio-ecology of the grey-crowned babbler: Population structure, unit size, and vegetation correlates. *Behav. Ecol. & Sociobiol.* 13:115–24.

Brown, J. L., and S. L. Pimm. 1985. The origin of helping: The role of variability in reproductive potential. *J. Theor. Biol.* 112:465–77.

Burley, N. 1982. Facultative sex-ratio manipulation. *Amer. Nat.* 120:81–107.

Bygott, J. D., B.C.R. Bertram, and J. P. Hanby. 1979. Male lions in large coalitions gain reproductive advantages. *Nature* 282:839–41.

Caraco, T. 1979a. Time budgeting and group size: A test of theory. *Ecology* 60:618–27.

———. 1979b. Time budgeting and group size: A theory. *Ecology* 60:611–17.

Caraco, T., and J. L. Brown. 1986. A game between communal breeders: when is food-sharing stable? *J. Theor. Biol.* 118:379–93.

Caraco, T., and L. L. Wolf. 1975. Ecological determinants of group sizes of foraging lions. *Amer. Nat.* 109:343–52.

Caughley, G. 1976. *Analysis of vertebrate populations*. New York: John Wiley & Sons.

Cavalli-Sforza, L. L., and W. F. Bodmer. 1961. *The genetics of human populations*. San Francisco: Freeman.

Charnov, E. L. 1982. *The theory of sex allocation*. Princeton: Princeton University Press.

Chase, I. D. 1980. Cooperative and noncooperative behavior in animals. *Amer. Nat.* 115:827–57.

Clark, A. B. 1978. Sex ratio and local resource competition in a prosimian primate. *Science* 201:163–65.

Clarke, M. F. 1984. Co-operative breeding by the Australian bell miner *Manorina melanophyrs* Latham: A test of kin selection theory. *Behav. Ecol. & Sociobiol.* 14:137–46.

Clutton-Brock, T. H., 1983. Selection in relation to sex. In *Evolution from molecules to man,* edited by D. S. Bendell, pp. 457–81. Cambridge: Cambridge University Press.

————, ed. In press. *Reproductive success.* Chicago: University of Chicago Press.

Cochran, W. G. 1947. Some consequences when the assumptions for the analysis of variance are not satisfied. *Biometrics* 3:22–38.

Counsilman, J. J. 1977a. Groups of the grey-crowned babbler in central southern Queensland. *Babbler* 1:14–22.

————. 1977b. A comparison of two populations of the grey-crowned babbler [pt. 1]. *Bird Behaviour* 1:43–82.

Craig, J. L. 1976. An interterritorial hierarchy: An advantage for a subordinate in a communal territory. *Z. Tierpsychol.* 42:200–205.

————. 1980. Pair and group breeding behaviour in a communal gallinule, the pukeko, *Porphyrio p. melanotus. Anim. Behav.* 32:23–32.

————. 1984. Are communal pukeko caught in the prisoner's dilemma? *Behav. Ecol. & Sociobiol.* 14:147–50.

Craig, J. L., and I. G. Jamieson. 1985. The relationship between presumed gamete contribution and parental investment in a communally breeding bird. *Behav. Ecol. & Sociobiol.* 17:207–211.

————. MS. Incestuous mating in a communal bird: a family affair.

Crook, J. H. 1965. The adaptive significance of avian social organisations. *Symp. Zool. Soc. Lond.* 14:181–218.

Crow, J. F., and M. Kimura. 1970. *An introduction to population genetics theory.* New York: Harper & Row.

Custer, T. W., and F. A. Pitelka. 1977. Demographic features of a lapland longspur population near Barrow, Alaska. *Auk* 94:505–525.

Darley-Hill, S., and W. C. Johnson. 1981. Acorn dispersal by the blue jay (*Cyanocitta cristata*). *Oecologia* 50:231–32.

Darwin, C. R. 1859. *On the origin of species*. London: Murray.

Davies, N. B. 1985. Cooperation and conflict among dunnocks, *Prunella modularis*, in a variable mating system. *Anim. Behav.* 33:628–48.

———. 1986. Reproductive success of dunnocks, *Prunella modularis*, in a variable mating system. I. Factors influencing provisioning rate, nestling weight and fledging success. *J. Anim. Ecol.* 55:123–38.

Davies, N. B., and A. I. Houston. 1986. Reproductive success of dunnocks, *Prunella modularis*, in a variable mating system. II. Conflicts of interest among breeding adults. *J. Anim. Ecol.* 55:139–54.

Davies, N. B., and A. Lundberg. 1984. Food distribution and a variable mating system in the dunnock, *Prunella modularis*. *J. Anim. Ecol.* 53:895–912.

Davis, D. E. 1940a. Social nesting habits of the smooth-billed ani. *Auk* 57:179–218.

———. 1940b. Social nesting habits of *Guira guira*. *Auk* 57:472–84.

———. 1941. Social nesting habits of *Crotophaga major*. *Auk* 58:179–83.

Dawkins, R. 1980. Good strategy or evolutionary stable strategy? In *Sociobiology: Beyond nature/nurture?*, edited by G. W. Barlow and J. Silverberg, pp. 331–67. Boulder, Co.: Westview Press.

Dawson, W. L. 1923. *The birds of California*. San Francisco: South Moulton Company.

Deevey, E. S., Jr. 1947. Life tables for natural populations of animals. *Quart. Rev. Biol.* 22:283–314.

Dixon, W. J. 1983. *BMDP. Biomedical computer programs*. Berkeley: University of California Press.

Donaldson, T. S. 1968. Robustness of the *F*-test to errors of both kinds and the correlation between the numerator and denominator of the *F*-ratio. *J. Amer. Stat. Assoc.* 63: 660–76.

Dow, D. D. 1982. Communal breeding. *Royal Austral. Ornith. Union Newsletter* (December): 6–7.

Dublin, L. I., and A. J. Lotka. 1925. On the true rate of natural increase. *J. Amer. Stat. Assoc.* 20: 305–339.

Eckstorm, F. H. 1901. *The woodpeckers.* Boston: Houghton Mifflin.

Ekman, J., G. Cederholm, and C. Askenmo. 1981. Spacing and survival in winter groups of willow tit *Parus montanus* Conrad and crested tit *P. cristatus* L.—a removal study. *J. Anim. Ecol.* 50: 1–9.

Elkin, R. G., W. R. Featherston, and J. C. Rogler. 1978. Investigations of leg abnormalities in chicks consuming high tannin sorghum grain diets. *Poultry Sci.* 57: 757–62.

Emigh, T. H., and E. Pollak. 1979. Fixation probabilities and effective population numbers in diploid populations with overlapping generations. *Theoret. Pop. Biol* 15: 86–107.

Emlen, S. T. 1981. Altruism, kinship, and reciprocity in the white-fronted bee-eater. In *Natural selection and social behavior: Recent research and new theory*, edited by R. D. Alexander and D. W. Tinkle, pp. 217–30. New York: Chiron Press.

———. 1982a. The evolution of helping. I. An ecological constraints model. *Amer. Nat.* 119: 29–39.

———. 1982b. The evolution of helping. II. The role of behavioral conflict. *Amer. Nat.* 119: 40–53.

———. 1984. Cooperative breeding in birds and mammals. In *Behavioural ecology: An evolutionary approach*, 2nd ed., edited by J. R. Krebs and N. B. Davies, pp. 305–339. Oxford: Blackwell.

Emlen, S. T., J. M. Emlen, and S. A. Levin. 1986. Sex-ratio selection in species with helpers-at-the-nest. *Amer. Nat.* 127: 1–8.

Emlen, S. T., and L. W. Oring. 1977. Ecology, sexual selection, and the evolution of mating systems. *Science* 197: 215–23.

Emlen, S. T., and S. L. Vehrencamp. 1983. Cooperative breeding strategies among birds. In *Perspectives in ornithology*, edited by A. H. Brush and G. H. Clark, Jr., pp. 93–120. New York: Cambridge University Press.

Emlen, S. T., and P. H. Wrege. 1986. Forced copulations and intra-specific parasitism: Two costs of social living in the white-fronted bee-eater. *Z. Tierpsychol.* 71:2–29.

Faaborg, J., T. de Vries, C. B. Patterson, and C. R. Griffin. 1980. Preliminary observations on the occurrence and evolution of polyandry in the Galápagos hawk (*Buteo galapagoensis*). *Auk* 97:581–90.

Faaborg, J., and C. B. Patterson. 1981. The characteristics and occurrence of cooperative polyandry. *Ibis* 123:477–84.

Feeny, P. 1970. Seasonal changes in the oak leaf. Tannins and nutrients as a cause of spring feeding by winter moth caterpillars. *Ecology* 51:565–81.

Felsenstein, J. 1975. A pain in the torus: Some difficulties with models of isolation by distance. *Amer. Nat.* 109:359–68.

Fisher, R. A. 1930. *The genetical theory of natural selection.* Oxford: Clarendon Press.

Fitzpatrick, J. W., and G. E. Woolfenden. 1981. Demography is a cornerstone of sociobiology. *Auk* 98:406–407.

———. 1986. Demographic routes to cooperative breeding in some New World jays. In *Evolution of animal behavior*, edited by M. H. Nitecki and J. A. Kitchell. Chicago: University of Chicago Press.

———. In press. Components of lifetime reproductive success in the Florida scrub jay. In *Reproductive success*, edited by T. H. Clutton-Brock. Chicago: University of Chicago Press.

Foster, M. S. 1978. Total frugivory in tropical passerines: A reappraisal. *Tropical Ecology* 19:131–54.

Frame, L. H., J. R. Malcolm, G. W. Frame, and H. van Lawick.

1979. Social organization of African wild dogs (*Lycaon pictus*) on the Serengeti Plains, Tanzania, 1967–1978. *Z. Tierpsychol.* 50:225–49.

Fry, C. H. 1972. The social organization of bee-eaters (Meropidae) and co-operative breeding in hot-climate birds. *Ibis* 114:1–14.

Gaston, A. J. 1973. The ecology and behaviour of the long-tailed tit. *Ibis* 115:330–51.

——. 1978a. Demography of the jungle babbler, *Turdoides striatus*. *J. Anim. Ecol.* 47:845–70.

——. 1978b. The evolution of group territorial behavior and cooperative breeding. *Amer. Nat.* 112:1091–1100.

Gessaman, J. A ., and P. R. Findell. 1979. Energy cost of incubation in the American kestrel. *Comp. Biochem. Physiol.* A63:57–62.

Gittleman, J. L. 1985. Functions of communal care in mammals. In *Evolution—essays in honour of John Maynard Smith*, edited by P. J. Greenwood, P. H. Harvey, and M. Slatkin, pp. 187–205. Cambridge: Cambridge University Press.

Gowaty, P. A. 1981. An extension of the Orians-Verner-Willson model to account for mating systems besides polygyny. *Amer. Nat.* 118:851–59.

——. 1985. Multiple parentage and apparent monogamy in birds. *Ornithol. Monogr.* 37:11–21.

Gowaty, P. A., and A. A. Karlin. 1984. Multiple paternity and maternity in single broods of apparently monogamous eastern bluebirds. *Behav. Ecol. & Sociobiol.* 15:91–95.

Gowaty, P. A., and M. R. Lennartz. 1985. Sex ratios of nestling and fledgling red-cockaded woodpeckers (*Picoides borealis*) favor males. *Amer. Nat.* 126:347–53.

Grafen, A. 1982. How not to measure inclusive fitness. *Nature* 298:425–26.

——. 1984. Natural selection, kin selection and group selection. In *Behavioural ecology: An evolutionary approach*, 2nd ed., edited by J. R. Krebs and N. B. Davies, pp. 62–84. Oxford: Blackwell.

Greenwood, P. J. 1980. Mating systems, philopatry and dispersal in birds and mammals. *Anim. Behav.* 28:1140–62.

Griffin, J. R. 1974. Botanical resources of the Hastings Reservation, Monterey County, California, *Madroño* 22:329–32.

———. 1983. Natural history of the Hastings Reservation. Unpublished MS (on file at Hastings Reservation).

Hamilton, W. D. 1964. The genetical evolution of social behaviour. I. II. *J. Theor. Biol.* 7:1–52.

Hannon, S. J., R. L. Mumme, W. D. Koenig, and F. A. Pitelka. 1985. Replacement of breeders and within-group conflict in the cooperatively breeding acorn woodpecker. *Behav. Ecol. & Sociobiol.* 17:303–12.

Hannon, S. J., R. L. Mumme, W. D. Koenig, S. Spon, and F. A. Pitelka. 1987. Acorn crop failure, dominance, and a decline in numbers in the cooperatively breeding acorn woodpecker. *J. Anim. Ecol.* 56: 197–207.

Hegner, R. E., S. T. Emlen, and N. J. Demong. 1982. Spatial organization of the white-fronted bee-eater. *Nature* 298:264–66.

Heinze, R. H., and A. E. Murneck. 1940. Comparative accuracy and efficiency in determination of carbohydrates in plant material. *Missouri Agr. Exp. Sta. Res. Bull.* 314.

Henshaw, H. W. 1876. Report on the ornithology of the portions of California visited during the field-season of 1875. In *Annual report upon the geographical surveys west of the one hundredth meridian, in California, Nevada, Utah, Colorado, Wyoming, New Mexico, Arizona, and Montana,* edited by G. M. Wheeler, pp. 224–78, appendix JJ.

Hensley, M. M., and J. B. Cope. 1951. Further data on removal and repopulation of the breeding birds in a spruce-fir community. *Auk* 68:483–93.

Herrera Tordesillas, A. de. 1601–1615. *Historia general de los hechos de los Castellaños en las Islas y Tiera Firme del Mar Oceaño.* Madrid: Emprenta Real.

Holmes, R. T. 1970. Differences in population density, territoriality, and food supply of Dunlin on arctic and subarctic tundra. In *Animal populations in relation to their food resources*, edited by A. Watson, pp. 303–317. Oxford: Blackwell.

Hoogland, J. L., and P. W. Sherman. 1976. Advantages and disadvantages of bank swallow (*Riparia riparia*) coloniality. *Ecol. Monogr.* 46: 33–58.

Horn, H. S. 1978. Optimal tactics of reproduction and life-history. In *Behavioral ecology: An evolutionary approach*, 1st ed., edited by J. R. Krebs and N. B. Davies, pp. 411–29. Oxford: Blackwell.

Houston, A. I., and N. B. Davies. 1985. The evolution of co-operation and life history in the dunnock, *Prunella modularis*. In *Behavioural ecology: The ecological consequences of adaptive behaviour*, edited by R. Sibly and R. H. Smith, pp. 471–87. Oxford: Blackwell.

Howell, T. R. 1972. Birds of the lowland pine savanna of north-eastern Nicaragua. *Condor* 74:316–40.

Hull, C. H., and N. H. Nie. 1981. SPSS Update 7–9. New York: McGraw-Hill.

Hunter, L. A. 1985. The effects of helpers in cooperatively breeding purple gallinules. *Behav. Ecol. & Sociobiol.* 18:147–53.

Jackson, C. T. 1866. [Untitled.] *Proc. Boston Soc. Nat. Hist.* 10:224–29.

Jackson, J. A. 1974. Gray rat snakes versus red-cockaded wood-peckers: Predator-prey adaptations. *Auk* 91:342–47.

———. 1977. Red-cockaded woodpeckers and red heart disease of pines. *Auk* 94:160–163.

James, F. C., and H. H. Shugart, Jr. 1970. A quantitative method of habitat description. *Amer. Birds* 24:727–36.

Jamieson, I. G. 1986. The functional approach to behavior: Is it useful? *Amer. Nat.* 127:195–208.

Jamieson, I. G., and J. L. Craig. In press. Critique of helping behaviour in birds: A departure from functional explana-tions. In *Perspectives in ethology*, vol 7., edited by P.P.G. Bateson and P. Klopfer. New York: Plenum.

Jarman, P. J. 1974. The social organisation of antelope in relation to their ecology. *Behaviour* 48:215–67.

Jeffreys, A. J., V. Wilson, and S. L. Thein. 1985. Hypervariable "minisatellite" regions in human DNA. *Nature* 314:67–73.

Jehl, J. R., Jr. 1979. Pine cones as granaries for acorn woodpeckers. *West. Birds* 10:219–20.

Jenkins, D., A. Watson, and G. R. Miller. 1963. Population studies on red grouse, *Lagopus lagopus scoticus* (Lath.) in northeast Scotland. *J. Anim. Ecol.* 32:317–76.

Joste, N. E., W. D. Koenig, R. L. Mumme, and F. A. Pitelka. 1982. Intragroup dynamics of a cooperative breeder: An analysis of reproductive roles in the acorn woodpecker. *Behav. Ecol. & Sociobiol.* 11:195–201.

Joste, N. E., J. D. Ligon, and P. B. Stacey. 1985. Shared paternity in the acorn woodpecker. *Behav. Ecol. & Sociobiol.* 17:39–41.

Kale, H. W., II. 1965. Ecology and bioenergetics of the long-billed marsh wren in Georgia salt marshes. *Publ. Nuttall Ornithol. Club* 5.

Kamil, A. C., and R. P. Balda. 1985. Cache recovery and spatial memory in Clark's nutcrackers (*Nucifraga columbiana*). *J. Exp. Psychol.: Anim. Behav. Proc.* 11:95–111.

Kattan, G. MS. Food habits, habitat and social organization of acorn woodpeckers in Colombia: Communal breeding without habitat saturation.

Kilham, L. 1958. Sealed-in winter stores of red-headed woodpeckers. *Wilson Bull.* 70:107–113.

———. 1963. Food storing of red-bellied woodpeckers. *Wilson Bull.* 75:227–34.

Kimura, M., and G. H. Weiss. 1964. The stepping stone model of population structure and the decrease of genetic correlation with distance. *Genetics* 49:561–76.

Kinniard, M. F., and P. R. Grant. 1982. Cooperative breeding by the Galápagos mockingbird, *Nesomimus parvulus*. *Behav. Ecol. & Sociobiol.* 10:65–73.

Kluijver, H. N. 1951. The population ecology of the great tit, *Parus m. major* L. *Ardea* 39:1–135.

Knapton, R. W., and J. R. Krebs. 1974. Settlement patterns, territory size, and breeding density in the song sparrow (*Melospiza melodia*). *Can. J. Zool.* 52:1413–20.

Knoke, D., and P. J. Burke. 1980. *Log-Linear models.* Beverly Hills, Calif: Sage Publications.

Koenig, W. D. 1980a. Variation and age determination in a population of acorn woodpeckers. *J. Field Ornith.* 51:10–16.

———. 1980b. The incidence of runt eggs in woodpeckers. *Wilson Bull.* 92:169–76.

———. 1981a. Space competition in the acorn woodpecker: Power struggles in a cooperative breeder. *Anim. Behav.* 29:396–409.

———. 1981b. Reproductive success, group size, and the evolution of cooperative breeding in the acorn woodpecker. *Amer. Nat.* 117:421–43.

———. 1981c. Coalitions of male lions: Making the best of a bad job? *Nature* 293:413.

———. 1982. Ecological and social factors affecting hatchability of eggs. *Auk* 99:526–36.

———. 1984. Geographic variation in clutch size in the northern flicker (*Colaptes auratus*): Support for Ashmole's hypothesis. *Auk* 101:698–706.

———. In press. Reciprocal altruism in birds: a critical review. *Ethol. & Sociobiol.*

Koenig, W. D., and S. S. Albano. 1987. Lifetime reproductive success, selection, and the opportunity for selection in the white-tailed skimmer, *Plathemis lydia. Evol.* 41:22–36.

Koenig, W. D., S. J. Hannon, R. L. Mumme, and F. A. Pitelka. In press. Parent-offspring conflict in the cooperatively breeding acorn woodpecker. *Proc. XIXth Int. Ornith. Congr.*

Koenig, W. D., R. L. Mumme, and F. A. Pitelka. 1983. Female roles in cooperatively breeding acorn woodpeckers. In

Social behavior of female vertebrates, edited by S. K. Wasser, pp. 235–61. New York: Academic Press.

———. 1984. The breeding system of the acorn woodpecker in central coastal California. *Z. Tierpsychol.* 65:289–308.

Koenig, W. D., and F. A. Pitelka. 1979. Relatedness and inbreeding avoidance: Counterploys in the communally nesting acorn woodpecker. *Science* 206:1103–1105.

———. 1981. Ecological factors and kin selection in the evolution of cooperative breeding in birds. In *Natural selection and social behavior: Recent research and new theory*, edited by R. D. Alexander and D. W. Tinkle, pp. 261–80. New York: Chiron Press.

Koenig, W. D., and P. L. Williams. 1979. Notes on the status of acorn woodpeckers in central Mexico. *Condor* 81:317–18.

Koford, R. R., B. S. Bowen, and S. L. Vehrencamp. 1986. Habitat saturation in groove-billed anis (*Crotophaga sulcirostris*). *Amer. Nat.* 127:317–37.

Krebs, J. R. 1971. Territory and breeding density in the great tit *Parus major* L. *Ecology* 52:2–22.

Krebs, J. R., and M. I. Avery. 1984. Chick growth and prey quality in the European bee-eater (*Merops apiaster*). *Oecologia* 64:363–68.

Lack, D. 1954. *The natural regulation of animal numbers*. Oxford: Clarendon Press.

———. 1966. *Population studies of birds*. Oxford: Clarendon Press.

———. 1968. *Ecological adaptations for breeding in birds*. London: Chapman and Hall.

Lawton, M. F., and C. F. Guindon. 1981. Flock composition, breeding success, and learning in the brown jay. *Condor* 83:27–33.

Lawton, M. F., and R. O. Lawton. 1985. The breeding biology of the brown jay in Monteverde, Costa Rica. *Condor* 87:192–204.

Leach, F. A. 1925. Communism in the California woodpecker. *Condor* 27:12–19.

413

Lewis, D. M. 1981. Determinants of reproductive success of the white-browed sparrow weaver, *Plocepasser mahali*. *Behav. Ecol. & Sociobiol.* 9:83–93.

———. 1982. Cooperative breeding in a population of white-browed sparrow weavers *Plocepasser mahali*. *Ibis* 124:511–22.

Ligon, J. D. 1970. Behavior and breeding biology of the red-cockaded woodpecker. *Auk* 87:255–78.

———. 1981. Demographic patterns and communal breeding in the green woodhoopoe, *Phoeniculus purpureus*. In *Natural selection and social behavior: Recent research and new theory*, edited by R. D. Alexander and D. W. Tinkle, pp. 231–43. New York: Chiron Press.

———. 1983. Cooperation and reciprocity in avian social systems. *Amer. Nat.* 121:366–84.

Ligon, J. D., and S. H. Ligon. 1978a. Communal breeding in green woodhoopoes as a case for reciprocity. *Nature* 276:496–98.

———. 1978b. The communal social system of the green wood-hoopoe in Kenya. *Living Bird* 17:159–98.

———. 1982. The cooperative breeding behavior of the green woodhoopoe. *Sci. Amer.* 247(1):126–34.

———. 1983. Reciprocity in the green woodhoopoe (*Phoeniculus purpureus*). *Anim. Behav.* 31:480–89.

———. MS. Embryo mortality and biased nestling sex ratio in the green woodhoopoe: Responses to an unpredictable environment?

Lotka, A. J. 1922. The stability of the normal age distribution. *Proc. Natl. Acad. Sci.* 8:339–45.

MacRoberts, M. H. 1970. Notes on the food habits and food defense of the acorn woodpecker. *Condor* 72:196–204.

———. 1974. Acorns, woodpeckers, grubs, and scientists. *Pacific Discovery* 27:9–15.

MacRoberts, M. H., and B. R. MacRoberts. 1976. Social organization and behavior of the acorn woodpecker in central coastal California. *Ornithol. Monogr.* 21:1–115.

Margolin, M. 1978. *The Ohlone way*. Berkeley, Calif.: Heyday Books.

Markl, H., ed. 1980. *Evolution of social behavior: Hypotheses and empirical tests*. Deerfield Beach, Fla.: Verlag Chemie.

Marquardt, R. R., and A. T. Ward. 1979. Chick performance as affected by autoclave treatment of tannin-containing and tannin-free cultivars of fababeans. *Can. J. Anim. Sci.* 59:781–89.

Martin, J. S., and M. M. Martin. 1982. Tannin assays in ecological studies: Lack of correlation between phenolics, proanthocyanidins and protein-precipitating constituents in mature foliage of six oak species. *Oecologia* 54:205–211.

Maynard Smith, J. 1964. Group selection and kin selection. *Nature* 201:1145–47.

Maynard Smith, J., and M. G. Ridpath. 1972. Wife sharing in the Tasmanian native hen, *Tribonyx mortierii:* A case of kin selection? *Amer. Nat.* 106:447–52.

Michael, E. 1927. Plurality of mates. *Yosemite Nature Notes* 6:78.

Michod, R. E. 1982. The theory of kin selection. *Ann. Rev. Ecol. & Syst.* 13:23–55.

Miller, A. H. 1963. Seasonal activity and ecology of the avifauna of an American equatorial cloud forest. *Univ. Calif. Publ. Zool.* 66:1–78.

Morton, E. S. 1973. On the evolutionary advantages and disadvantages of fruit eating in tropical birds. *Amer. Nat.* 107:8–22.

Mumme, R. L. 1984. Competition and cooperation in the communally breeding acorn woodpecker. Ph.D. diss. University of California, Berkeley.

Mumme, R. L., and A. de Queiroz. 1985. Individual contributions to cooperative behaviour in the acorn woodpecker: Effects of reproductive status, sex, and group size. *Behaviour* 95:290–313.

Mumme, R. L., and W. D. Koenig. MS. Calculation of the future component of indirect fitness in cooperative breeders.

Mumme, R. L., W. D. Koenig, and F. A. Pitelka. 1983a. Mate guarding in the acorn woodpecker: Within-group reproductive competition in a cooperative breeder. *Anim. Behav.* 31:1094–1106.

———. 1983b. Reproductive competition in the communal acorn woodpecker: Sisters destroy each other's eggs. *Nature* 306:583–84.

Mumme, R. L., W. D. Koenig, R. M. Zink, and J. A. Marten. 1985. Genetic variation and parentage in a California population of acorn woodpeckers. *Auk* 102:312–20.

Murray, B. G., Jr., and L. Gårding. 1984. On the meaning of parameter x of Lotka'a equations. *Oikos* 42:323–26.

Myers, H. W. 1915. A late nesting record for the California woodpecker. *Condor* 17:183–85.

Nie, N. H., C. H. Hull, J. G. Jenkins, K. Steinbrenner, and D. H. Brent. 1975. *Statistical package for the social sciences.* San Francisco: McGraw-Hill.

Nilsson, S. G. 1984. The evolution of nest-site selection among hole-nesting birds: The importance of nest predation and competition. *Ornis. Scandinavica* 15:167–75.

———. 1986. Evolution of hole-nesting in birds: On balancing selection pressures. *Auk* 103:432–35.

Nudds, T. D. 1978. Convergence of group size strategies by mammalian social carnivores. *Amer. Nat.* 112:957–60.

O'Conner, R. J. 1984. *The growth and development of birds.* New York: John Wiley & Sons.

Perrins, C. M. 1976. Possible effects of qualitative changes in the insect diet of avian predators. *Ibis* 118:580–84.

Price, T., S. Millington, and P. Grant. 1983. Helping at the nest in Darwin's finches as misdirected parental care. *Auk* 100:192–94.

Pulliam, H. R., and T. Caraco. 1984. Living in groups: Is there an optimal group size? In *Behavioural ecology: An evolutionary approach*, 2nd ed., edited by J. R. Krebs and N. B. Davies, pp. 122–47. Oxford: Blackwell.

Pynnönen, A. 1939. Beiträge zur Kenntnis der Biologie Finnisher Speckte. Pt. 1. *Ann. Zool. Soc. Zoologicae-Botanicae-Fennicae Vanamo* 7:1–66.

Rabenold, K. N. 1984. Cooperative enhancement of reproductive success in tropical wren societies. *Ecology* 63:871–85.

———. 1985. Cooperation in breeding by nonreproductive wrens: Kinship, reciprocity, and demography. *Behav. Ecol. & Sociobiol.* 17:1–17.

Rabenold, K. N., and C. R. Christensen. 1979. Effects of aggregation on feeding and survival in a communal wren. *Behav. Ecol. & Sociobiol.* 6:39–44.

Reyer, H.-U. 1980. Flexible helper structure as an ecological adaptation in the pied kingfisher (*Ceryle rudis rudis* L.). *Behav. Ecol. & Sociobiol.* 6:219–27.

———. 1984. Investment and relatedness: A cost/benefit analysis of breeding and helping in the pied kingfisher (*Ceryle rudis*). *Anim. Behav.* 32:1163–78.

———. 1986. Breeder-helper-interactions in the pied kingfisher reflect the costs and benefits of cooperative breeding. *Behaviour* 96:277–303.

Ricklefs, R. E. 1969. An analysis of nesting mortality in birds. *Smith. Contrib. Zool.* 9:1–48.

———. 1973. Fecundity, mortality, and avian demography. In *Breeding biology of birds*, edited by D. S. Farner, pp. 366–435. Washington, D.C.: National Academy of Sciences.

———. 1974. Energetics of reproduction in birds. In *Avian energetics*, edited by R. A. Paynter, Jr. *Publ. Nuttall Ornithol. Club* 15.

———. 1975. The evolution of co-operative breeding in birds. *Ibis* 117:531–34.

———. 1980. Geographic variation in clutch size among passerine birds: Ashmole's hypothesis. *Auk* 97:38–49.

Riedman, M. L. 1982. The evolution of alloparental care and adoption in mammals and birds. *Quart. Rev. Biol.* 57:405–35.

Ritter, W. E. 1921. Acorn-storing by the California woodpecker. *Condor* 23:3–14.

———. 1922. Further observations on the activities of the California woodpecker. *Condor* 24:109–122.

———. 1929. The nutritional activities of the California woodpecker (*Balanosphyra formicivora*). *Quart. Rev. Biol.* 4:455–83.

———. 1938. *The California woodpecker and I.* Berkeley: University of California Press.

Roberts, R. C. 1979a. Habitat and resource relationships in acorn woodpeckers. *Condor* 81:1–8.

———. 1979b. The evolution of avian food-storing behavior. *Amer. Nat.* 114:418–38.

Rossi, R. S. 1979. The history of cultural influences on the distribution and reproduction of oaks, northern San Luis Obispo County, California. Ph.D. diss., University of California, Berkeley.

Rowley, I. 1965. The life history of the superb blue wren, *Malurus cyaneus. Emu* 64:251–97.

———. 1978. Communal activities among white-winged choughs *Corcorax melanorhamphus. Ibis* 120:178–97.

———. 1981. The communal way of life in the splendid wren, *Malurus splendens. Z. Tierpsychol.* 55:228–67.

Sappington, J. N. 1977. Breeding biology of house sparrows in north Mississippi. *Wilson Bull.* 89:300–309.

Selander, R. K. 1964. Speciation in wrens of the genus *Campylorhynchus. Univ. Calif. Publ. Zool.* 74:1–224.

Sherman, P. W. 1981. Electrophoresis and avian genealogical analyses. *Auk* 98:419–21.

Shields, W. M. 1987. Dispersal and mating systems: Investigating their causal connections. In *Patterns of dispersal among mammals and their effects on the genetic structure of populations,* edited by B. D. Chepko-Sade and Z. Halpin. Chicago: University of Chicago Press.

Short, L. L., Jr. 1970. Notes on the habits of some Argentine and Peruvian woodpeckers (Aves, Picidae). *Amer. Mus. Novitates* 2413:1–37.

————. 1982. *Woodpeckers of the world*. Delaware Museum of Natural History Monographs 4.

Silk, J. B. 1984. Measurement of the relative importance of individual selection and kin selection among females of the genus *Macaca*. *Evol.* 38:553–59.

Skutch, A. F. 1935. Helpers at the nest. *Auk* 52:257–73.

————. 1961. Helpers among birds. *Condor* 63:198–226.

————. 1969. Life histories of Central American birds, III. *Pacific Coast Avifauna* 35.

————. 1985. *Life of the woodpecker*. Santa Monica, Calif. Ibis Publishing Co.

Slatkin, M. 1981. Populational heritability. *Evol.* 35:859–71.

Smith, D., G. N. Paulsen, and C. A. Raguse. 1964. Extraction of total available carbohydrates from grass and legume tissue. *Plant Physiol.* 39:960–62.

Smith, S. M. 1978. The "underworld" in a territorial sparrow: Adaptive strategy for floaters. *Amer. Nat.* 112:571–82.

————. 1984. Flock switching in chickadees: Why be a winter floater? *Amer. Nat.* 123:81–98.

Sokal, R. R., and F. J. Rohlf. 1981. *Biometry*, 2nd ed. San Francisco: Freeman.

Sork, V. L., P. Stacey, and J. E. Averett. 1983. Utilization of red oak acorns in a non-bumper crop year. *Oecologia* 59: 49–53.

Southwood, T.R.E. 1978. *Ecological methods*, 2nd ed. London: Chapman and Hall.

Spray, C., and M. H. MacRoberts. 1975. Notes on molt and juvenal plumage in the acorn woodpecker. *Condor* 77:342–45.

Stacey, P. B. 1979a. Habitat saturation and communal breeding in the acorn woodpecker. *Anim. Behav.* 27:1153–66.

————. 1979b. Kinship, promiscuity, and communal breeding in the acorn woodpecker. *Behav. Ecol. & Sociobiol.* 6: 53–66.

————. 1981. Foraging behavior of the acorn woodpecker in Belize, Central America. *Condor* 83:336–39.

———— 1982. Female promiscuity and male reproductive success in birds and mammals. *Amer. Nat.* 120:51–64.

Stacey, P. B., and C. E. Bock. 1978. Social plasticity in the acorn woodpecker. *Science* 202:1298–1300.

Stacey, P. B., and T. C. Edwards. 1983. Possible cases of infanticide by immigrant females in a group-breeding bird. *Auk* 100:731–33.

Stacey, P. B., and R. Jansma. 1977. Storage of piñon nuts by the acorn woodpecker in New Mexico. *Wilson Bull.* 89:150–51.

Stacey, P. B., and W. D. Koenig. 1984. Cooperative breeding in the acorn woodpecker. *Sci. Amer.* 251(2):114–21.

Stacey, P. B., and J. D. Ligon. 1987. Helping in the acorn woodpecker: Habitat saturation or mutualism? *Amer. Nat.* (in press).

Steinbeck, J. 1952. *East of Eden.* New York: Viking.

Stewart, R. E., and J. W. Aldrich. 1951. Removal and repopulation of breeding birds in a spruce-fir community. *Auk* 68:471–82.

Sullivan, K. A. 1984. Advantages of social foraging in downy woodpeckers. *Anim. Behav.* 32:16–22.

Sumner, F. B. 1944. William Emerson Ritter: Naturalist and philosopher. *Science* 99:335–38.

Swanberg, P. O. 1951. Food storage, territory and song in the thick-billed nutcracker. *Proc. X Int. Ornith. Congr.*, pp. 545–54.

Swearingen, E. M. 1977. Group size, sex ratio, reproductive success and territory size in acorn woodpeckers. *West. Birds* 8:21–24.

Tanner, J. T. 1978. *Guide to the study of animal populations.* Knoxville: University of Tennessee Press.

Tempel, A. S. 1981. Field studies of the relationship between herbivore damage and tannin concentration in bracken (*Pteridium aquilinum* Kuhn). *Oecologia* 51:97–106.

Thompson, R. L., ed. 1971. *The ecology and management of the red-cockaded woodpecker.* Washington, D.C.: Bureau of Sport Fish and Wildl., U.S. Dept. of Interior.

Time. 1984. For the birds. January 10, p. 24.

Tomback, D. F. 1978. Foraging strategies of Clark's nutcracker. *Living Bird* 16:123–61.

Trail, P. W. 1980. Ecological correlates of social organization in a communally breeding bird, the acorn woodpeaker, *Melanerpes formicivorus. Behav. Ecol. & Sociobiol.* 7:83–92.

Trail, P. W., S. D. Strahl, and J. L. Brown. 1981. Infanticide in relation to individual and flock histories in a communally breeding bird, the Mexican jay (*Aphelocoma ultramarina*). *Amer. Nat.* 118:72–82.

Trivers, R. L. 1971. The evolution of reciprocal altruism. *Quart. Rev. Biol.* 46: 35–57.

Trivers, R. L., and D. E. Willard. 1973. Natural selection of parental ability to vary the sex ratio of offspring. *Science* 179:90–92.

Troetschler, R. G. 1976. Acorn woodpecker breeding strategy as affected by starling nest-hole competition. *Condor* 78:151–65.

van Balen, J. H. 1980. Population fluctuations of the great tit and feeding conditions in winter. *Ardea* 68:143–64.

Vander Wall, S. B., 1982. An experimental analysis of cache recovery in Clark's nutcracker. *Anim. Behav.* 30:84–94.

Vander Wall, S. B., and R. P. Balda. 1977. Coadaptations of the Clark's nutcracker and the piñon pine for efficient seed harvest and dispersal. *Ecol. Monogr.* 47:89–111.

Vehrencamp, S. L. 1977. Relative fecundity and parental effort in communally nesting anis, *Crotophaga sulcirostris. Science* 197:403–405.

———. 1978. The adaptive significance of communal nesting in groove-billed anis (*Crotophaga sulcirostris*). *Behav. Ecol. & Sociobiol.* 4:1–33.

———. 1979. The roles of individual, kin, and group selection in the evolution of sociality. In *Social behavior and communication,* edited by P. Marler and J. Vandenburgh, pp. 351–94. New York: Plenum.

———. 1983. A model for the evolution of despotic versus egalitarian societies. *Anim. Behav.* 31:667–82.

Verbeek, N.A.M. 1973. The exploitation system of the yellow-billed magpie. *Univ. Calif. Publ. Zool.* 99:1–58.

Vleck, C. M. 1981. Energetic cost of incubation in the zebra finch. *Condor* 83:229–37.

von Schantz, T. MS. Territory economics and population stability—can populations be socially regulated?

Wade, M. J. 1979. The evolution of social interactions by family selection. *Amer. Nat.* 113:399–417.

———. 1980. Kin selection: Its components. *Science* 210:665–67.

Walsberg, G. E. 1983. Avian ecological energetics. In *Avian biology,* vol. 7, edited by D. S. Farner, J. R. King, and K. C. Parkes, pp. 161–220. New York: Academic Press.

Waser, P. M. 1981. Sociality or territorial defense: The influence of resource renewal. *Behav. Ecol. & Sociobiol.* 8:231–37.

Watson, A., and R. Moss. 1970. Dominance, spacing behaviour and aggression in relation to population limitation in vertebrates. In *Animal populations in relation to their food resources,* edited by A. Watson, pp. 167–218. Oxford: Blackwell.

West Eberhard, M. J. 1975. The evolution of social behavior by kin selection *Quart. Rev. Biol.* 50:1–33.

Wiley, R. H. 1974. Effects of delayed reproduction on survival, fecundity, and the rate of population increase. *Amer. Nat.* 108:705–709.

Wiley, R. H., and K. N. Rabenold. 1984. The evolution of cooperative breeding by delayed reciprocity and queuing for favorable social positions. *Evol.* 38:609–621.

Wilkinson, R. 1982. Social organization and communal breeding in the chestnut-bellied starling (*Spreo pulcher*). *Anim. Behav.* 30:1118–28.

Wilkinson, R., and A. E. Brown. 1984. Effects of helpers on the feeding rates of nestlings in the chestnut-bellied starling *Spreo pulcher. J. Anim. Ecol.* 53:301–310.

Williams, G. C. 1966. *Adaptation and natural selection.* Princeton: Princeton University Press.

Williams, P. L. 1982. A comparison of colonial and non-colonial nesting by northern orioles in central coastal California. Masters thesis, University of California, Berkeley.

Williams, P. L., and W. D. Koenig. 1980. Water dependence of birds in a temperate oak woodland. *Auk* 97:339–50.

Wilson, D. S., 1975. A theory of group selection. *Proc. Natl. Acad. Sci.* 72:143–46.

———. 1977. Structured demes and the evolution of group-advantageous traits. *Amer. Nat.* 111:157–85.

———. 1980. *The natural selection of populations and communities.* Menlo Park, Calif.: Benjamin/Cummings.

Wilson, E. O. 1975. *Sociobiology: The new synthesis.* Cambridge, Mass: Harvard University Press.

Winer, B. J. 1971. *Statistical principles in experimental design*, 2d ed. New York: McGraw-Hill.

Woolfenden, G. E., and J. W. Fitzpatrick. 1978. The inheritance of territory in group-breeding birds. *BioScience* 28:104–108.

———. 1984. *The Florida scrub jay: Demography of a cooperative-breeding bird.* Princeton: Princeton University Press.

———. 1986. Sexual asymmetries in the life history of the Florida scrub jay. In *Ecological aspects of social evolution: Birds and mammals,* edited by D. I. Rubenstein and R. W. Wrangham. pp. 87–107. Princeton: Princeton University Press.

Wright, S. 1931. Evolution in Mendelian populations. *Genetics* 16:97–159.

———. 1943. Isolation by distance. *Genetics* 28:114–38.

———. 1946. Isolation by distance under diverse systems of mating. *Genetics* 31:39–59.

———. 1951. The genetical structure of populations. *Ann. Eugenics* 15:323–54.

———. 1969. *The theory of gene frequencies.* Chicago: University of Chicago Press.

Wynne-Edwards, V. C. 1962. *Animal dispersion in relation to social behaviour.* Edinburgh: Oliver and Boyd.

Yapar, Z., and D. R. Clandinin. 1972. Effect of tannins in rapeseed meal on its nutritional value for chicks. *Poultry Sci.* 51:222–28.

Yom-Tov, Y. 1980. Intraspecific nest parasitism in birds. *Biol. Rev.* 55: 93–108.

Zack, S., and J. D. Ligon. 1985a. Cooperative breeding in Lanius shrikes. I. Habitat and demography of two sympatric species. *Auk* 102:754–65.

———. 1985b. Cooperative breeding in Lanius shrikes. II. Maintenance of group-living in a nonsaturated habitat. *Auk* 102:766–73.

Zahavi, A. 1974. Communal nesting by the Arabian babbler. *Ibis* 116:84–87.

———. 1976. Co-operative nesting in Eurasian birds. *Proc. XVIth Int. Ornith. Congr.*, pp. 685–93.

Zucker, W. V. 1983. Tannins: Does structure determine function? An ecological perspective. *Amer. Nat.* 121:335–65.

Author Index

425

Subject Index

Library of Congress Cataloging-in-Publication Data

Koenig, Walter D., 1950–
   Population ecology of the cooperatively breeding
acorn woodpecker.

   (Monographs in population biology; 24)
   Bibliography: p.
   Includes index.
   1. California woodpecker—Ecology.   2. California
woodpecker—Reproduction.   3. Population biology.
I. Mumme, Ronald L., 1954–   .   II. Title.
III. Series.
QL696.P56K64   1987   598′.72   87-2385
ISBN 0-691-08422-X
ISBN 0-691-08464-5 (pbk.)